Interpretation and Explanation in the Study of Animal Behavior

Volume II: Explanation, Evolution, and Adaptation

Interpretation and Explanation in the Study of Animal Behavior

Volume II: Explanation, Evolution, and Adaptation

EDITED BY
Marc Bekoff and Dale Jamieson

Westview Press
BOULDER, SAN FRANCISCO, & OXFORD

This Westview softcover edition is printed on acid-free paper and bound in library-quality, coated covers that carry the highest rating of the National Association of State Textbook Administrators, in consultation with the Association of American Publishers and the Book Manufacturers' Institute.

Published in 1990 in the United States of America by Westview Press, Inc., 5500 Central Avenue, Boulder, Colorado 80301, and in the United Kingdom by Westview Press, Inc., 36 Lonsdale Road, Summertown, Oxford OX2 7EW

Library of Congress Cataloging-in-Publication Data
Interpretation and explanation in the study of animal behavior /
edited by Marc Bekoff and Dale Jamieson.
p. cm.
Bibliography: p.
Includes index.
Contents: v. 1. Interpretation, intentionality, and communication—
v. 2. Explanation, evolution, and adaptation.
ISBN 0-8133-7704-8 (v. 1)—ISBN 0-8133-7979-2 (v. 2)
1. Animal behavior. I. Bekoff, Marc. II. Jamieson, Dale.
QL751.I57 1990
591.5′1—dc19 88-28019
CIP

Printed and bound in the United States of America

The paper used in this publication meets the requirements of the American National Standard for Permanence of Paper for Printed Library Materials Z39.48-1984.

10 9 8 7 6 5 4 3 2 1

These volumes are dedicated to the memory of Milton E. Lipetz, who strongly supported interdisciplinary behavioral research when he was a senior administrator at the University of Colorado, Boulder, and to Everly B. Fleischer, who, as Dean of the College of Arts and Sciences, introduced us to each other, supported our initial efforts to teach together, and provided the initial support that helped get this project off the ground.

Contents

VOLUME II
Explanation, Evolution, and Adaptation

VOLUME I
Interpretation, Intentionality, and Communication

Foreword

Donald R. Griffin

The stimulating chapters in these two volumes are a highly heterogeneous assembly, ranging from critical analyses of how investigators of animal behavior can best minimize the hazards of erroneous interpretations to thoughtful philosophical discussions. The latter include arguments based on deeply held convictions that many animals deserve better treatment by members of our species. A significant development reflected in several chapters is the increased concern with both cognition and mental experiences in nonhuman animals. For example, Patrick Bateson tentatively concludes that animals have preferences and make choices, Robert Mitchell infers that animals sometimes act intentionally, and W. John Smith accepts the presence of expectations. This relatively recent trend to reopen the old and long neglected questions of animal mentality presents us with extremely difficult scientific challenges; but many contributors to these volumes share at least a tentative belief that these questions are important and that they can be studied profitably by scientific methods. This is a welcome change from the puritanical behaviorism that used to repress all consideration of animal minds.

Yet despite this liberalization of viewpoint, stubborn residues of positivistic behaviorism are still evident. They seem to center on an excessive fear of anthropomorphism, apprehension that one may be accused of uncritical sentimentality if one suggests that any nonhuman animal might experience subjective emotions such as fear, or think consciously in even the simplest terms, such as believing that food is located in a certain place. This widespread taboo might be called by the tongue-twisting name of "*anthropomorphophobia*," or *A-phobia* for short. The "A word" has had frightening connotations among scientists, and we have customarily taken it more or less for granted that only uncritical thinkers would suggest that nonhuman animals might experience subjective emotions or simple conscious thoughts.

But as John Fisher points out, there is no solid logical basis for the belief that anthropomorphism is a fallacy. Only if one assumes in advance that our species has an exclusive monopoly on mental experiences is it unsound to consider to what degree members of other species may think and feel, as I have discussed elsewhere (Griffin 1977, 1984, 1990). It may therefore be helpful to outline an alternative cognitive perspective that is equally compatible with critical scientific inquiry and analysis:

A. As materialists who accept the fact of biological evolution, we take it for granted that mental experiences, that is, subjective feelings and conscious thoughts, result from the functioning of central nervous systems.

B. All central nervous systems consist of quite similar neurons and synapses; and, as far as we know, they function in much the same way.

C. We know firsthand that mental experiences occur in one species, although they accompany only a small fraction of the information processing that takes place in the central nervous system.

D. It is therefore parsimonious to infer that the central nervous systems of all animals give rise at times to some sort of subjective feelings and conscious thought.

E. Inference D does not require, or even imply, that the mental experiences of all species are alike or equal in complexity. Indeed they differ to some extent among members of the one species where we know they occur, for instance, between children and grown-ups, men and women, and members of different cultural groups. Interspecific differences are almost certainly greater than intraspecific differences. Thus recognizing the likelihood that animals experience simple thoughts and feelings need not detract from a full recognition of the enormous superiority of human mentality and its many potent ramifications.

F. Because the mental experiences of one group differ from those of others does not mean that those of either group are nonexistent.

These statements are quite generally accepted by scientists concerned with neurophysiology and behavior, except for inference D, probably because it conflicts with deep-seated beliefs that human mentality is qualitatively different in kind from the results of nonhuman brain function. Of course human thinking is astronomically more complex and versatile than that of other species, but even this enormous quantitative superiority does not, in principle, require that there must be basic differences in kind between the mental experiences resulting from the functioning of human and nonhuman brains.

This brings us back to the "A word," and its baggage of negative connotations among scientists. If we recognize that the views outlined above are at least plausible and should not be rejected out of hand, it follows that there is no valid reason to dismiss as fallaciously anthropomorphic the suggestion that nonhuman animals may experience simple feelings or conscious thoughts. There are often good reasons to doubt that a particular mental experience could occur to a given animal, for example because it lacks the necessary information or is incapable of organizing such information into the appropriate pattern. Finding answers to such challenging questions as those discussed in this book is not facilitated by dogmatically negative prejudgments. Yet this basic consideration tends to be overlooked. For example, behavioral scientists often qualify any cognitive interpretations of animal behavior with disclaimers that they are suggesting the occurrence of *human* fear, belief, intention or the like. This customary wording implies that human emotions or thoughts are the only possible kind.

Just how similar or different are the mental experiences of other species? Recognizing that they may differ greatly across species makes it more difficult to identify them than would be the case if they were necessarily the same as ours. But it would probably be an overreaction to jump to the conclusion that fear, affection or, say, the belief that a certain animal is a dangerous predator differ substantially between various species. Insofar as they occur, such mental experiences must play an equivalent functional role in a wide variety of animals; and their emotional impact is presumably much the same regardless of the specific sensory or perceptual processes by which they are established. Thus it would be prematurely pessimistic to assume in advance that the ingenuity of scientists cannot eventually reduce significantly our present ignorance about these important matters.

Another important aspect of several chapters in these volumes, as well as similar discussions of animal behavior elsewhere, is a demand for premature perfectionism. Scientists pondering the questions of animal mentality yearn for clear-cut formulations by which cognitive processes and mental experience can be unambiguously identified. Ideally we would like such identifications to permit predictions that could then be confirmed or disconfirmed to provide empirical tests of the validity of our interpretations. Because such ideal litmus tests are not immediately available, many behavioral scientists tend to throw up their hands, figuratively speaking, and dismiss the whole enterprise as hopelessly beyond the reach of scientific inquiry. This may result, in part, from a deep-seated nostalgia for a satisfyingly simple philosophical outlook from which qualitative human uniqueness and inherent superiority are taken for granted.

In this situation it is helpful to remind ourselves that many if not most major scientific advances have begun by tentative explorations of possibilities that could initially be little more than shrewd guesses. The investigation of animal mentality is in a very early stage of development, where tidy theories and definitive evidence are not yet available. Therefore cognitive ethology must grope inquisitively through tangled thickets of uncertainty, and gradually reduce our current ignorance by gathering and evaluating evidence that can increase or decrease the plausibility of specific hypotheses rather than settling definitive questions once and for all by simple predictions that are then confirmed or not. Several chapters exemplify this constructive endeavor. To insist at the outset on tidy packages of totally consistent and universally applicable theories would be at best unrealistic, and at worst a repressive inhibition of a significant area of scientific investigation.

An additional impediment to scientific investigation of animal mentality is a political or quasi-religious sort of claim that narrowing the perceived gulf between human and animal mentality would threaten fundamental human values. Adler (1967) argued at length that if people were persuaded that animals differ from men only in degree and not radically in kind, this would destroy our moral basis for holding that all men deserve equal treatment in matters of ethics and law. His view is summed up in the following quotation (Adler 1967: 263):

> If in the future we should discover that man differs from other animals *only in degree*, the line that divides the realm of persons from the realm of things would be rubbed out, and with its disappearance would go the basis in fact for a principled policy of treating men differently from the way in which we now treat other animals and machines.

Recently Allen (1987: 158-160) has reiterated a similar argument, adding the charge that cognitive ethologists are proto-fascists:

> Sociobiologists and students of animal awareness, though coming from different directions, arrive at the same end: they blur the distinction between animals and humans by setting up an evolutionary continuum...students of animal awareness see fully developed human awareness existing in rudimentary form in lower animals....[T]o blur the distinction between animal and human especially by distorting the biological reality (or by claiming for the biological reality more than it can offer), is to play into the hands of a political mood that leads ultimately to fascism....[B]lurring the distinction between humans and other animals, whether by evolutionary, genetic, or neurobiological arguments, paves the way for relegating some people to the sub-human category on the basis of their biology. Once there, the usual moral restraints and considerations cease to apply, and fascism has arrived.

These political views entail a belief that certain scientific subjects must not be investigated, because if the truth should turn out in one conceivable way, our moral standards would be seriously undermined. Such arguments are highly reminiscent of the outrage that greeted Darwin's conclusion that animals have evolved over geological time, and that our species was part of this evolutionary continuum. Biological continuity is now generally accepted (even Adler and Allen speak of human and nonhuman animals), but mentality is still viewed by some as a sort of sacred area where evolutionary continuity must be denied at all costs. Just as the suggestion that genetic background exerts some influence on behavior was greeted with quasi-religious outrage, so now a few ardent advocates of radical human uniqueness see cognitive ethology as a threat to human morality, following in the footsteps of Bishop Wilberforce and other religious opponents of evolution.

Fortunately these imagined threats are remote and absurd. Religion and ethics have survived the Copernican and Darwinian revolutions, strengthened rather than weakened by the correction of factual errors. The astronomical differences in degree of mental capability is quite sufficient to distinguish our species from others. As Whitehead (1938: 27) put it succinctly, "The distinction between men and animals is in one sense only a difference of degree. But the extent of the degree makes all the difference." It would surely be tragic to base our morals and ethics on erroneous scientific conclusions. And although scientists have no special license to advocate ethical standards, it is appropriate to question the soundness of basing them on exclusion of supposedly inferior creatures. In any event the increasing realization that the extent and nature of animal mentality has fundamental philosophical implications has stimulated many philosophers, including those represented in this book and others such as Midgley (1983) and Radner & Radner (1989), to reexamine the ramifications of cognitive ethology. Such thoughtful consideration is now limited by our lack of knowledge of the actual mental experiences of nonhuman animals, and it will be greatly facilitated by learning as much as we can about what various animals think and feel. We have a very long way to go in this endeavor, but several chapters in these volumes represent important milestones at the beginning of this road.

LITERATURE CITED

Adler, M.J. 1967. *The Difference of Man and the Difference It Makes*. New York: World Publishing.

Allen, G.E. 1987. Materialism and reductionism in the study of animal consciousness. In: *Cognition, Language, and Consciousness: Integrative Levels* (ed. by G. Greenberg & E. Tobach), pp. 137-160. Hillsdale, New Jersey: Lawrence Erlbaum Associates.

Griffin, D.R. 1977. Anthropomorphism. *BioScience* 27, 445-446.

_____. 1984. *Animal Thinking*. Cambridge, Massachusetts: Harvard University Press.

_____. 1990. *Animal Thinking*. Revised Edition. Cambridge, Massachusetts: Harvard University Press.

Midgley, M. 1983. *Animals and Why They Matter*. Athens, Georgia: University of Georgia Press.

Radner, D. & M. Radner. 1989. *Animal Consciousness*. Buffalo, New York: Prometheus Books.

Whitehead, A.N. 1938. *Modes of Thought*. New York: Macmillan.

Preface

In these two volumes, numerous and previously unacquainted thinkers meet one another in settings that probably none would or could have predicted - Paul Grice meets Gregory Bateson, Charles Lloyd Morgan meets Daniel Dennett, Charles Darwin meets Ludwig Wittgenstein, and Aristotle meets Niko Tinbergen. The original essays in this collection, written by biologists, psychologists, philosophers, an anthropologist, and a historian, are each concerned in their own way with the interpretation and explanation of animal behavior.

Many books and articles have been written on interpretation and explanation (see, for example, Cummins 1983; Miller 1987; Dretske 1988; Giere 1988; Pitt 1988; Humphreys 1989; Kitcher & Salmon 1989; Packer & Addison 1989), and we would not be so foolish as to attempt to provide full analyses of these concepts here. And anyway, different contributors have different conceptions of these notions. However, we can say this: Problems of interpretation arise in the neighborhood of observation, while those of explanation arise in the vicinity of theory. Both interpretation and explanation involve empirical and conceptual dimensions. In order to understand better nonhuman animals, we need to know more about what they do and about our own concepts and the grounds for attributing them.

Problems of interpretation generally concern *how* we know what an animal is doing; whether, for example, two dogs are playing or fighting, whether a ground squirrel recognizes another individual, whether a Japanese quail chooses a specific individual with whom to mate, whether an evening grosbeak selects a particular area in which to breed, or whether a dog goes to a door because it wants a human to let it out (and perhaps intends to accomplish this by going to the door). These and other problems of interpretation are taken up in different chapters in this collection.

Problems of explanation generally concern *why* something occurs. Traditionally, explanation has been thought to be logically related to prediction (Hempel & Oppenheim 1948); both involve the deduction of a particular observable event from a universal covering law. Without endorsing any particular view, we can say that an explanation makes a behavior intelligible to us by assimilating it to a larger pattern. Contributors to this collection are concerned with such questions as why birds respond differently to various song dialects, why helping behavior has evolved, and why there appears to be a predictable relationship between social group size and antipredatory (vigilance) behavior in some birds and mammals. One recurring theme concerns whether an animal's behavior can ever be explained by reference to its mental states.

Our own interests in these areas of inquiry originate from different perspectives on animal behavior and from different experiences with animals. Yet it was obvious to us from the beginning that these interests and experiences complement and reinforce each other and that it was quite natural for an ethologist and a philosopher to converge on problems of interpretation and explanation in animal behavior (Bekoff & Jamieson 1990a,b). Since our graduate school days, in different fields in different universities, we have both been fascinated by questions about "what it is like" to be a nonhuman animal (Nagel 1974).

In recent years there has been a revival of interest in the study of animal minds (e.g. Mitchell & Thompson 1986; Blakemore & Greenfield 1987; Richards 1987; Byrne & Whiten 1988; Montefiore & Noble 1989; Radner & Radner 1989; Robinson 1989), an area pioneered by Charles Darwin (1871/1946, 1872/1965) and his immediate followers (see Burghardt 1985). Zoologists such as Griffin (1976, 1984, 1990), psychologists such as Walker (1983) and Kamil (1988), and historians such as Haraway (1989) have all made contributions. Philosophers have been mainly concerned with producing general theories of mind, but in many cases these theories have implications for animal minds. For example, some philosophers have argued that "folk psychological" notions such as "believe," "desire," and "intend" apply to both humans and some nonhumans (e.g. Dennett 1983, 1987; Searle 1983; Fodor 1987; Rollin 1989). Other philosophers have claimed that such notions apply to neither humans nor nonhumans (P. M. Churchland 1981; Stitch 1983; P. S. Churchland 1986). Still others have held that such notions apply to some humans but not to languageless creatures such as nonhumans (Davidson 1985).

This growing interest in animal minds is not merely academic. In Western societies, generally, a reevaluation is occurring with respect to the cognitive and affective capacities of animals, and what this may mean for our relations with them. Just one example of this is the May 23, 1988, issue of *Newsweek* magazine. Pictured on its cover was a "thinking dog," and the caption boldly stated: "How smart are animals? They know more than you think."

Despite this flurry of activity there has been a great deal of insulation (both social and intellectual) between different scientific disciplines, different research programs, and those studying different animals. Moreover, very little of the philosophical work (Daniel Dennett's being the main exception) has had much effect on behavioral biologists and comparative psychologists, and, similarly, very little work in these latter fields has had much impact on philosophers (the work of Donald Griffin being the main exception). This lack of communication, both between and among disciplines, has inhibited progress in the study of animal minds and behavior, and this may explain the surprising fact that there was no comprehensive, truly interdisciplinary collection that addressed the issues of interpretation and explanation in the study of animal behavior. When we discovered that there was no such collection, we decided to invite a few friends and colleagues to help us fill this gap. This two-volume work, which borders dangerously on 1,000 pages in length, is the result.

The main purpose of this collection is to make accessible to a wide audience diverse and interdisciplinary views of the study of animal behavior. We wish to show how various concerns in different fields can be brought together to provide a more coherent picture of animal lives and our relation to them. Despite the varying backgrounds and views of the contributors, some common themes and concerns emerge. These include questions about how behavior is categorized; anthropomorphism; the role of values in behavioral and other types of research; levels and methods of analysis; intervening and confounding variables; critical experiments; units of selection; perceptual worlds, or *umwelts*; the meaning of key terms such as "social relationship," "recognition," "choice," and "play"; communication and language; questions about adaptation and optimality; cognitive skills and affective states; intentionality; mental continuity; the scientific legitimacy of animal mentation; and ethical issues about animal welfare. Our belief is that views on these matters are interrelated - sometimes by logic, sometimes by culture, psychology, and temperament. What we do in our analyses of animal behavior, how we do it, and how information is interpreted, explained, and disseminated all hang together. The picture that we have of a prairie dog's cognitive and affective states influences our views about how prairie dogs ought to be treated. Theoretical concerns about what a dog is doing, or believes, desires, or intends when it goes to a door are only artificially separated from more practical issues concerning our responsibilities to our animal companions (Regan 1983; Bekoff & Jamieson 1990a,b; Singer 1990).

In this collection we have tried to encourage, and sometimes almost compel, communication among the contributors. Rather than just stapling together contributions from different authors, we have circulated manuscripts for review among scholars from many different fields. We also have critically read each manuscript several times in the belief that if an author can communicate with both of us, her or his chances of reaching a broad, diverse audience are pretty good.

This is not the sort of project that two people can accomplish on their own. Indeed, in retrospect, we would probably rather try invading Russia in winter or riding mountain bikes across Antarctica than edit another camera-ready anthology with 37 chapters and 51 authors, who assaulted us with discs of various sizes containing manuscripts prepared on 34 different word-processing programs. We have had a lot of help along the way, and although defects remain, there are many fewer due to the support and advice of many good people.

When we first floated the idea of this anthology, Kellie Masterson of Westview Press was immediately receptive. Spencer Carr, Lynn Arts, and Lindsay Schumacher, also at Westview, provided helpful advice at various crucial stages. Perhaps most importantly, the folks at Westview managed to retain their composure, at least in our presence, when the manuscript weighed in at twice its anticipated length.

We also have been very fortunate that our university has permitted us to teach together a unique course on philosophical issues concerning animal behavior. We have even been provided with financial support to bring some of our contributors to Boulder to take part in our seminar. For this we thank the President of the University of Colorado, E. Gordon Gee, Chancellor James Corbridge, Associate Vice-Chancellor Mark Dubin, Dean Everly Fleischer (now Vice-Chancellor at the University of California, Riverside), Dean Charles Middleton, and our department chairs, Michael Breed and John Andrew Fisher. We would also like to thank Oscar and Beatrice Bekoff for financial support.

The following people reviewed various chapters, and we are grateful to all of them: Mark Anderson, Steve Austad, Chris Barnard, Anne Bekoff, Gordon Burghardt, John Byers, Carol

Cleland, Thomas Daniels, Randolf DiDomenico, John Dupré, Margaret Dussault, Robert Eaton, John Fentress, Sandra Mitchell, Allen Moore, W. John Smith, Randy Thornhill, and Hugh Wilder. Students in our seminars also contributed comments on drafts of many of these essays, and Eric Ervin also helped track down references. For somehow transforming discs into drafts we are grateful to Phyllis O'Connell. Laura Heigl also helped with processing unmanageable manuscripts. Michael Breed and Elizabeth Owen graciously permitted their staff to spend an inordinate amount of time helping us to complete these volumes.

Anne Bekoff and Toby Jacober put up with our self-indulgence as we worked (and worked and worked) on this project. We hope they'll continue to put up with us in the future. Finally Sasha Bekoff and Gretta Jacober-Jamieson - two prototype examples of canid cognitive machinery - provided invaluable inspiration and insights. Why, they asked us, do we question not only the existence of their minds, but also their higher-order intentions? Unfortunately, however, not everyone has been able to hear their questions and even fewer have been able to understand them (see John Dupré's chapter in section IV, Volume I).

We have divided this collection into two complementary volumes. The first deals mainly with problems of *interpretation, intentionality, and communication.* Issues that are discussed include how behavior is categorized; the role of anthropomorphism in comparative analyses of behavior; gender-related issues in the study of behavior; how social relationships are studied; recognition; choice; play; animal communication and cognition; "language" in nonhuman primates; animal mentation; psychological continuity; what it means to speak of animal minds; and why animal minds seemed to disappear from behavioral science from the latter part of the nineteenth century to the middle of the twentieth century. In the second volume *explanation, evolution, and adaptation* are highlighted. It includes discussions of early natural history studies; adaptation; the nature of evolutionary explanations; methodology in neuroethology; the application of artificial intelligence to studies of animal behavior; comparative analyses of behavior and phylogeny; and methodological issues in various areas of research including bird-song, vigilance (antipredatory) behavior, helping behavior, and life-history studies. Ethical issues are also discussed, including those relating to domestication, genetics research, and the use of animals in behavioral and other types of research.

The study of animal behavior requires careful observation and sampling in a wide range of conditions and also the application of rigorous methods of analysis. Detailed studies of nonhumans will allow us to assess not only what they do when we ask them to do things in conditions that we control, but also will expose what they *can* do when permitted to express themselves in more permissive environments. However, the reliability of the information we collect must be assessed carefully and critically so that we do not fool ourselves into thinking that we know something when we do not. Questions about how behavior is categorized, how to study animals, where to do it and when all demand careful consideration, for these matters color how we interpret and explain the behavior of our nonhuman counterparts. Animal minds, animal cognitive abilities and affective states, and animal behavior in general are difficult to study. Too often we have studied what is easy and denied the existence of what is difficult. We have been like the man who searches for his keys under the streetlight because that is where the light is good.

Taken together, these volumes present original discussions of key issues in the study of animal behavior. While each volume can be read on its own, each informs the other. Collectively, these essays serve as a broad interdisciplinary introduction to wide-ranging questions about the animals with whom we share the planet. We hope that these volumes will set the standard for future research.

Marc Bekoff
Dale Jamieson
Boulder, Colorado

LITERATURE CITED

Bekoff, M. & Jamieson, D. 1990a. Reflective ethology, applied philosophy, and the moral status of animals. *Perspectives in Ethology* 9, in press.

_____. 1990b. Cognitive ethology and applied philosophy: The significance of an evolutionary biology of mind. *Trends in Ecology & Evolution* 5, 156-159.

Blakemore, C. & Greenfield, S. (eds.) 1987. *Mindwaves: Thoughts on Intelligence, Identity, and Consciousness.* New York: Basil Blackwell.

Burghardt, G.M. (ed.) 1985. *Foundations of Comparative Ethology.* New York: Van Nostrand Reinhold.

Byrne, R. & Whiten, A. (eds.) 1988. *Machiavellian Intelligence: Social Expertise and the Evolution of Intellect in Monkeys, Apes, and Humans.* New York: Oxford University Press.

Churchland, P.M. 1981. Eliminative materialism and propositional attitudes. *Journal of Philosophy* 78, 67-90.

Churchland, P.S. 1986. *Neurophilosophy: Toward a Unified Science of the Mind/Brain.* Cambridge, Massachusetts: MIT Press.

Cummins, R. 1983. *Psychological Explanation.* Cambridge, Massachusetts: MIT Press.

Darwin, C. 1871/1946. *The Descent of Man and Selection in Relation to Sex.* London: Watts & Company.

_____. 1872/1965. *The Expression of the Emotions in Man and Animals.* Chicago, Illinois: University of Chicago Press.

Davidson, D. 1985. Rational animals. In: *Actions and Events: Perspectives on the Philosophy of Donald Davidson* (ed. by E. LePore & P. McLaughlin), pp. 473-480. London: Basil Blackwell.

Dennett, D.C. 1983. Intentional systems in cognitive ethology: The 'Panglossian paradigm' defended. *Behavioral and Brain Sciences* 6, 343-390.

_____. 1987. *The Intentional Stance.* Cambridge, Massachusetts: MIT Press.

Dretske, F. 1988. *Explaining Behavior: Reasons in a World of Causes.* Cambridge, Massachusetts: MIT Press.

Dupré, J. (ed.) 1987. *The Latest on the Best: Essays on Evolution and Optimality.* Cambridge, Massachusetts: MIT Press.

Fodor, G. 1987. *Psychosemantics: The Problem of Meaning in the Philosophy of Mind.* Cambridge, Massachusetts: MIT Press.

Giere, R.N. 1988. *Explaining Science: A Cognitive Approach.* Chicago, Illinois: University of Chicago Press.

Griffin, D.R. 1976. *The Question of Animal Awareness.* New York: Rockefeller University Press.

_____. 1984. *Animal Thinking.* Cambridge, Massachusetts: Harvard University Press.

_____. 1990. *Animal Thinking.* Revised Edition. Cambridge, Massachusetts: Harvard University Press.

Haraway, D. 1989. *Primate Visions: Gender, Race, and Nature in the World of Modern Science.* New York: Routledge.

Hempel, C.G. & Oppenheim, P. 1948. Studies in the logic of explanation. *Philosophy of Science* 15, 567-579.

Humphreys, P. 1989. *The Chances of Explanation: Causal Explanation in the Social, Medical, and Physical Sciences.* Princeton, New Jersey: Princeton University Press.

Kamil, A.C. 1988. A synthetic approach to the study of animal intelligence. *Nebraska Symposium on Motivation* 1987: 257-308.

Kitcher, P. & Salmon, W. C. (eds.) 1989. *Scientific Explanation.* Minneapolis, Minnesota: University of Minnesota Press.

Miller, R. W. 1987. *Fact and Method: Explanation, Confirmation and Reality in the Natural and the Social Sciences.* Princeton, New Jersey: Princeton University Press.

Mitchell, R.W. & Thompson, N.S. (eds.) 1986. *Deception: Perspectives on Human and Nonhuman Deceit.* Albany, New York: SUNY Press.

Montefiore, A. & Noble, D. (eds.) 1989. *Goals, No-Goals and Own Goals.* London: Unwin Hyman.

Nagel, T. 1974. What is it like to be a bat? *Philosophical Review* 83, 435-450.

Packer, M.J. & Addison, R.B. (eds.). 1989. *Entering the Circle: Hermeneutic Investigation in Psychology.* Albany, New York: SUNY Press.

Pitt, J.C. (ed.). 1988. *Theories of Explanation.* New York: Oxford University Press.

Radner, D. & Radner, M. 1989. *Animal Consciousness.* Buffalo, New York: Prometheus Books.

Regan, T. 1983. *The Case for Animal Rights.* Berkeley, California: University of California Press.

Richards, R.J. 1987. *Darwin and the Emergence of Evolutionary Theories of Mind and Behavior.* Chicago, Illinois: University of Chicago Press.

Robinson, D.N. 1989. *Aristotle's Psychology.* New York: Columbia University Press.

Rollin, B. 1989. *The Unheeded Cry: Animal Consciousness, Animal Pain and Science.* New York: Oxford University Press.

Searle, J.R. 1983. *Intentionality: An Essay in the Philosophy of Mind.* New York: Cambridge University Press.

Singer, P. 1990. *Animal Liberation,* 2nd Edition. New York: New York Review of Books.

Stitch, S. 1983. *From Folk Psychology to Cognitive Science: The Case Against Belief.* Cambridge, Massachusetts: MIT Press.

Interpretation and Explanation in the Study of Animal Behavior

Volume II: Explanation, Evolution, and Adaptation

I
Explanation and Confirmation

Introduction

Animal behavior has a rich and interesting history. Part of that history is reviewed by Richard Burkhardt in his contribution to this section. Perhaps what is most fascinating is that many of the disputes and divisions that currently exist in the field were foreshadowed in the eighteenth and nineteenth centuries. During that period there were differences in opinion about whether studying wild animals or domestic animals provided the most reliable information; about whether the behavior of animals could be explained by appeals to "instinct" or whether their behavior is modifiable and can overcome inherited predispositions; and about the importance of studying living animals versus the role of dissecting prepared specimens. Edmund Selous emerges from this review as an especially interesting, if irascible, character. He called zoologists "murderers" and "thanatologists," and referred to natural history museums as "mausoleums."

There was much discussion during this period about the fundamental differences between humans and nonhumans. Although the influential French scientist, Buffon, claimed to have scorn for this idea, the investigation of animal behavior was seen by many as a branch of natural theology. The remarkable fit between a species' needs and its abilities was seen as evidence of the perfection of God's creation.

Richard Levins & Richard Lewontin (1985) have argued that contemporary adaptationist views are related to these older theological doctrines. In association with Stephen Jay Gould they have claimed that biologists often see teleology where there is randomness, and order where there is drift. These themes are taken up in the chapters by John Byers & Marc Bekoff, Randy Thornhill, and Sandra Mitchell.

Byers & Bekoff argue that studies of social evolution often commit the logical fallacy of affirming the consequent. This fallacy involves supposing that a hypothesis is true if it implies true predictions. This is a fallacy because distinct hypotheses can imply the same predictions, and the fact that these predictions turn out to be true provides no support for one of the hypotheses to the exclusion of others. Byers & Bekoff relate this fallacy to "facile adaptationism." All too often predictions are made about what would be optimal behavior, and if an animal's behaviorconforms to these predictions then it is supposed that the animal is "optimally designed."

While clearly within the adaptationist paradigm, broadly construed, Thornhill does not commit simple fallacies. Indeed he identifies one fallacy sometimes committed by adaptationists: that of inferring evolutionary function on the basis of strength of selection on an adaptation in a current environment. Thornhill's chapter is a careful, sober discussion of how to study adaptation.

Mitchell is not sympathetic to the adaptationist program, especially when it is applied to animals that have the capacity for cultural learning. She focuses on problems with individuating evolutionarily significant behaviors and in recognizing similarities across species. As her target she takes sociobiological explanations of rape. Her point is that in order to explain human rape on the basis of the behavior of other animals, such as scorpionflies, we must be certain that the same behavior has been identified. But since human rape is "essentially intentional" there is no reason to believe that the same behavior is being compared in these explanations. For this reason she finds sociobiological explanations of such human behaviors as rape unconvincing.

Mitchell's concerns are reminiscent of those of Alexander Rosenberg (section II of Volume I) and Sarah Stebbins (section III of Volume I). Rosenberg denies that there can be an evolutionary biology of play on the ground that play is an intentional concept and there can be no evolutionary theory of intentionality. Stebbins believes that other animals could only be doing what we do when we talk if the behaviors are homologous. Mitchell isn't committed to Rosenberg's views about evolutionary theory but, like him, she is bothered by the extrapolation across biological categories of behaviors that are intentional. For Mitchell, like Stebbins, a trait's "causal history" is important when trying to identify similar behaviors in different species.

Thomas Wynn's chapter concerns tool-use, and some of the same themes explored by Burkhardt and Mitchell are also discussed here. For many years it was believed that tool-use served to distinguish humans from nonhumans. Like so many other supposed differences between humans and nonhumans, this one too has fallen by the wayside. However Wynn points out that only humans seem to imbue tools with cultural significance that manifests itself in nonfunctional design choices. We may wonder whether this supposed difference will go the way of the others. Is this apparent difference a feature of nature or a feature of our ignorance? If this apparent difference is a real difference, we may wonder whether it is true after all to say that nonhumans use *tools*.

The essays in this section open up many questions about explanation and confirmation. They also make clear that this is an area in which much more work needs to be done. Although there are important exceptions, philosophers of science have traditionally focused on confirmation and explanation in the physical sciences. It is plain that the life sciences in general and the study of behavior in particular confront us with different problems than the physical sciences, and also poses different questions. These chapters also make clear that students of behavior, if they are to avoid simple fallacies and unthinking allegiance to ideologies such as adaptationism, are going to have to bring greater reflection to bear on the logic of their science.

LITERATURE CITED

Levins, R., & Lewontin, R. 1985. *The Dialectical Biologist.* Cambridge, Massachusetts: Harvard University Press.

1. Theory and Practice in Naturalistic Studies of Behavior Prior to Ethology's Establishment as a Scientific Discipline

Richard W. Burkhardt, Jr.

The sustained scientific study of the behavior of animals, and especially that of animals in the wild, is a relatively recent phenomenon. It has been only in the twentieth century, and then primarily in the period since the Second World War, that a significant number of biological researchers have pursued field studies of animal behavior as their primary scientific activity. The conceptual, methodological, institutional, and professional dimensions of this modern development all deserve careful scrutiny.

It is also the case, however, that as early as the eighteenth century a number of prominent naturalists underlined the importance of studying the behavior of free-ranging, wild animals - as opposed to that of domesticated animals or animals in captivity. In 1771, for example, the famous French naturalist Georges Louis Leclerc, Comte de Buffon, observed that neither the remains of animals preserved in museums nor the live animals held captive in menageries could provide anything but an imperfect view of Nature. Museums, he said, displayed a Nature that was "dead, inanimate, and superficial." Menageries contained animals whose behavior was "altered, constrained, and scarcely worthy of the consideration of a philosopher." The only Nature worthy of philosophical contemplation, Buffon said, was a Nature that was "free, independent, and -if you will - wild" (Buffon 1771: 3-4). Similar views had recently been expressed by Hermann Samuel Reimarus and Charles-Georges Le Roy, and the same theme was to recur explicitly or implicitly in the writings of a sizable number of naturalists from the end of the eighteenth century to the beginning of the twentieth, including Gilbert White, John Blackwall, Charles Darwin, J.-H. Fabre, Bernard Altum, George and Elizabeth Peckham, Charles Otis Whitman, Edmund Selous, and Julian Huxley - to name only a few (see for example

Reimarus 1760; Le Roy 1802; Blackwall 1834; Darwin 1859; Fabre 1879-1907; Altum 1868; Peckham & Peckham 1898; Whitman 1898; Selous 1905b; Huxley 1914).

Among the issues that these investigators confronted are many that continue to attract the attention of scientists and philosophers today - as evidenced by the present volume. These include the topics of whether animals have feelings, consciousness, language, or the capacity to choose; how bird song develops; the adaptiveness of behavior; and the role of sexual selection in organic change. However, the historical interest of the work of these earlier researchers is not to be measured solely in terms of the extent to which their concerns correspond to those of present-day biologists and philosophers. If we are to understand the aims and methods of these earlier investigators of animal behavior, we must reconstruct the contexts in which these investigators themselves conceived and conducted their researches. By considering how the questions addressed and practices adopted by the early students of behavior reflected and were adapted to their respective scientific, institutional, social, and local settings, we may begin to appreciate the different contexts that made the study of behavior meaningful to them and to gain at least some sense of why the study of animal behavior did not emerge as a science in its own right until well into the twentieth century.

The present survey must necessarily be selective in order to be of a suitable length for the present volume. The period covered will be from the mid-eighteenth century to the first quarter of the twentieth century - stopping short, in other words, of the 1930s, when Konrad Lorenz and Nikolaas Tinbergen began constructing the conceptual and methodological foundations on which the discipline of ethology was erected.

The primary motives for studying animal behavior in the eighteenth century were essentially the same as those for studying natural history and science in general. The glory of God and the relief of man's estate were the two main justifications for science offered by Francis Bacon, John Ray, and other promoters of science in the seventeenth century, and in the eighteenth century these themes were voiced with even greater confidence (Ray 1691; Bacon 1905). Within such a framework, the study of the habits and manners of animals fit quite comfortably. While the remarkable correlation between the instincts and needs of the respective animal species was hailed as compelling evidence of God's care in designing the creation, a knowledge of the habits and ruses of animals was seen as essential to humankind's exploitation of useful creatures and avoidance of the ravages of harmful creatures.

Field work was essential, of course, to the naturalist's enterprise. If the final assessments of how individual species should be classified were characteristically made in urban science centers (where the major museums and herbarium collections were assembled), the source of the specimens for these collections was necessarily the field, and the field remained the place where many naturalists made their reputations. Not only were the reports and collections from voyagers to distant lands greeted with enthusiasm in the eighteenth century, less strenuous one-day excursions into the field also achieved a certain vogue. In Uppsala, the citizens were treated to the ceremony of Linnaeus and his students, bedecked in special botanical uniforms, marching out from town early in the morning for a day in the field and then marching back in the evening, their return announced by the triumphal sound of French horns (Lindroth 1983). In France, excursions to observe plants in their natural habitats - "*herborisations*" - were promoted in Rousseau's *Lettres elementaires sur la botanique* as a way of providing the body with exercise while giving the soul the opportunity to commune with nature.

It remained the case, however, that gathering detailed information on the behavior of animals in the wild was quite different from observing plants in their natural surroundings. Travelers engaged in a general reconnaissance of the unfamiliar plants, animals, and peoples of a given region were unlikely to have the time or occasion to learn a great deal about the habits of individual species, and wild creatures were by nature disinclined to reveal themselves to groups of nature-enthusiasts invading the countryside on popular outings. Furthermore, the major systematic treatises of the period had little room for descriptions of animal behavior. Linnaeus's goal in his *Systema naturae* was to catalogue the whole of the natural world, providing a name, classification, and *brief* description for every plant, animal, and mineral species known to humankind.

In contrast to Linnaeus' *Systema naturae*, Buffon's *Histoire naturelle, générale et particulière* (1749-67) eschewed systematics and promised to provide detailed "histories" of the different animals, starting with the mammals. Buffon's goal was to describe, at least in general terms, the natural behavior of each mammal species. Rather than recounting anecdotes about the special feats of individual animals, he limited himself, with but few exceptions, to describing species-typical behavior. Though necessarily forced to rely on the accounts of others, he prided himself on his ability to recognize which of these accounts to trust (Burkhardt 1990a).

Despite Buffon's fine statement about wild Nature being the only Nature worthy of the philosopher's contemplation, however, his *Histoire naturelle*, both in its general structure and its specific details, was a continuous testimony to man's dominion over the rest of the animate world. In treating the different animals of the animal kingdom, Buffon proceeded according to what the historian Jacques Roger has called "an order of decreasing dignity" (Roger 1963: 531). After first treating man, Buffon turned next to the horse, as "man's greatest conquest." He then moved on to consider the other mammals that man had succeeded in domesticating. Not until he was done with them did he take up the animals of the wild.

Throughout the *Histoire naturelle*, Buffon maintained that "an essential and infinite distance" (1753: 30) separated the human species from the rest of the animal kingdom. Animals, in his view, were purely material beings, moved by wholly mechanical causes which depended in each case on the animal's organization. He readily acknowledged that animals had feelings - he was indeed prepared to credit them with such feelings as pleasure and pain, fear and anger, and love and jealousy, as well as the ability to learn - but he insisted that animals lacked souls, and that they were therefore, unlike humans, incapable of thinking and reflection. However, it was not a metaphysical argument so much as an argument involving actual historical achievement that formed the basis of his distinction between humans and animals. Where Descartes (1637/1963: 629) had maintained that the essential difference between man and animal could be seen in the inability of animals to communicate with their own kind by words or signs, Buffon maintained that what set man apart from the other animals was that he represented the only species that had brought other species under its dominion by domesticating them.

Pushing this claim further, Buffon (1760: 282) observed that "In proportion as man has risen above the state of nature, the other animals have sunk below it." This, he believed, was particularly evident in the case of the beaver. Beaver societies, he maintained, were not comparable to insect societies. While the latter arose simply as the result of physical necessity, beaver societies were the result of "a kind of choice" (Buffon 1760: 285) on the part of the beavers. Acknowledging that this implied that beavers possessed "a glimmer of intelligence," Buffon was careful to assert that this intelligence was "very different in principle from that of man." (1760: 285) He suggested, nonetheless, that the achievements of beavers in constructing their dams and lodges could be properly compared to the social accomplishments of human savages. His additional claim was that beavers refrained from constructing their common works in areas where human society was a threat to them. Where civilized man was present, beavers lived solitary, asocial lives. For Buffon, the beaver was living testimony to how the actions of man had caused other species to decline in capabilities from their ancient, natural state (Buffon 1760).

Buffon was eager to insist that studying animal nature was essential for learning what it was that made man specifically human. Significantly enough, however, he found fault with those of his contemporaries who believed that in the smallest details of nature they could find evidence of God's handiwork. Disdainful of things small, unimpressed by the natural theologians' argument from design, and reluctant as well to credit the achievements of a rival naturalist, Buffon had little good to say about Réaumur's remarkable studies of the lives of insects (Réaumur 1734-42). Condescending only to acknowledge that the detailed description of insect generation, multiplication, and metamorphosis might occupy the *leisure* of a naturalist, Buffon made clear his own opinion that serious work in natural history involved the establishment of *broad* views of nature operations, not the assembly of minute details about minute organisms. To the idea of God promoted in the writings of Réaumur and others he objected:

> Who, in fact, has the greatest idea of the Supreme Being, he who envisions him creating the universe, setting in order the forms of life, founding nature on invariable and perpetual laws; or he who looks for him and wants to find him intent upon managing a republic of flies, and very much occupied with the way in which a beetle must fold its wing? (Buffon 1753: 95)

Despite Buffon's scorn for natural theology, the seeking of God's handiwork in nature continued to play a central role in eighteenth-century natural history. It lay behind what was arguably the most perceptive work on animal behavior published in the eighteenth century, that of Hermann Samuel Reimarus.

Professor of Oriental Languages at the Hamburg Academic Gymnasium Reimarus was a pious proponent of rational rather than revealed religion. Having published one treatise in which animal behavior was enlisted as one of several kinds of evidence revealing "The principal Truths of Natural Religion" (1754), he devoted a second, major work specifically to the instincts or drives (*Triebe*) of animals and what these seemed to demonstrate regarding God's relation to humankind (Reimarus 1760 [French translation 1770]; Jaynes & Woodward, 1974a,b).

In the second of these works, Reimarus surveyed the behavior of a wide range of animal species for the purpose of showing how the various instincts of these species were perfectly proportioned to the animals' specific needs. He also surveyed the major authors who had addressed the question of animal behavior before him, identifying the various inadequacies he found in their respective treatments. He explicitly rejected both Descartes' idea that animals were automata and Condillac's idea that animal instincts developed as the result of reason and habit. In his view, instincts or drives were implanted in each species by God at the creation as part of their innate physiological organization. This gave the members of each species the innate competence to perform the particular actions necessary for the species' welfare and conservation. The individual animal performed these actions without prior experience - and without knowledge of their purpose.

Like Buffon, Reimarus found it necessary to base his analysis of animal behavior on facts that had been related by others. His avowed strategy was to accept the reports only of those authors who were known for the exactitude of their observations and who specified the conditions under which these observations were made. He placed much more credit in Réaumur's observations than in Buffon's. By his own account, he could not help laughing when he read Buffon's assertions that it was not a matter of natural necessity that caused beavers to assemble and work together, but rather a matter of choice and convenience, and that beavers were solitary and inactive in areas inhabited by man. Beavers, Reimarus maintained, continued to live and work in society, constructing their lodges even in the vicinity of man, because this corresponded to their specific, natural needs (Reimarus 1770: 1, 198-199).

Not only Reimarus but also a number of Buffon's own countrymen concluded that Buffon's authority was not incontestable in the area of animal behavior. The *philosophe*

Condillac, for example, took quick exception to Buffon's general view of animals as unthinking, Cartesian automata. Condillac maintained that the animal mind was formed through principles of sensation and association in basically the same way as the human mind. His method in his *Traité des animaux,* in fact, was to observe the faculties of humans and then to judge of the faculties of animals by analogy. He himself was not a close observer of animals. The same cannot be said, however, of his disciple, Charles-Georges Le Roy. Le Roy was at one and the same time a natural philosopher and a hunter, holding the official position of keeper of the king's game at Versailles and Marly. The idea that animals are unthinking machines was inconsistent with his knowledge of the behavior of wild game. Animals, he was convinced, were able to respond intelligently to novel circumstances. He wrote:

> I would like, in order to have the complete history of an animal, that after having rendered an account of its essential character, its natural appetites, its way of life, etc., one should seek to observe it in all the circumstances which could put obstacles in the way of the satisfaction of its needs: circumstances the variety of which would interrupt the ordinary uniformity of its conduct, and force it to invent new means (Le Roy 1768/1802: 7).

Choosing common, wild animals as his subjects - animals such as deer, hares, and wolves - Le Roy argued that within the limits established by their respective constitutions, these animals had the capacity to feel, to compare, to judge, to reflect, and to choose. He maintained that if there were such a thing as instinct - "a principle which directs the beasts in their actions" - it was something which was developed and perfected through experience.

Le Roy's views on instinct were thus just the opposite of Reimarus'. Reimarus understood instincts to be innate, the products of the animals' God-given natural forces, and not perfectible through experience. Le Roy, on the other hand, regarded instincts as modifiable and perfectible. That no clear choice was made in the eighteenth century between the views of Le Roy and those of Reimarus is strikingly demonstrated by the article on "Instinct" that appeared in the second edition of Diderot's famous *Encyclopédie*. There the views of the two men were simply presented one after the other, with no acknowledgment that they were even contradictory (Burkhardt 1990b).

Interestingly enough, however, if Le Roy and Reimarus offered markedly different interpretations of animal behavior, they provided very similar suggestions regarding how animal behavior ought to be studied. The major methodological message of each was that if one wished to understand animal behavior, one had to study animals in the wild. Reimarus, who noted that savants living in towns were less likely to know certain animals than were hunters, voyagers, birdcatchers, and fisherman, wrote:

> The histories of tame or locked up animals offer only doubts and uncertainties. One judges badly of particular events when one cannot know exactly the circumstances of them. It is not on the basis of the actions of individuals that have lost their natural liberty that one can judge of their true instincts; in leading a kind of life that is not natural to them, these instincts vary, degenerate, or die out entirely. It seems to me much more certain to speak only about animals that enjoy their natural state of liberty; discomfort and constraint have no influence on their actions (Reimarus 1770: xvii-xviii).

Le Roy represented himself as "a dedicated hunter" who had "taken his course of philosophy in the woods." He urged that the naturalist, after conducting the proper anatomical observations on animals, "should give up his scalpel, abandon his cabinet, and plunge into the woods to study the ways of these conscious beings [*êres sentans*] ..." (Leroy 1768/1802: xvi, 4-5).

These exhortations notwithstanding, however, the practice of the majority of naturalists of the late eighteenth century remained a practice centered on naming and classifying plant and animal species, and it was an activity conducted primarily indoors. In Paris in the 1790s, for example, the professors at the new Muséum d'Histoire Naturelle (transformed from the Jardin du Roi during the French Revolution) had at their disposal the richest natural history collections in the world. B.-G.-E. Lacépède, J.-B. Lamarck, Etienne Geoffroy Saint-Hilare, and Georges Cuvier all made their names by identifying, describing, and classifying the materials available to them. The individual at the Museum who contributed most to the study of animal behavior was Georges Cuvier's younger brother, Frédéric Cuvier, who in 1804 was given the junior, non-professorial position of keeper of the museum's menagerie. Frédéric Cuvier sought to make the most of his situation by claiming that menageries could be for zoologists what the chemist's laboratory was to the student of inanimate matter: places where one could study what *could* happen in nature, given the proper conditions. (Cuvier 1807: 119). Despite his efforts, however, two decades later Cuvier had to acknowledge that "the science of animal intelligence" was "still only in its infancy," with its basic principles yet to be established (Cuvier 1822: 529). Cuvier intended to write a general work on the origin of animal actions (Cuvier 1823: 241), but he died before completing it.

In Great Britain, in the meantime, where science remained less professionalized than in France, a model of the field naturalist emerged that was to prove of long-standing appeal to British sensibilities. This model was provided by that perceptive chronicler of "*parochial history*," the Reverend Gilbert White, in his *Natural History and Antiquities of Selborne* (1789). Defining himself as "an *out-door naturalist*, one that takes his observations from the subject itself, and not from the writings of others" (White 1789: 115), White was ever attentive to what he called "the life and conversation of animals." While he granted the utility of the efforts of the systematists - "without system the field of Nature would be a pathless wilderness" - he insisted that "system should be subservient to, not the main object of, pursuit." (White 1789: 232) In his words:

> *Faunists*...are too apt to acquiesce in bare descriptions, and a few synonyms: the reason is plain; because all that may be done at home in a man's study, but the investigation of the life and conversation of animals, is a concern of much more trouble and difficulty, and is not to be attained but by the active and inquisitive, and by those that reside much in the country (White 1789: 144).

Contending that "every kingdom, every province, should have its own *monographer*" (White 1789: 132), White became the classic example of the field naturalist who over the course of decades developed an extraordinary familiarity with the natural history of his own district. He came to know the local birds by their songs and mannerisms as well as by their physical appearance. In studying bird song, he carried with him a list of the birds he expected to see as he rode or walked about his business, and he recorded which birds continued to sing and which did not. On the basis of their notes alone, for example, he asserted that his district harbored three distinct species of willow wrens.

White fashioned an approach to the study of animal behavior that was admirably suited to the role of the country clergyman. The point is not simply that White found ample opportunity in the field to reflect on the "methods of Providence." It is that in field natural history, much more than in the related activities of hunting, horticulture, or the studies of the cabinet naturalist, the social opportunities for the clergyman were unexcelled. As explained years later by J. C. Loudon:

> The naturalist is abroad in the fields, investigating the habits and searching out the habitats of birds, insects, or plants, not only invigorating his health, but affording ample opportunity for frequent intercourse with his parishioners. In this way their reciprocal acquaintance is cultivated, and the clergyman at last becomes an adviser and friend, as well as a spiritual teacher (cited by Allen 1978: 23).

White's *Natural History and Antiquities of Selborne* had a major influence on British natural history. The book went through many editions, with its successive editors and annotators including such naturalists as Sir William Jardine, James Rennie, Edward Blyth, Leonard Jenyns, J. G. Wood, Frank Buckland, and J. E. Harting. In an important paper on bird song presented to the Manchester Literary and Philosophical Society in 1822, the naturalist John Blackwall wrote that "[White's] example in investigating nature cannot be too highly recommended" (Blackwall 1824: 291), and he echoed previous complaints about those who studied nature only by studying books and preserved specimens, stating:

> To those whom business or inclination leads to reside chiefly in large towns, such are almost the only means of information that offer themselves; but who, that enjoys the opportunity of observing the free denizens of the fields and woods in their native haunts, would exchange their lively and unrestrained activity, their curious domestic economy, their mysterious migrations, and their wild but delightful melody, for

the fixed glassy eye and the mute tongue of the inanimate forms that are crowded together in melancholy groups in the museum? (Blackwall 1824: 289-290)

Blackwall, a man of sufficient means to be able to enjoy "the advantages of leisure and locality" unencumbered by "the ties of professional business" (Pickard-Cambridge 1881: 145), recommended the careful study of bird song as a means of making fresh discoveries about birds "without placing the observer under the painful necessity of destroying life." Believing "that the calls of birds, which seem to be the simplest expressions of their sensations, are natural, not acquired," he proceeded to test this supposition experimentally, raising baby birds under conditions where they were unable to hear the calls of other birds. Finding that these birds developed the songs of their species independent of any instruction, he argued that bird song in some cases provided ornithologists with a better means of distinguishing species than could be achieved by the examination of specimens. He also maintained that the songs of birds could be properly termed their "language" - a "melodious interchange of thought and feeling; which, though very limited and imperfect, still answers many important purposes, and contributes materially to the happiness and preservation of species." (Blackwall 1824: 293-294)

Among others impressed by White's *Selborne* was Charles Darwin. Darwin recalled in his autobiography that as a youth, after reading White's *Selborne*, he "took much pleasure in watching the habits of birds, and even made notes on the subject," an activity that left him wondering "why every gentleman did not become an ornithologist" (Darwin 1958: 45). In May of 1843, having moved his family from London to Down House in the Kent countryside, Darwin began an "Account of Down." His son Francis later identified this as an attempt on his father's part "to write a natural history diary after the manner of Gilbert White" (Moore 1985: 460).

Whatever Darwin's thoughts may have entertained in 1843 with regard to emulating White as the monographer of Down, his own intellectual aspirations extended much further. By 1843, he had already for half a dozen years been embracing and developing a framework of analysis markedly different from that of his predecessors and contemporaries - a framework that had fundamental consequences for understanding the whole of organic life, behavior included. He had come to regard the behavior of animals as part of the broader question of organic evolution. If he saw the explanation of complex instincts as a serious test for his idea of evolution by natural selection (Richards 1981, 1987), he also saw behavioral evidence as providing a wealth of evidence for demonstrating the continuity between humans and higher animals with respect to their mental faculties. In the case of sexual selection, furthermore, he identified behavior as having a special role to play as an evolutionary mechanism. Throughout his career, Darwin was fascinated with behavioral questions, and his thoughts on behavior and evolution enriched each other in a continuing interaction (Burkhardt 1983, 1985).

Nonetheless, Darwin's incorporation of behavioral phenomena into his concerns did not serve to establish the study of animal behavior as a prominent enterprise in late nineteenth century zoology. In 1890 E. B. Poulton astutely remarked that the reason that Darwin's theory of sexual selection was still a matter of debate was that there were "comparatively few true naturalists - men who would devote much time and the closest study to watching living animals amid their natural surroundings, and who would value a fresh observation more than a beautiful dissection or a rare specimen." (Poulton 1890: 287) It was, Poulton said,

> a very remarkable fact that the great impetus given to biological inquiry by the teachings of Darwin has chiefly manifested itself in the domain of Comparative Anatomy, and especially in that of Embryology, rather than in questions which concern the living animal as a whole and its relations to the organic world. And yet these were the questions in which Darwin himself was principally interested. (Poulton 1890: 286)

Though Poulton's characterization of Darwin's way of looking at things was in some measure self-serving, Poulton's assessment of the fields of zoology that had flourished most in the aftermath of the Darwinian revolution was essentially accurate. Professional zoologists of the period focused for the most part on comparative anatomy and embryology, carrying their research out in museums and laboratories. The study of animal behavior in the wild at this time was promoted not so much by professionals as by amateurs, at least in England, where the amateur tradition remained quite strong, particularly among field ornithologists. As it became increasingly clear to amateur ornithologists that little remained to be accomplished with respect to identifying and classifying the birds of Britain, and as popular sentiment began to stir against the shooting of birds and the over-zealous collecting of bird eggs, a new enthusiasm arose among the amateurs for studying the lives and *minds* of birds in the birds' native haunts. Of the two English amateurs who did the most to promote the scientific watching of birds - Edmund Selous and H. Eliot Howard - it is Selous that we will concentrate upon here.

Edmund Selous (1857-1934) was an ardent Darwinian who insisted upon the importance of field work for resolving major issues in animal behavior and evolutionary theory (Nice 1935; Lack 1959; Simmons 1984; Burkhardt 1990c). Though trained as a barrister, Selous supported himself not by the practice of law but instead by a small private income and by writing articles and books on natural history. An ardent exponent of the preservation of animal life, he laced his writings with diatribes against the zoological establishment of his day, calling zoologists "murderers" or "thanatologists" and referring to their museums as "mausoleums." Irascible and anti-social, he belonged to no scientific societies, preferring the solitary watching of bird life to any participation in scientific assemblies. He described the motivations of his field work as "Joy in all wild life and its surroundings, with another joy in Darwin and a social-shunning disposition - an intellectual love of truth too - (for in any other way I mostly hate her)" (Selous 1927: v).

Selous' painstaking studies of what he called the "domestic habits" of birds - their behavior in courtship, mating, nest-building, rearing young, interacting with conspecifics, and so forth - were unprecedented in their approach and attention to detail. He made it his practice to record his observations on the spot and as soon as the actions he was observing took place, noting precisely the date and time the observations were made. His most significant observations were on display behavior and the role of female choice in sexual selection.

Selous insisted on the importance of making one's own observations instead of relying on the authority of others and on observing not just what was expected of animals but also how they deviated from that expectation. Deviations from the type, he told his readers, were the material upon which organic evolution was based. He was attentive both to the intraspecific variations that made natural and sexual selection possible and to particular behavior patterns which he believed represented behavior in the course of evolutionary change. He hypothesized, for example, that the frenzied motions of birds when sexually excited were the basis from which the courting displays and nest-building habits of birds gradually evolved.

Among Selous' most important papers were his studies of sexual selection in the Ruff (1906-1907) and the Blackcock (1909-1910). In his paper on the Ruff he provided a powerful confirmation of Darwin's view of the importance of female choice in sexual selection, a view which had few proponents among the scientists of the day. He found that the females of this lek-breeding species played an active rather than a passive role in the mating process, that they did indeed choose among the males, and that this choice was not simply a reflection of male "vigor." His subsequent studies on the Blackcock provided further support for the role of female choice in sexual selection. His papers are of interest not only for their observations on animal behavior but also for their comments on the pains of fieldwork on the one hand and the complacency of armchair professionals on the other. Thus as he watched the courtship of the Ruff very early one "bright, but bitterly cold morning," he could not help complaining of the "learned ornithologists all over Europe" who lay "sleeping in their pleasant beds" but who would "come down all the fresher to breakfast" and then from the comfort of their studies issue "bulls against sexual selection" (Selous 1907: 163). Writing on the general difficulties of field studies of bird courtship, he lamented: "All, or at least the greater part - unless, perhaps, if one is rich, as an enthusiast ought to be - is wretchedness, cold, and discomfort; such, upon close acquaintanceship, are the charms of early spring in north temperate Europe" (Selous 1906: 216).

Selous was scheduled to be a major contributor to F. B. Kirkman's four-volume *The British Bird Book* (1910-1913), a work aimed especially at bringing together all the available knowledge on the *habits* of British Birds. The need for such a work, as its editor explained, was great:

> It is not possible at present to give, in the case even of many of our commoner birds, a detailed reliable description of the differences in the nuptial displays that occur at the beginning of the breeding season; yet one has only to turn over the pages of Darwin, Wallace, and their successors to realise how important is the evidence that the ornithologists might bring to the solution of the vexed question of sexual selection (Kirkman 1910: I, iii).

Selous' attacks on the other ornithologists of the day, however, caused him to be dropped from Kirkman's project after contributing to the first volume. If Selous' strong sense of alienation from the scientists and society of his day suited him well to the solitary vigils that his field studies demanded, it also militated against him playing a more central role than he did in the conceptual and methodological reformation of behavioral field studies at the beginning of the twentieth century.

While Selous was an isolated amateur, identifying himself as "as much of a hermit as I am mercifully permitted to be" (Selous 1905b: 114), professional biologists who were well-connected with their scientific communities also had difficulty making animal behavior an integral part of the concerns of their fellow biologists. Two different cases will be mentioned here: that of the American biologist Charles Otis Whitman and that of the British biologist Julian Huxley.

In 1898, when he delivered at the Marine Biological Laboratory (MBL) at Woods Hole his now classic lecture on animal behavior, Charles Otis Whitman (1843-1910) was perhaps the most influential biologist in America. As Director of the MBL, Head of the Department of Zoology at the University of Chicago, and founder of the *Journal of Morphology*, he was in a powerful position to speak to the needs of contemporary biology. Significantly, he believed that biology needed special institutions if it was to flourish as a broadly-conceived, well-organized enterprise. Neither the laboratory nor the field, in his view, was perfectly suited to his approach, which he referred to as "experimental natural history." He called for the establishment of a special experiment station - a "biological farm" - where "the study of life-histories, habits, instincts and intelligence" would be conducted along with "the experimental investigation of heredity, variation, and evolution." As he explained, "the laboratory is too narrow, and the world too wide for the continuous study of living organisms, under conditions that can be definitely known and controlled" (Whitman 1898, 1902; Burkhardt 1988).

Whitman's idea of proper biological research was to take a single species or a group of closely related species and study it in all aspects of its existence. In his 1898 paper on animal behavior he referred to three different kinds of animals that he had studied extensively: the leech, *Clepsine*; the fresh-water salamander, *Necturus*; and a number of doves and pigeons. In the paper, Whitman set forth his views on the proper methods of studying behavior, the nature of instinct, the importance of studying instinct from a phylogenetic standpoint, the means by which behavior has evolved, and the relations - both ontogenetic and phylogenetic - between instinct and intelligence. He addressed himself to questions of evolutionary theory, pointing out among other things that instinctive behavior could not have arisen as a sort of "lapsed intelligence," as the neo-Lamarckians believed. At the same time he chastized those neo-Darwinians who, in believing that instincts could in effect be "rubbed into the ready-made organism," paid insufficient attention to the pre-existing, underlying organization that was necessarily the basis of any organic change. In maintaining that "instinct and structure are to be studied from the common standpoint of phyletic descent" (Whitman 1898: 328) - the statement that modern ethologists are especially fond of quoting - he had in mind not only that instincts could be treated as organs, but also that instincts and organs both had to be traced back to their "more or less remote origin" (Whitman 1898: 308-309).

A lover of animals from his youth - as most ethologists today have also claimed themselves to have been - Whitman was particularly enchanted with pigeons. Beginning in about 1895, after he had already established his scientific reputation through his embryological researches and his founding of biological enterprises, Whitman undertook a monumental study of of pigeons. This study was aimed at reconstructing the evolutionary history of pigeons through a careful study of their variation, heredity, and development, looking in particular at their color patterns, their instinctive behavior, the interfertility of their hybrids, and so forth.

Whitman was not quick, however, to publish the results of his pigeon studies. The secret to Darwin's success, he assured himself, was that Darwin was "a cool, patient, indefatigable investigator, counting not the yeas devoted to preliminary work, but weighing rather the facts collected by his tireless industry, and testing his thoughts and inferences over and over again, until well-assured that they would stand." The trouble with modern students of animal behavior, Whitman caustically observed, was that they were "ambitious to reach the heights of comparative psychology through a few hours of parlor diversion with caged animals, or by a few experiments on domestic animals" (Whitman 1899: 524-525).

Needing a major facility for his pigeon studies, and failing to secure the support to establish a biological farm, Whitman put all his efforts into developing a pigeon facility of his own in and around his home in Chicago. Eventually his collection consisted of some 550 individual birds representing about thirty species. He and his wife cashed in all their life insurance to pay for improvements to the plant, and his wife's wealthy brother contributed $20,000 to the cause. Tragically, Whitman died before his research could be completed (Burkhardt 1988).

Julian Huxley's case is different from Whitman's, for Huxley came to his behavioral work early in his career rather than late (and while he was worrying about how such work corresponded to the lofty scientific goals to which he, as the grandson of T. H. Huxley, Charles Darwin's greatest champion, was aspiring). Years later, in his autobiography, Huxley described his paper of 1914 on the Great Crested Grebe as "a turning point in the study of bird courtship, and indeed of vertebrate ethology in general," and he also credited himself with "having made field natural history scientifically respectable" (Huxley 1970: 79, 83). These claims, however, need to be qualified. Huxley played a significant role in the study of animal behavior, but it was ultimately not as great a one as he might have played, nor was it as decisive a one as his autobiography indicates (Burkhardt 1990d).

Huxley's early love for bird-watching and his concern with the problems of Darwinian evolution led him to produce some excellent field studies of bird behavior early in his career. His first bird behavior paper, on the courtship of the redshank (1912), resulted from observations he made while on a reading holiday in Wales, where his primary scientific intention had evidently been to master Butschli's massive work on protozoology. His conclusion from watching the mating behavior of the redshank was that the action of the birds leading up to pairing were "explicable only on the Darwinian theory of Sexual Selection, or on some modification of that theory." There was obvious display on the part of the male, he claimed, and "an equally marked power of choice" on the part of the female (Huxley 1912: 651, 654). Two years later Huxley published his more famous paper on the great crested grebe paper, which is especially noteworthy for Huxley's accounts of the "nuptial displays" of these birds and for his development of the idea of "mutual selection." Also deserving of note, however, is the fact that Huxley offered this paper to his readers as an example of what a clever and enjoyable time one could have with "a spare fortnight in the spring." In other words, he was not staking his professional reputation on it. He seems to have recognized from the outset that field studies did not represent the most prestigious area in which to make a contribution to biology. At the same time he worried that he did not have the ability to do good research in physiology (Burkhardt 1990d).

Huxley seems to have recognized soon enough that in addition to not having the prestige of laboratory work, field studies of behavior did not represent an area in which novel and major contributions could be made *in a hurry*. Selous, who had chosen to devote his personal resources to watching birds almost full time, spent hour after hour of patient watching, often seeing nothing. Huxley, with his professional aspirations and responsibilities, did not feel he could afford to do the same. Huxley's major papers on bird courtship, written over the period 1911-1925, were the product of a total of at most forty to fifty days of field work - and partial days at that - snatched from his vacation time. He did serve to bridge the gap, however, between the amateurs Selous and Howard at the beginning of the century and the scientific birdwatchers of the 1930s (Lack 1959). He made the observations and ideas of writers Selous and Howard more available to the scientific community by presenting them in a coherent biological framework and publishing them in accessible places.

If in terms of his field methods, his observations, and his general theoretical framework, Huxley's behavioral work was largely continuous with that of Selous and Howard, he was still capable of much that Selous and Howard were not. He had both the training and breadth of perspective necessary to place behavior studies squarely in a broad biological context. In 1925, believing that the time had come to gather the data from "field observation, animal psychology & behavior, genetics, & comparative psychology ... [and consider] the problem [of behavior] from a truly broad & unitary biological standpoint," he started to sketch out the manuscript for a book on bird courtship (Burkhardt 1990d). Had he completed this work, it would have been a singular contribution to the study of animal behavior. Neither Charles Otis Whitman nor Oskar Heinroth, the two other professionals interested in bird behavior in the early years of the century, was a *field* zoologist, however much each of them stressed the importance of observing animal behavior under natural rather than laboratory conditions. Furthermore, while Whitman died without ever consolidating his unrivalled knowledge of the behavior of pigeons, Heinroth was disinclined to be a generalist. Huxley, however, failed to carry through on the bird courtship book. In addition, in the broad, synthetic book that he eventually did write, *Evolution, the Modern Synthesis,* he neither made behavior part of the synthesis nor offered guidelines regarding how that might be accomplished. Though Huxley was a leading spokesmen for field studies of animal behavior in the early twentieth century, and though he was enthusiastic in his support of other ornithologists interested in behavior, his own career could not serve as a wholly reassuring model for someone interested in taking up a career as a field biologist.

CONCLUSION

From the eighteenth century to the twentieth, an appreciable number of amateur and professional naturalists and biologists advocated studying the naturally occurring behavior of animals in the wild. Motivated initially by utilitarian and natural theological concerns, as well as by an interest in assessing the similarities and differences between humans and the rest of the animal kingdom, these investigators treated animal behavior as an important part of natural history. Additional incentives for behavior study were provided when Charles Darwin argued that behavioral evidence could be highly instructive with regard to both the mechanisms and the historical course of organic evolution. In the decades after Darwin's death, various scientists brought behavioral evidence to bear on such issues as the role of sexual selection or the inheritance of acquired characters in the evolutionary process, while others, most notably Charles Otis Whitman and Oskar Heinroth, found instinctive behavior patterns to be characters of special value in reconstructing phylogenies.

Nonetheless, in the early years of the twentieth century, the study of the naturally-occurring behavior of animals - particularly as it occurred in the wild - did not enjoy a secure place amongst the research concerns of professional zoologists and biologists. While physiologically-oriented studies - studies of tropisms, sensory capabilities, and motor coordination - thrived in the new laboratories that were reshaping academic biology, field studies in contrast tended to appear old-fashioned, inconclusive, and professionally unpromising. A young biologist seeking to establish a scientific career at this time was better advised to engage in the kind of experimental work that produced tangible results than to give himself or herself up to those long periods of "watching and wondering" that ethologists have deemed are essential for understanding the behavior of animals in the wild.

At the beginning of the century, Edmund Selous urged that "The habits of animals are really as scientific as their anatomies, and professors of them, when once made, would be as good as their brothers" (Selous 1905a: 49-50). Animal behavior studies have since become professionalized, and the awarding of the Nobel Prize in Physiology or Medicine for 1973 to the founders of ethology can perhaps be interpreted as the ultimate realization of Selous' words. How the biological study of animal behavior achieved disciplinary status in the twentieth century, however, is a story that remains to be told (though see Thorpe 1979; Burkhardt 1981, 1983; Durant 1981, 1986).

A full historical analysis of the development of animal behavior studies in the twentieth century will require attention not only to the theories and practices of twentieth-century investigators but also to the institutional, social, and technological factors that have been a part of that development. The expansion of university biology and psychology departments and the creation of new, specialized research institutes have provided career opportunities for professional students of animal behavior where no such positions existed before. Specialized societies and journals have given an institutional focus to animal behavior studies that they lacked previously, even if the breadth of questions encompassed by ethology and its sister disciplines is such that coherence for the field as a whole remains more of an ideal than a fact of existence. New techniques and technologies, from bird-banding and other means of marking or identifying individual animals to the use of cinematography and computer analysis, have made possible behavioral studies that were impractical or even inconceivable at the beginning of the century. New theoretical formulations, such as those of kin selection and optimality theory, have, together with a revived interest in sexual selection, sent more investigators into the field.

The special value of looking at the aims, methods, and historical situations of early investigators of behavior, from Buffon and others in the eighteenth century to Selous and Whitman and Huxley at the beginning of the twentieth, is not only that it alerts us to long-standing issues in the study of behavior. It is instructive, to be sure, to discover that the debate over what is learned and what is innate in animal behavior long predates Lorenz's concern with that subject, or that the question of animal choice figured in discussions of animal behavior well before Darwin raised the issue by identifying "female choice" as one of the two bases of sexual selection. It is furthermore highly important to recognize how a concern with animal mind, common through the nineteenth century and still basic to the formulations of a Whitman or a Huxley at the beginning of the twentieth century, was studiously avoided in Tinbergen's programmatic writings as he sought to define his and Lorenz's approach as an "objectivistic Ethology ... applying physiological methods to the objects of animal Psychology" (Tinbergen 1942: 40). It is similarly interesting to see that many of the early proponents of field studies of behavior - among them Gilbert White, John Blackwall, Edmund Selous, and Julian Huxley - found it important to call for the preservation rather than the destruction of animal life. I would like to suggest, however, that the historical significance of these early students of animal behavior is to be found not so much in the extent to which their views correspond to (or depart from) the views of biologists of the present, but rather in how their own cases, taken in their respective historical contexts, provide us with a rich set of examples of the adaptive behavior of investigators engaged in the ongoing process of evaluating, reevaluating, and negotiating what is scientifically interesting, socially important, and institutionally or methodologically feasible.

LITERATURE CITED

Allen, D.C. 1978. *The Naturalist in Britain. A Social History.* London: Penguin.

Altum, B. 1968. *Der Vogel und sein Leben.* Munster: Wilhelm Riemann.

Bacon, F. 1905. *The Works of Francis Bacon.* 2 volumes. London: G. Routledge & Sons.

Blackwall, J. 1824. Observations on the notes of birds, including an enquiry whether or not they are instinctive. *Memoirs of the Literary and Philosophical Society of Manchester* (second series) 4, 289-323.

_____. 1834. *Researches in Zoology, Illustrative of the Manners and Economy of Animals.* London: Simpkin and Marshall.

Buffon, G.-L.L., Comte de. 1749-1767. *Histoire naturelle, générale et particulière,* 15 volumes. Paris: Imprimérie royale.

_____. 1753. Discours sur la nature des animaux. *Histoire naturelle, générale et particulière* 4, 3-110.

_____. 1760. Le Castor. *Histoire naturelle, générale et particulière* 8, 282-306.

_____. 1770-1783. *Histoire naturelle des oiseaux,* 7 volumes. Paris: Imprimérie royale.

_____. 1771. L'Outarde. *Histoire naturelle des oiseaux* Volume 2, 1-39.

Burkhardt, R.W., Jr. 1981. On the emergence of ethology as a scientific discipline. *Conspectus of History* 1 (no. 7), 62-81.

_____. 1983. The development of an evolutionary ethology. In: *Evolution from Molecules to Men* (Ed. by. D.S. Bendall), pp. 429-444. Cambridge: Cambridge University Press.

_____. 1985. Darwin on animal behavior and evolution. In: *The Darwinian Heritage* (ed. by D. Kohn), pp. 435-481. Princeton, New Jersey: Princeton University Press.

_____. 1988. Charles Otis Whitman, Wallace Craig, and the biological study of behavior in America, 1898-1925. In: *The American Development of Biology* (ed. by R. Rainger, K. Benson & J. Maienschein), pp. 185-218. Philadelphia, Pennsylvania: University of Pennsylvania Press.

_____. 1990a. Le comportement animal et l'idéologie de domestication chez Buffon et les éthologistes modernes. In press.

_____. 1990b. Animal behavior and organic mutability in the age of Lamarck. In press.

_____. 1990c. Edmund Selous. *Dictionary of Scientific Biography.* Supplement II. In press.

_____. 1990d. Julian Huxley and the rise of ethology. In press.

Cuvier, F. 1807. Du rut. *Annales du Muséum d'Histoire Naturelle* 9, 118-130.

_____. 1822. Instinct. *Dictionnaire des Sciences Naturelles* 23, 528-544.

_____. 1823. Examen de quelques observations de M. Dugald-Stewart, qui tendent à détruire l'analogie des phénomènes de l'instinct avec ceux de l'habitude. *Mémoires du Muséum d'Histoire Naturelle* 10, 241-260.

Darwin, C. 1859. *On the Origin of Species by Means of Natural Selection.* London: Murray.

_____. 1958. *The Autobiography of Charles Darwin 1809-1882, with the Original Omissions Restored* (ed. by N. Barlow). London: Collins.

Descartes, R. 1637/1963. *Oeuvres philosophiques* Volume 1 (ed. by F. Alquie). Paris: Garnier Frères.

Durant, J. 1981. Innate character in animals and man: a perspective on the origins of ethology. In: *Biology, Medicine and Society, 1840-1940* (ed. by C. Webster), pp. 157-192. Cambridge: Cambridge University Press.

Durant, J. 1986. The making of ethology: the Association for the Study of Animal Behaviour, 1936-1986. *Animal Behavior* 34, 1601-1616.

Fabre, J.H. 1879-1907. *Souvenirs entomologiques, études sur l'instinct et les moeurs des insectes.* 10 vol. Paris: C. Delagrave.

Huxley, J. 1912. A first account of the courtship of the redshank (*Totanus calidris* L.). *Proceedings of the Zoological Society of London,* 647-655.

_____. 1914. The courtship habits of the great crested grebe (*Podiceps cristatus*); with an addition to the theory of sexual selection. *Proceedings of the Zoological Society of London,* 491-562.

_____. 1970. *Memories I.* London: Allen and Unwin.

Jaynes, J. & Woodward, W. 1974a. In the shadow of the Enlightenment: I. Reimarus against the Epicureans. *Journal of the History of the Behavioral Sciences* 10, 3-15.

_____. 1974b. In the shadow of the Enlightenment: II. Reimarus and his theory of drives. *Journal of the History of the Behavioral Sciences* 10, 144-159.

Kirkman, F.B. (ed.) 1910-1913. *The British Bird Book* 4 Volumes. London: T. C. E. C. Jacks.

Le Roy, C.-G. 1768/1802. *Lettres philosophiques sur l'intelligence et la perfectibilité des animaux, avec quelques lettres sur l'homme.* Paris: Bossange, Masson & Besson.

Lack, D. 1959. Some British pioneers in ornithological research, 1859-1939. *Ibis* 101, 71-81.

Lindroth, S. 1983. The two faces of Linnaeus. In: *Linnaeus: The Man and His Work* (ed. by T. Frangsmyr), pp. 1-62. Berkeley, California: University of California Press.

Moore, J.R. 1985. Darwin of Down: the evolutionist as squarson-naturalist. In: *The Darwinian Heritage* (ed. by D. Kohn), pp. 435-481. Princeton, New Jersey: Princeton University Press.

Nice, M.M. 1935. Edmund Selous - an appreciation. *Bird-Banding* 6, 90-96.

Peckham, G.W. & Peckham, E.G. 1898. *On the Instincts and Habits of the Solitary Wasps.* Madison, Wisconsin: Wisconsin Geological and Natural History Survey.

Pickard-Cambridge, O. 1881. John Blackwall, F.L.S. *The Entomologist* 14, 145-150.

Poulton, E.B. 1890. *The Coulours of Animals, Their Meaning and Use, Especially Considered in the Case of Insects.* New York: D. Appleton and Company.

Réaumur, R.A.F. de. 1734-1742. *Mémoires pour servir à l'histoire des insectes.* 6 vols. Paris: Imprimérie royale.

Reimarus, H.S. 1760. *Allgemeine Betrachtungen über die Triebe der Thiere, hauptsächlich über ihre Kunsttriebe, zur Erkenntnis des Zusammenshanges zwischen dem Schöpfer und uns selbst.* Hamburg: J. C. Bohn.

_____. 1770. *Observations physiques et morales sur l'instinct des animaux, leur industrie & leurs moeurs.* Amsterdam: D. J. Changuion. (Translation of the 1762 German edition)

Richards, R.J. 1981. Instinct and intelligence in British natural theology: some contributions to Darwin's theory of the evolution of behavior. *Journal of the History of Biology* 14, 193-230.

_____. 1987. *Darwin and the Emergence of Evolutionary Theories of Mind and Behavior.* Chicago, Illinois: University of Chicago Press.

Roger, J. 1963. *Les sciences de la vie dans la pensée française du xviii siècle.* Paris: Armand Colin.

Selous, E. 1899. An observational diary of the habits of nightjars (*Caprimulgus europaeus*), mostly of a sitting pair. Notes taken at time and on spot. *The Zoologist* 3, 388-402, 486-505.

_____. 1901. *Bird Watching.* London: J. M. Dent.

_____. 1905a. *Bird Life Glimpses.* London: George Allen.

_____. 1905b. *The Bird-Watcher in the Shetlands: with Some Notes on Seals - and Digressions.* London: J. M. Dent.

_____. 1906-1907. Observations tending to throw light on the question of sexual selection in birds, including a day-to-day diary on the breeding habits of the Ruff (*Machetes pugnax*). *The Zoologist* 10, 209-219, 285-294, 419-428; 11, 60-65, 161-182, 367-381.

_____. 1909-1910. An observational diary on the nuptial habits of the Blackcock (*Tetrao tetrix*) in Scandinavia and England. *The Zoologist* 13, 401-413; 14, 23-29, 51-56, 176-182, 248-265.

_____. 1927. *Realities of Bird Life, Being Extracts from the Diaries of a Life-Loving Naturalist.* London: Constable.

Simmons, K.E.L. 1984. Edmund Selous (1857-1934): fragments for a biography. *Ibis* 126, 595-596.

White, G. 1789. *The Natural History and Antiquities of Selborne.* London: Bensley.

Whitman, C.O. 1898. Animal behavior. *Biological Lectures, Woods Hole 1898* (1899), 285-338.

_____. 1899. Myths in animal psychology. *The Monist* 9, 524-537.

_____. 1902. A biological farm, for the experimental investigation of heredity, variation and evolution and for the study of life-histories, habits, instincts and intelligence. *Biological Bulletin* 3, 214-224.

2. The Study of Adaptation

Randy Thornhill

The central problem in evolutionary biology is to elucidate the long-term process of evolution because it is this process that has produced the diversity of life. Thus it is essential to distinguish procedures that can provide direct evidence about the workings of long-term evolution from those that cannot. This essay treats teleonomy, the study of the purposeful or functional design of living systems and the directional selection pressures that have designed adaptations during long-term evolution. The study of adaptation is of fundamental importance in evolutionary biology because adaptations are information about how the long-term evolutionary process actually works; adaptations are the long-term consequences of evolution by selection and thus understanding the functional design of an adaptation is synonymous with understanding how evolution by directional selection worked over the frame of geological time to produce the adaptation. Some popular methods of analyzing adaptations cannot in themselves provide evidence useful for testing hypotheses about long-term evolution. These methods of analyzing adaptations provide information about the microevolutionary process and the action of current selection, and they yield hypotheses about long-term evolution, but when used alone they cannot elucidate the nature of long-term evolution.

THE GENERAL METHOD OF TELEONOMY

The scientific study of adaptation has been called teleonomy (Pittendrigh 1958; Williams 1966) and the adaptationist program (Gould & Lewontin 1979; Symons in press). Regardless of the name attached to the study of adaptation, the field is based on the explicit recognition, elucidation and analysis of functional design in living systems. It is the complexity of functional organization of living systems and the apparent purposiveness and goal-directed nature of adaptations that attracts the attention of teleonomists. Evolutionists typically view an adaptation as any feature of an organism that performs a function or purpose "with sufficient precision, economy, efficiency, etc. to rule out pure chance as an adequate explanation" (Williams 1966: 10; also see Darwin 1859, 1874; Curio 1973; Alexander 1979; Dawkins 1982, 1986; Burian 1983; Mayr 1983; Symons 1989, in press). Natural and sexual selection are the only scientific explanations available for the initial production and maintenance of purposeful phenotypic features, that is, adaptation. The teleonomist thus assumes that the phenotypic design that he is interested in is the long-term consequence of some type of directional selection. The hypotheses the teleonomist generates to explore the phenotypic design pertain to the nature of directional selection that may have produced it.

Teleonomic studies do not typically include analysis of the evolutionary origin of adaptations - that is, the phenotypic precursors that were modified by directional selection over long-term evolution into complex features with identifiable purposeful designs. Instead teleonomy focuses on the understanding of phenotypic design and thus the forces of selection responsible for adaptation. Both the origin and selection history of an adaptation are important to understand, but origin and selective history are different questions; the two types of questions deal with different historical causes. Many teleonomists recognize that the evolutionary purpose/function of an adaptation can be studied productively without any reference to or understanding of the adaptation's origin. However, Sherman's (1988) treatment reveals that confusion about origin versus selective history of adaptation is still a significant problem in the study of adaptation by evolutionary biologists.

The two ways by which teleonomists typically apply the hypothetico-deductive method of science (Hempel 1966) in the study of adaptation are:

1. The initial step in a very common application of teleonomic analysis is awareness of an apparent adaptation that is not understood. Recognition of an adaptation involves identification of a feature of an organism that is too complexly organized to be due to chance. The complex organization implies that the feature is not merely a byproduct of an adaptation or the product of drift; the feature is likely the product of a long evolutionary history of directional selection. (Criteria for recognizing adaptations are discussed by Sommerhoff 1950; Williams 1966; Curio 1973; Dawkins 1982, 1986; Burian 1983; Mayr 1983; Symons 1987, in press). The recognition of an adaptation leads the teleonomist to the question "What is the

adaptation's evolutionary function?" This question in turn leads to alternative hypotheses. Often detecting adaptation necessarily entails the simultaneous generation of a preliminary functional hypothesis which then leads to more specific functional questions and alternative hypotheses. The recognition of adaptation and the perception of possible function often are simultaneous events because the application of the criterion of "too complexly organized to be due to chance" frequently is based on a trait's possible purpose.

A hypothesis is a statement about presumed or possible causation. Teleonomic hypotheses attempt to identify the nature of selection that may have built an adaptation. The predictions of a hypothesis are necessary consequences of the hypothesis. By necessary consequences I mean the phenomena that must exist if the hypothesis is true, and thus if they do not exist the hypothesis is false. The predictions of a teleonomic hypothesis pertain to the functional design of an adaptation. In teleonomy each hypothesized cause, that is, each hypothesized form of directional selection (e.g. a predator of a particular type) is viewed as a potential designer of the adaptation. Functional analysis of the actual design of an adaptation tells the teleonomist which designer is most likely to have caused the evolution of the adaptation. By study of the purposeful design of an adaptation one can eliminate hypothesized designers, i.e. hypothesized selective agents. With sufficient information about the details of functional design strong conclusions can be made about the nature of historical directional selection responsible for the adaptation.

2. Another typical approach in teleonomy is similar to the first, but with the following modifications. In many cases, the investigation of phenotypic design does not begin with the observation of an adaptation that has not been explained. Instead the teleonomic study begins with the recognition of the need for an adaptation to deal with some significant ecological problem that the investigator feels is likely to have been part of the study-organism's environment for an evolutionarily relevant period of time - e.g. locating dispersed or clumped mates, or food of a particular type. The recognition of such a problem faced by an organism leads to the question, "How will the organism solve the problem?" This in turn leads to alternative hypotheses about mechanisms that might solve the ecological problem. The predictions of each hypothesis pertain to the phenotypic design that must exist if the hypothesis is correct. Tests of the predictions of alternative hypotheses result in support or

falsification of the hypotheses. In this way adaptations are identified and characterized that were originally hypothetical. This form of teleonomy is typical of studies of optimal foraging (e.g. Stephens & Krebs 1986), mate searching (Parker 1984; Thornhill 1984a), sex allocation (Charnov 1982), human psychological design (Daly & Wilson 1988; Cosmides & Tooby 1989; Symons in press) and in many other analyses of the rules organisms have evolved in solution to ecological problems that would have impinged on fitness in evolutionary history.

The long-term consequences of evolution - e.g. species, subspecies, and adaptations - hold the real facts about the pathways taken and the processes involved in evolution over geological time. If we are to understand long-term evolution, we must use Darwin's method of historical science, which he applied to problems of evolutionary causation as well as other historical *causal* issues such as the formation of coral reefs and of soil by earthworms (see Ghiselin 1969). Throughout his work Darwin used retrospective analysis of causation in order to study history scientifically. Long-term historical causes cannot be observed directly. However, the consequences of historical causation are all around us. The predictions of a hypothesis about historical cause pertain to the long-term consequences that must exist currently if the historical cause was actually in operation. Such hypotheses are falsified when their predictions are not met and are supported when their predictions are met. Application of this approach eliminates some hypothesized historical causes and provides evidence for others. Unquestionably, the actual *evidence* of how long-term evolution works can be illuminated only by examination of the long-term consequences of evolution.

The strength of teleonomic analysis, as outlined above, is that it can lead to strong inferences about the long-term process of evolution acting in nature. By "strong inference" here I mean that the two methods of teleonomy outlined above can provide actual evidence of the kind of selection that was important in long-term evolution and data that can falsify hypotheses about the kind of selection that acted over long-term evolution.

It is important to distinguish the study of the long-term outcomes of the process of natural evolution from studies of microevolution and of variation in the reproductive success of individuals. Studies of reproductive success and of microevolutionary studies may be done in the lab, the field, or in agricultural systems. By microevolutionary analysis I mean the study of phenomena such as 1) changes in the frequencies of known genes in populations, 2) genetic parameters in populations (e.g. heritabilities of traits) and 3) artificial selection. Results from studies of microevolution and of reproductive success may provide *hypotheses* about long-term evolution (also see Betzig 1989), but results from these studies do not yield direct evidence of how the evolutionary process actually worked over the long-term to produce the diversity of life. This is why the study of adaptation is of fundamental importance in biology: teleonomic analysis provides direct evidence of how long-term evolution works because adaptations contain actual information about long-term evolution. The knowledge that we have about the kinds of directional selection that have produced adaptations represents most of our understanding of long-term evolution. I agree with Pittendrigh's (1958: 395) assessment. As he put it, "The study of adaptation is not an optional preoccupation with fascinating fragments of natural history, it is the core of biological study".

Although the two versions of teleonomy I outlined above are the typical and most productive ways to study adaptations, they are not the only ways in which adaptations are studied. I will treat other methods of teleonomy later in the paper and discuss their relationship to the two approaches outlined above. Now, I will discuss an aspect of my research in which I have applied the method of teleonomy I have emphasized. The portion of my research I have selected will serve to illustrate how the first version of the approach I have discussed is applied, and also will allow me to contrast my approach with an inappropriate teleonomic approach offered by Wade (1987).

THE ABDOMINAL CLAMP OF MALE SCORPIONFLIES

My analysis of functional design will focus on my study of a clamp-like structure on the dorsum of the abdomen of male, but not female, *Panorpa* scorpionflies (Mecoptera: Panorpidae) (Thornhill 1980, 1984b). The clamp is composed of two parts: a specialized section of the posterior edge of the top of abdominal segment three and a highly sclerotized spine from the anterior edge of abdominal segment four (Figure 1). The anterior edge of one of the female's forewings is placed in the clamp before mating. Intersegmental muscles between the third and fourth abdominal segments contract to bring the two parts of the clamp together and securely hold the female's wing during mating. Obviously, the clamp is an adaptation. It is too complexly organized for the purpose of clamping the female's wing to be a byproduct of an

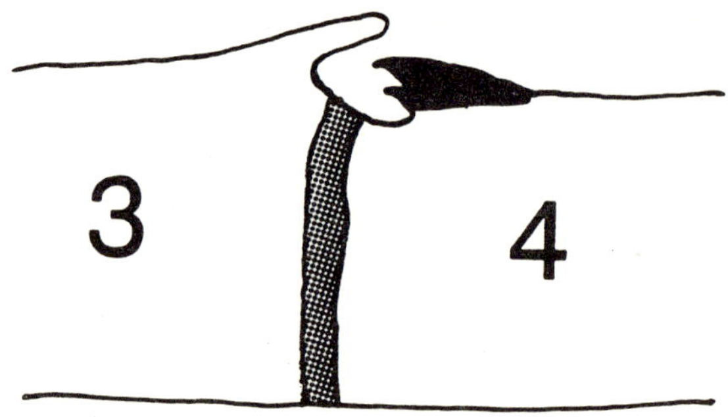

Figure 1. Diagram of the clamp on the top of the abdomen of male *P. latipennis*. Numbers refer to abdominal segments. The posterior portion of the clamp (inked) is more sclerotized than the anterior portion.

adaptation or the product of drift. Thus it must be the product of directional selection over the long-term of evolution. The puzzle is, "What kind of selection made it?"

Alternative hypotheses that might explain the kind of selection that produced the clamp were tested and will be discussed below. The tests favor the hypothesis that the evolutionary function of the clamp is forced copulation - i.e. the clamp is designed for forced mating with unwilling females. Said differently, the favored hypothesis says that variation in the clamp covaried with the fitness component male mating success in a specific way during the evolution of the clamp - the way being that the clamp increased male mating success via increasing the number of matings achieved by force.

Alternative Mating Tactics

Male scorpionflies have three alternative mating tactics (Thornhill 1979, 1980, 1981, 1986, 1987). Two tactics involve the feeding of the female by the male. Nuptial offerings of dead arthropods or hardened salivary masses are employed. Males with nuptial gifts release pheromone, which attracts females from a distance. The third tactic is forced copulation. In forced copulation, a male without a nuptial gift grabs a female with his genital claspers and then secures the anterior edge of one of her forewings in the abdominal clamp. The male then forcibly mates with the female. Forced copulation is not preceded by pheromone release by the male. All males are capable of using all three tactics, and adoption of each tactic is condition-dependent. Females prefer gift-giving males as mates and attempt to avoid males without gifts that will attempt forced copulation. Males must feed on dead arthropods in order to produce salivary masses. Large males have the advantage in aggressive competition for dead arthropods and therefore tend to use the alternative tactics of arthropod- and saliva-presentation more than small males. Small males primarily use forced copulation. Body size does not change after the attainment of adulthood in scorpionflies.

Tests of Hypotheses for the Clamp's Design

In a series of laboratory experiments using *Panorpa latipennis*, I studied the role of the clamp in forced and unforced copulation (Thornhill 1980, 1984b, 1987). One experiment determined whether the clamp is necessary for forced copulation (Thornhill 1980). The clamps of forty males were rendered nonfunctional by applying warmed beeswax to them. The clamps of forty control males were unaltered. All males were starved prior to the experiment so that they could not secrete saliva during the experiment. The males were not given dead arthropods to use as nuptial gifts. Thus all males were conditionally manipulated to be forced copulators. The experiment consisted of four replicates, each consisting of ten control males, ten males whose clamps were covered with beeswax, and ten receptive virgin females. The scorpionflies were individually marked. Each replicate was observed for four hours.

The results revealed that treated and control males grasped females with their genital claspers with similar frequency. However, none of the treated males secured the wing of a female in the clamp and none mated. Treated males tried to position females so as to secure their forewing in the clamp, but the females always escaped by struggling. About one-half of the control males secured the forewing of at least one female in their clamps. Eight of the forty control males succeeded in forcing copulation. Four of the eight copulating females were inseminated. (My studies indicate that females can often prevent insemination in forced copulation.) A second experiment, essentially identical to the first, provided very similar results (Thornhill 1984b). These two experiments indicate that forced copulation cannot be accomplished without an operative clamp.

Two additional experiments revealed that the clamp is not necessary for mating and insemination when males provide nuptial gifts of a salivary mass or a very small arthropod (Thornhill 1984b). The four experiments combined suggest that the clamp is designed for forced copulation.

I studied natural morphological variation in the clamp of *P. latipennis* in order to determine if males that engage in more forced copulation have clamps of a different morphology compared to males that engage in less forced copulation (Thornhill unpublished). Although males of all body sizes conditionally will adopt the tactic of forced copulation, small males primarily do so (Thornhill 1981, 1986, 1987). I measured the body sizes (forewing length) and the length, width and height of the posterior spine of the clamp of 60 males collected from one population over a period of a few days. The results reveal that small males have larger clamps: there are significant negative correlations between male body size and each of the three measures of size of the posterior spine of the clamp. At this time I do not know how clamp size relates functionally to the success of forced copulation by males of different size categories. Also, small males, compared to large males, possess a tooth-like structure at the base of the posterior spine (see Figure 1), but the role of the tooth in mating is unknown. Thus males that primarily use forced copulation, i.e. small males, have significantly different clamps than males that use forced copulation less often. This analysis of natural variation in the size and structure of the clamp provides additional evidence to that of the experiments that the clamp's evolutionary function is forced copulation.

I examined four additional hypotheses that might explain the reason the clamp evolved. I have suggested that the clamp functions to prevent disruption of copulating pairs and the insemination of the female of the pair by an intruding male (Thornhill 1974). A fundamental prediction about the functional design of the clamp from this hypothesis was not supported: In a lab experiment, copulating males with treated clamps and copulating males with normal clamps experienced equal rates of copulation disruption and mate take over by intruding males (Thornhill 1984b). Apparently, the clamp does not reduce the probability of a copulating pair being disrupted by intruders.

The clamp is not a source of a male pheromone, as suggested by Felt (1895). Histological studies I have conducted reveal no glandular tissue in the vicinity of the clamp (Thornhill 1984b). It might be suggested that the clamp functions in sexual or species visual recognition. However, the clamp does not show the design features that the sexual or species recognition hypotheses require. The clamp typically is not visible to females during courtship, because the males' wings are held roof-like over the abdomen. Also, in addition to the male-produced pheromone, male scorpionflies have numerous sexually differentiated and very visible morphological features that are species-specific and thus could serve as features that identify sex and species (see Thornhill 1984b).

The results of my studies of the clamp support the hypothesis that its evolutionary purpose is forced copulation.

Wade's Alternative Method of Teleonomy

Wade (1987) recently has criticized my interpretation of the results from the first experiment (Thornhill 1980) I described above in which I concluded that the clamp is essential for forced copulation and may be designed for forced copulation. His critique also applies to my repetition of the experiment (Thornhill 1984b) because techniques and results were virtually identical. Using techniques for measuring current selection acting in populations that he and his colleagues have developed (Wade & Arnold 1980; Lande & Arnold 1983; Arnold & Wade 1984), Wade calculated from the data of the first experiment that only 11% of "the total opportunity for selection" on males is focused on the presence or absence of the abdominal clamp. He concluded that there is weak selection on the presence vs. the absence of the clamp in *Panorpa*. Wade's calculation of weak selection derives from the fact that there was great variation in copulation-success among the males with functionally normal clamps. If the data from the two experiments are combined only nineteen of eighty males (24%) with normal clamps mated. Wade (1987: 204) uses this example to emphasize that, "Hypothesis testing [what I did] and parameter measurement [measuring the strength of current selection] on the same data set do not always lead to the same evolutionary inferences." Apparently, Wade feels that the magnitude of current selection on a trait determines the inferences that can be made about evolutionary function.

The research Wade criticizes was done as part of a research program to determine the functional design of the clamp. Specifically, the part he criticized was done to determine if males without a functional clamp could mate when they did not have resources to offer females, but employed forced copulation behavior. The answer is no. In thirty-two hours (the results of the two experiments combined) not a single male of eighty males with nonfunctional clamp's mated; however, nineteen of eighty normal males mated. Clearly, the clamp is necessary for mating when males do not possess nuptial gifts.

For the following reasons, I feel the tentative conclusion that the clamp's evolutionary function is forced copulation is warranted: a) the results showing that the clamp *per se* (and not a correlated trait) is necessary for mating when males do not provide a resource and use forced copulation behavior, b) the results showing that the clamp is unnecessary for mating and insemination when males provide resources to mates, c) the results revealing the correlation between use of the clamp and the clamp's structure within a species and d) the results that seriously question the alternative functional hypotheses not involving forced copulation.

Note that a) demonstrates the existence of current selection on the clamp in the context of the artificial variation in the lab experiments. That is, the presence versus absence of the clamp causally influences male mating success. The demonstration of current selection associated with artificial experimental variation was only an incidental aspect of my study of the design of the clamp.

Wade's approach of inferring evolutionary function from the *strength* of current selection is problematic. The strength of current selection on the clamp (and other adaptations) will be variable, sometimes relatively weak or nonexistent and sometimes relatively strong. My research on the mating system of *Panorpa* has revealed that small females are the primary targets of forced copulations (Thornhill 1987), and the abundance of small females is variable in space and time. Also, the success of forced copulation, compared to unforced copulation, depends on the extent of competition within and between *Panorpa* species for dead arthropods (Thornhill 1979, 1981, 1986, 1987), which varies considerably within seasons and between seasons and populations. Wade's approach would only allow forced copulation as the evolutionary function of the clamp when current selection on the clamp is strong in some arbitrarily defined sense. If it were shown that the current selection on visual ability in humans in the USA is weak, because of the availability of eyeglasses and other ophthalmic technology, it would be ludicrous to conclude that the evolutionary function of the human eye in the USA is not vision.

My most fundamental criticism of Wade's approach is that it totally misunderstands the problem in the study of adaptation. The problem that the teleonomist strives to solve is the evolutionary purpose of complexly organized traits of individual organisms. The problem of the evolutionary purpose of an adaptation is solved only when its true functional design is elucidated. Such elucidation demonstrates how the trait covaried with fitness in the environment of evolutionary adaptation and thus the nature of directional selection that produced the adaptation. It is usually inappropriate to make conclusions about evolutionary function solely on the basis of information about the mere presence or nature of current selection on an adaptation. Because of changed current environments, compared to the environment in which an adaptation evolved, the variation (natural or experimentally-induced) of an adaptation may be unrelated to variation in either total fitness or any fitness component, or the adaptation may be related to an evolutionarily novel fitness component. Even if human visual ability in the USA were unrelated to survival and all other fitness components, the physiology, biochemistry and morphology of the human eye would demonstrate that its evolutionary purpose is vision. (Incidentally, this means that the presence or absence of current selection on an adaptation cannot be used to distinguish effects of adaptations [byproducts of adaptations, Williams 1966] from actual adaptations.) It is never appropriate to make inferences about evolutionary function on the basis of the strength of selection on an adaptation. In current environments the variation in an adaptation may show a correlation with a fitness component that was not the important fitness correlation during the evolution of the adaptation. Thus, regardless of how strong a form of current selection on an adaptation may be, that form of selection may have had nothing to do with the production of the adaptation during evolutionary history.

The study of the current selection on an adaptation and the study of the evolutionary function of the adaptation are philosophically different endeavors in terms of how results from each bear on the elucidation of long-term evolution. I will explore this in some detail later in this essay. Here I want to emphasize that the kind of directional selection that designed an adaptation can only be addressed directly via tests of alternative hypotheses about the functional design of the adaptation-elucidating by hypothesis testing the details of how the adaptation works to allow its bearers to cope with an environmental problem. The nature of current selection on an adaptation (but never the magnitude of selection) can be an important piece of evidence about an adaptation's design, but more information than current selection is needed because of the problems in interpretation of data on current selection mentioned in the previous paragraph.

This framework for studying historical directional selection is especially productive when coupled with analysis of functional design using the comparative method to test hypotheses about the evolved diversity in adaptation. (For discussion of the strengths and weaknesses of the comparative method, and other methods such as experimentation, see Thornhill 1984c; also for discussion of the comparative method see Ridley 1983; Pagel & Harvey 1988.) Both the analysis of the phenotypic design of an adaptation and the study of the diversity of adaptation yield actual evidence about historical selection, because these approaches characterize the long-term phenotypic consequences of selection.

DIRECT METHODS OF TELEONOMIC ANALYSIS

Teleonomy is the study of evolutionary purpose. Each of the methods discussed in this section can falsify hypotheses about long-term evolution and provide actual evidence about the nature of directional selection that has been important in long-term evolution because the methods are focused on illuminating phenotypic design. I refer to such methods as direct.

Modern teleonomists realize the important role of selection theory in directing them to observations that would be overlooked in the absence of the theory. Some understanding of sexual selection theory, for example, is essential for deriving the hypothesis that the clamp of male *Panorpa* is an adaptation to forced copulation. This hypothesis has led to some unique discoveries about the clamp. The primary value of theory in evolutionary science, as in any other science, is to render phenomena discoverable that otherwise would not be. Thus the usefulness of a scientific theory can be measured by the discoveries it provides. Evolution by selection is a very useful theory by this criterion: It continually leads biologists to new findings about life (e.g. see recent issues of *The American Naturalist, Animal Behaviour, Ethology, or Evolution*).

In general the theory of evolution by selection is used in the study of adaptations in the two ways outlined earlier in the essay, coupled with comparative analysis of the distribution of adaptations across taxa in relation to hypothesized selective agents. These ways of using selection theory to study adaptations might be called the standard approach of teleonomy, because it is the most commonly employed form of teleonomic analysis. For example, about 75% of the papers in the issue of the journal *Animal Behaviour* that happens to be beside my computer (Volume 36, 6, Nov./Dec. 1988) use what I have called the standard teleonomic approach. This is the approach that C. Darwin (e.g. 1859, 1874) developed: It involves analysis of phenotypic design in order to elucidate a predicted evolutionary function and thus support the expected selection responsible for the design. Darwin's approach combined testing hypotheses about the functional design of adaptations with comparative analysis of the taxonomic distribution of adaptations. (See for example Darwin's [1874] discussion of evidence that male ornamental features function in sexual competition among males; also see Ghiselin's 1969 discussion of Darwin's approach in studying evolution.)

This standard approach to teleonomy has been used as the basis for developing other very important direct methods for analyzing adaptations. As mentioned earlier, the study of current selection can provide evidence of functional design and thus evidence of the historical selection responsible for the design. The evidence that identifies causal current selection on an adaptation is the evidence for the adaptation's functional design. For example, the experimental evidence that the clamp of male scorpionflies itself causes variation in male mating success in the context of forced copulation is part of the evidence for the clamp's design for forced copulation. However, for reasons mentioned above, more information than current causal selection on a trait is necessary to identify adaptation and to elucidate the evolutionary purpose of adaptation. Furthermore, the identification of causal selection on a trait can be extremely difficult. Most commonly, the results of studies of current selection demonstrate selection but do not demonstrate the trait upon which selection is focused (e.g. most chapters in Clutton-Brock 1988); such studies do not provide evidence of functional design. I will treat the study of current selection in general in the next section, which deals with indirect methods of teleonomic analysis - that is, methods that in themselves do not provide direct evidence of functional design. I delay the discussion because a somewhat lengthy evaluation is needed to distinguish the analysis of causal selection on a trait, a direct method of teleonomic analysis, from the study of variation in individual reproduction in general.

The evolutionarily stable strategy (ESS) approach (reviewed in Maynard Smith 1982) is a direct method for study of adaptation. It is specifically formulated to provide hypotheses for the evolution of adaptations by frequency-dependent selection. Said differently, ESS hypotheses attempt to identify and characterize the phenotypic design that has arisen due to long-term frequency-dependent selection. A set of strategies (phenotypes) is envisioned and their relative fitness depends on the use of the strategies in the population. This approach is based on optimality. It determines the theoretical solution, the evolutionarily stable strategy or mixture of stable strategies, i.e. the "optimum" or best, under frequency-dependent selection.

The ESS approach in studies of adaptations may or may not involve mathematics as an accessory tool. Fisher's (1958) hypothesis about the evolution by selection of the equal investment by parents in sons and daughters is an example of a nonmathematical evolutionary hypothesis based on ESS reasoning (see Parker 1984). Fisher envisioned the hypothetical circumstance of numerical disparity of the sexes of offspring as a phenotype of parental reproduction that was an alternative to the parental strategy of production of equal numbers of sons and daughters. Later Hamilton (1967) used ESS reasoning to formulate his successful hypothesis for the evolution of female-biased sex ratios under inbreeding. The ESS approach was made highly mathematical in the 1970s by the application of game theory to evolutionary problems involving frequency-dependent selection (Maynard Smith & Price 1973). The ESS approach has been especially valuable in providing useful hypotheses about phenotypic design associated with contesting resources (e.g. Austad 1983) and mating (see Parker 1984).

Another important direct method of teleonomy also incorporates a mathematical optimality approach in the study of adaptations (e.g. MacArthur & Pianka 1966; Charnov 1976; Stephens & Krebs 1986). This approach, sometimes called optimality theory, provides hypotheses that attempt to elucidate the kind of selection that has produced adaptations for coping with fixed environmental problems, that is, when the fitness of hypothetical phenotypes under selection is not frequency-dependent. It has proven its value, especially in successful prediction of foraging adaptations (Stephens & Krebs 1986).

The standard teleonomic approach, the analysis of current selection on an adaptation, and the two versions of optimality mentioned above typically ignore the nature of genetic control of phenotypic design. Although the ignoring of the genetic control of adaptations by teleonomists has been criticized by Lewontin (1978), this is not a general problem in teleonomy (e.g. see Grafen 1984; Stephens & Krebs 1986). Cases where genetics might make a difference in the teleonomist's hypothesizing (e.g. overdominance) probably are rare (Maynard Smith 1982; Grafen 1984). Of course, any time a phenotypic feature meets the criteria of an adaptation (see above), the very existence of the feature demonstrates that there was heritability in the trait in the evolutionary past.

The final direct teleonomic approach is based explicitly on genetics. This approach provides hypotheses based on the dynamics of gene frequencies that may be brought about by selection. Genetical modeling of the evolution of phenotypes continues to make important contributions to the understanding of the evolved design of living systems. For example, Hamilton & Zuk's (1982) parasite hypothesis of sexual selection, which is based on the dynamics of gene frequencies in parasites and their hosts under frequency-dependent selection, has led to a number of novel findings about the signal content and thus the design of sexual displays of birds (Hamilton & Zuk 1982; Read 1987; Zuk in press; Zuk et. al. in press) and fish (Ward 1988; McMinn in press).

In sum, each of the teleonomic procedures discussed above are direct methods for study of evolutionary purpose, because each can characterize phenotypic design and thereby can yield actual evidence that can be used to refute or support hypotheses about the kind of selection that was important in long-term evolution. Unquestionably, these methods work in the sense that matters in science - that is, they lead to discoveries about phenotypic design and thus about the selection responsible for the design. Just because the procedures work, it does not follow that they are perfect and not in need of refinement. As Williams (1966: 260) pointed out over 20 years ago:

> "How, ultimately, does one ascertain the function of a biological mechanism?...I have assumed, as in customary, that functional design is something that can be intuitively comprehended by an investigator and convincingly communicated to others. Although this

may often be true, I suspect that progress in teleonomy will soon demand a standardization of criteria for demonstrating adaptation, and a formal terminology for its description."

It would seem that a formal system for recognizing and describing phenotypic design is urgently needed in biology given that the concept of adaptation is fundamental for understanding long-term evolution.

INDIRECT METHODS OF TELEONOMY

There are other methods used in the study of adaptation. I emphasize that I am not saying that the methods discussed in this section are unscientific. They are used with the hypothetico-deductive model and thus meet scientific standards. Nor am I saying that the methods do not provide interesting and valuable information about organisms. Our understanding of natural history has been advanced significantly by each of the methods discussed below. Nor am I saying that the methods below do not yield hypotheses that lead to better understanding of long-term natural evolution; they do. My point about the so-called indirect methods is that, without the assistance of the direct methods discussed above, they do not allow quality inferences about long-term evolution. In contrast, each of the direct methods can provide actual evidence for or against a hypothesis about long-term evolution. All the indirect teleonomic methods comprise studies of microevolution or reproductive success.

Artificial Selection

As Williams (1985) has pointed out with respect to studies of artificial selection, the findings of such studies - Williams mentioned Wade's (1976) lab research on artificial group selection in flour beetles as an example - can provide only hypotheses of how evolution might have worked over geological time but never actual evidence of the action of long-term evolution in nature. Williams' point applies to Wade's results because Wade studied artificial selection and not because the research was done in the lab. Evolution in nature proceeds in an open system over the long-term. The openness of gene pools means that the events that would be necessary for foretelling the future long-term course of evolution are unpredictable. Which mutations will occur is not predictable; nor is the nature of the environment that will impinge on any population in the long-term future predictable. Thus we cannot know the future availability of genetic raw material or future adaptations and evolutionary constraints that would be necessary for predicting future long-term evolution. Similarly, because evolution proceeds in an open system it is impossible to determine if the specific events that contribute to the occurrence of a particular form of current evolution (e.g. evolution by group selection in flour beetles in the lab) held in the evolutionary past.

Wade's (1976) results suggest how long-term evolution might work-provide a hypothesis about long-term evolution. The hypothesis that group selection has been effective in producing adaptations during long-term evolution is testable only by determination of whether the long-term phenotypic consequences of such a process are present in nature. That is, do we see the stamp of a history of group selection in the phenotypes of organisms? Examination of organisms for evidence of phenotypic design indicative of effective group selection reveals that the process has not been important in shaping adaptation (see especially Williams 1985; also Williams 1966; Dawkins 1982, 1986; Alexander 1979, 1987; Thornhill & Alcock 1983; and many others).

I'm not arguing that knowledge about artificial selection is unimportant for our general understanding of long-term evolution. For example, knowledge of artificial selection apparently helped Darwin clarify his ideas about the role of natural and sexual selection in long-term evolution. Darwin argued that the phenotypic consequences of artificial selection are analogous to the phenotypic consequences of selection generated by nonhuman agents during long-term evolution. Also, the results from studies of artificial selection since Darwin's time demonstrate that microevolution occurs by selection. (Darwin's "demonstration" of selection was by deduction from the fact of high reproductive output and the inevitability of nonrandom mortality of individuals that must follow given the stability of populations.) Thus work on artificial selection has been essential in clarifying the general conceptual connection between the microevolutionary process of evolution and the long-term historical process of evolution.

I am arguing that results from studies of artificial selection that demonstrate or falsify a particular microevolutionary process cannot be viewed as evidence for or against the same process in long-term evolution. However, such results may lead to valuable hypotheses about long-term evolution.

The work on senescence in *Drosophila* by Rose & Charlesworth (1980) can be used to clarify points about artificial selection touched on above. Williams (1957) hypothesized that senescence is the inevitable evolutionary consequence of the pleiotropic effects of alleles with positive fitness effects early in life and negative effects late in life. Genes may have many effects, some positive and some negative. Selection favors alleles whose net effect is more positive than alternative alleles. Williams argued that senescence is the maladaptive result of selection of alleles whose negative effects are expressed late in life. The *Drosophila* study revealed an evolutionary response to artificial selection for later high rates of reproduction and a correlated evolutionary response in delay of onset of senescence. This result suggests that the kind of pleiotropic genes Williams' hypothesis requires do indeed exist. There is a genetic correlation, apparently due to pleiotropy, between late reproduction and delayed senescence. However, the evolutionary response in the study was a microevolutionary response, and therefore cannot be viewed as providing evidence for William's view of the evolution of senescence in long-term evolution. The evolutionary response in the study however does suggest that the hypothetical scenario envisioned by Williams possibly could apply to long-term evolution.

Williams' hypothesis can best be tested by deriving the predicted long-term phenotypic consequences of the evolutionary process that the hypothesis contains and then determining if these consequences are actually manifested in senescence. Senescence is not an adaptation but it is a product of long-term evolution, and therefore the evolutionary process that produced it can be understood by detailed analysis of the nature of the phenotypic responses that occur during senescence (for discussion of the kind of evidence most useful for testing Williams' hypothesis see Williams 1957; Hamilton 1966; Bell 1984; Alexander 1987).

Reproductive Success

A currently popular form of evolutionary study involves measurement of the extent of current variation in reproductive performance among individuals. This approach was the basis of Wade's analysis of the selection acting on the abdominal clamp of male scorpionflies discussed above. Another example is the collection of papers, edited by Clutton-Brock (1988), reporting measurement of variation in reproductive success of individuals of numerous animal species over a large part of their lives. Grafen (1988) pointed out in his paper in Clutton-Brock's edited collection that the contributors to the book are interested to an important extent in understanding adaptations of the animals whose lives they are studying. Grafen's paper provides a detailed critique of analyzing adaptation by study of individual variation in reproduction. Symons (1989; in press) has provided important papers dealing with the erroneous notion that human adaptations can best be understood by study of current variation in individual reproductive success. Williams (1966) and Stephens & Krebs (1986) also have criticized briefly this approach in the study of adaptation. I will briefly cover some of the points discussed in these critiques and raise some additional problems associated with teleonomic studies that are based on individual variation in reproduction.

Studies of individual variation in reproduction take two forms:

1. Mere differential reproduction among individuals does not demonstrate the action of current selection, because selection is *nonrandom* differences in reproduction (see below). Reproductive variance among individuals may be random with regard to phenotypic differences and thus due to drift rather than selection. Sutherland (1985) has shown that the data from Bateman's well known study of variation in reproductive success in *Drosophila melanogaster* cannot be distinguished from a random pattern. Also, studies of mere differentials in reproduction say nothing about phenotypic design and therefore nothing about adaptation and long-term evolution. As Williams (1966: 159) so appropriately put it, "measuring reproductive success focuses attention on the rather trivial problem of the degree to which an organism actually achieves reproductive survival. The central biological problem is not survival as such, but design for survival."

2. Data on individual variance in reproductive performance may identify selection in action - that is, the occurrence of nonrandom differential reproduction of individuals. Note that my definition of selection does not include an evolutionary response to selection. Evolution typically is defined as any change in the frequency of genes in a population. Evolution may result from any agent of evolution (drift, mutation, migration, or selection). Defining selection in terms of changes in gene frequency resulting from nonrandom individual reproduction (e.g. Endler 1986)

confuses selection, a phenotypic event, with evolution, a genetic event. As pointed out by Fisher (1958) selection is not evolution (also see Arnold 1983; Lande & Arnold 1983; Wade 1987).

The action of current selection is demonstrated: 1) by a correlation between trait variation and some fitness-related measure such as mating success or survival, or 2) when there is repeatability in the performance of individuals in some fitness-related measure in the absence of a detectable correlation between phenotypic trait variation and individual performance. An example of 2) is McVey's (1988) demonstration that males of a dragonfly species when experimentally forced to reestablish a territory obtain a territory that yields a positive correlation with their original territory in terms of reproductive success. Thus variance in male reproductive success was due to variance in male competitive ability, even though McVey could not determine which aspects of male phenotype were important. Either 1) or 2) is sufficient to demonstrate nonrandom variation in the reproductive performance of individuals.

A demonstration of selection by 1), however, does not identify the trait that correlates with fitness-related variation as the actual focus of selection, because other traits that may correlate with the trait may be the causal source of variation in nonrandom individual reproduction. Grafen's (1988) paper mentioned above deals in detail with the important problem of correlated traits in the study of current selection (also see Arnold 1983; Lande & Arnold 1983; Endler 1986: 162-165; Mitchell-Olds & Shaw 1987; Crespi & Bookstein 1989). Grafen emphasizes the important role of experimentation in untangling a trait that is thought to correlate with fitness from known or unknown correlations with other traits.

This is why I created experimentally a category of male scorpionflies without functional clamps. This manipulation allowed examination of the role of the clamp itself in forced copulation, uncomplicated by variation in other features (behavioral, morphological, physiological, etc.) that might correlate with natural variation in the clamp and confound my results on the functional design of the clamp for use in forced mating. Note that my reason for manipulating the clamp was to provide evidence of its evolutionary function, and in the course of the experiment, causal sexual selection on the clamp was demonstrated incidentally; that is, variation in the clamp (functional versus nonfunctional in the experiments) caused a difference in male mating success. Indeed any time an adaptation is experimentally manipulated the real question is evolutionary purpose/functional design and not current selection.

Some traits are very hard to manipulate experimentally. In these cases, only by detailed functional analysis, characteristic of the field of functional morphology (e.g. Altenbach 1979), can it be demonstrated that a trait is causally related to selection. The primary question in such analyses pertains to phenotypic design and not the action of current selection.

The demonstration of causal current selection on an adaptation, by experimental manipulation or detailed functional analysis, can identify phenotypic design and thus results from such studies bear directly on the kind of selection that produced the adaptation. Studies of current selection become relevant to the issue of adaptation only when functional design is analyzed-that is, only when causal current selection is detected. Otherwise, a study of current selection has limited significance: such information alone does not contribute to a better understanding of the long-term process of evolution in nature although it may lead to hypotheses that can be tested. As Symons (1989: 132) put it "In modern evolutionary biology, an adaptation is usually considered to be an aspect of a phenotype...that was designed by natural selection to serve a specific function...hence, to the extent that one fails to describe or characterize phenotypic design, one fails to describe or characterize adaptation."

The issue in teleonomy is not the existence of current selection in natural populations; nor the magnitude of current selection in natural populations; nor the extent of causal current selection on adaptations; nor whether adaptations are related causally to current selection. Clearly, evidence of a causal relationship of a trait to selection may provide evidence of phenotypic design and thus evidence of historical directional selection. However, for reasons mentioned above, this evidence is not the only evidence of interest, nor is it the most informative of historical selection. Other information about the adaptation's design beyond the existence and nature of current selection on the adaptation is essential for determining whether the adaptation evolved in the selective context indicated by data on its current covariation with fitness.

Thus I would disagree with the emphasis in Arnold's (1983: 347) interesting paper on the measurement of current selection acting on adaptations. He states:

> My thesis in this paper is that it is possible to measure adaptive significance directly. In particular it is possible to characterize statistically the relationship between fitness and morphology in natural populations. One can argue that this statistical approach constitutes the highest grade of evidence for selection and adaptation.

I have argued that methods other than measurement of selection on adaptations address adaptive significance directly. Also, I have argued that measurement of selection on adaptations does not "constitute the highest grade of evidence for...adaptation." Finally, I have argued that measurement of selection on adaptations may not provide any evidence at all for adaptations.

I would agree completely however with the following point made by Arnold (1983: 347):

> Despite its virtues measurements of selection should not be considered a substitute for other modes of attack on adaptive significance. Direct analysis of selection will be most valuable when it is combined with analytical studies of function and with comparative studies that describe the scope of evolution. Likewise, inferences from functional and comparative studies will be strengthened by companion studies of selection...

The fundamental issue in teleonomy is, "How did an adaptation of interest relate to fitness during the evolution of the adaptation?" This question is the same as "What is the evolutionary purpose of an adaptation?" The answer can be determined directly only by analysis of the functional design of an adaptation, and such analyses may include measurement of current selection on an adaptation.

Although evidence of causal current selection on an adaptation can provide evidence of the adaptation's evolutionary purpose, I suggest that studies in evolutionary biology that emphasize the identification and characterization of current selection are important primarily for the following two reasons. First, the demonstration of current selection of a particular form suggests the possibility that the same selection may have operated in long-term evolution. Such a demonstration then may give some credibility to a hypothesis about long-term evolution resulting from the same selection. This is useful because the relative credibility of alternative hypotheses about the selection that acted in long-term evolution is important to assess in determining research priorities for testing alternative hypotheses. For example, many hypotheses have been proposed to explain how sexual selection works in long-term evolution (review in Bradbury & Andersson 1987). Studies of the current action of sexual selection may be helpful in determining how evolutionary biologists should proceed in prioritizing for testing the numerous hypotheses for long-term evolution by sexual selection. Second, studies of current selection sometimes yield hypotheses about the effective selection that may have operated in long-term evolution that might not be realized otherwise (see e.g. Howard 1988; also Chapter 29 by Clutton-Brock in Clutton-Brock 1988).

Genetic Parameters

The final category of indirect analysis of adaptation I will discuss is the empirical evaluation of genetic parameters of populations. Recently this approach has focused on the measurement of the genetic basis of life history characters (e.g. Rose & Charlesworth 1980; Dingle & Hegmann 1982) and traits related to sexual selection (e.g. Majerus et al. 1982). I will not argue here that evolutionary ideas that are explicitly genetical are all indirect analyses of adaptation. Recall that I emphasized earlier that hypotheses for long-term evolution that are based on the dynamics of hypothetical genetic alternatives represent a direct approach in teleonomy. I do feel that although the measurement of genetic parameters of populations is a useful tool in the overall effort to solve the problems of long-term evolution, it is not the panacea. Thus I disagree with Lande's (1987: 83) statement that "Salient features of sexual selection and the evolution of sexual dimorphism can be understood *only* through the study of genetic mechanisms." (emphasis added) Genetic correlations between the sexes are the genetic mechanisms that Lande argues will elucidate sexual selection and sexual dimorphism. Lande is a geneticist and it is not surprising that he is enthusiastic about genetics. Lewontin is also a geneticist, and he (1978) apparently feels that it is essential for teleonomists to become geneticists. Enthusiasm about one's own field usually is a good thing. However, the view that the field of genetics is synonymous with modern evolutionary science is erroneous and ignores many excellent evolutionary studies.

I suggest that the major empirical contribution of geneticists to the understanding of evolution will not derive from measuring genetic parameters such as genetic correlations or heritabilities of traits or from measuring changes in the frequencies of genes in populations. Such studies do provide interesting information about the natural history of organisms, and we are much better off with this information than without it. The same can be said about all facts of natural history. Information about the genetic parameters of traits and microevolution in a species is as interesting as information about what the species eats, season of breeding, etc.; any fact of natural history may lead to useful hypotheses about long-term natural evolution and can falsify or support these hypotheses if studied to test their predictions.

My guess is that the major empirical contribution of genetics to the understanding of long-term evolution will derive from teleonomic analysis of the functional design of genetic adaptations. Genetic adaptations are as important to understand as other categories of adaptations, because all adaptations are the long-term consequences of selection in nature. The only way to understand the way selection has worked on genetic systems is by discovery of the nature of the design of genetic systems.

Consider the following example. Lande (1987) argues that variation in sexual dimorphism in sexually-selected traits across species should coincide with the degree of pleiotropic genetic correlation of the traits. He offers this as an explanation for the pattern of "transference" of male secondary sexual characters to females discovered by Darwin (1874). Empirically this idea can be tested by measuring, in a group like the Phasianidae (pheasants), the degree of genetic correlation between the sexes in features such as morphological ornaments that may be expressed in both sexes. Lande would expect that the magnitude of the genetic correlation will correlate positively with the similarity of the sexes in ornamentation, because his hypothesis for transference assumes limited or ineffective selection on the postulated genetic mechanisms. The view that genetic systems reflect fine-tuned adaptation to the same degree as the morphological and physiological features of individuals (also see Trivers 1988: 271-272) would suggest a different and more encompassing line of research on sexual dimorphism than Lande's perspective: If the empirical finding was against Lande's hypothesis, more research would be needed to determine how the genetic systems across species of phasianids proximately regulate the expression of genes for ornaments in relation to the sex of the individual. If the empirical finding favored Lande's hypothesis, the same research just mentioned would be needed. That is, viewing genomes as highly evolved adaptive systems would lead to questions about the design of the genetic adaptations for sexual dimorphism.

The diversity in the phenotypic expression of features in both sexes in groups like the phasianids ranges from sexual uniformity to restricted to only one sex. Even within single species certain features are "transferred" and others are not. The question is why? (For example, the red jungle fowl, *Gallus gallus,* hen has a small comb compared to the large comb of the rooster. The hen lacks the other ornaments of males.) I suggest that this diversity in sex-limited control of traits reflects the diversity of adaptations of the genetic systems involved.

Genetic correlations are often viewed by geneticists as mere constraints on long-term evolution (e.g. Lande 1987). In its most naive form this view confuses proximate and ultimate levels of causation. Genetic correlations, indeed all properties of living systems, are constraints on long-term evolution in only one sense: Selection will act on what evolution in the past has produced. Genetic correlations and other complex genetic properties may be most appropriately viewed as an aspect of the evolved design of genetic systems, a perspective that suggests very different questions about genetic features than would be asked in descriptive quantitative genetics.

I will make one final comment on the recent flurry of interest in quantitative genetics among evolutionary biologists. Some evolutionary biologists that I have discussed the matter with apparently feel that if all the genetic parameters and current selective forces in a natural system are empirically documented, future evolution in the system should be predictable. It is conceivable that such an approach may sometimes work over the short-term, although the thorough understanding of all the forces of selection, heritabilities, genetic correlations, etc. that would be necessary is mind-boggling. However, it would not be wise to attempt to predict long-term future evolution from knowledge of current genetic parameters and selection for the reasons mentioned earlier.

CONCLUSIONS

The central problem in evolutionary biology is to understand the long-term process of evolution in nature. Thus it is essential to distinguish the direct from the indirect methods of analyzing long-term evolution. There are several methods used in teleonomy, the study of adaptations. The functional design of an adaptation provides the relevant evidence of the kind of directional selection responsible for the adaptation. Methods that do not elucidate the purposeful design of phenotypic traits do not provide direct evidence of the ultimate causation of adaptation. The standard approach used in teleonomy today, the Darwinian approach, has been modified to form the basis of optimality theory, evolutionarily stable strategy theory, genetic modelling of the evolution of adaptation, and the study of causal current selection acting on adaptation. All these approaches analyze phenotypic design and thus yield evidence directly relevant to illuminating long-term evolution, specifically the selection that produced phenotypic design.

Some popular methods of analyzing adaptation do not focus on phenotypic design and therefore do not actually address adaptation directly. These indirect methods of analyzing adaptation focus on microevolution and on variation in reproductive performance of individuals, as opposed to long-term outcomes of evolution. Microevolutionary questions are addressed by studies involving: artificial selection; tallies of changes in the frequencies of identifiable genes in populations; and measures of the genetic parameters of populations. Microevolutionary analyses and studies characterizing the current selection acting in populations may lead to testable hypotheses about long-term evolution, but alone they cannot provide evidence of how evolution actually works over the long-term.

I suggest that the major empirical contribution of the field of genetics to the understanding of long-term evolution will derive from teleonomic analysis of the functional design of genetic adaptations.

ACKNOWLEDGEMENTS

I am grateful to M. Daly, M. Bekoff, L. Betzig, D. Jamieson, W. Kuipers, R. Pierotti, P. Stacey, D. Symons, F. Taylor and an anonymous reviewer for useful comments on the manuscript. Discussion about methodology with M. Daly, W. Kuipers, P. Stacey, D. Symons, F. Taylor, P. Watson, M. Wilson, M. Zuk, and especially N.W. Thornhill were helpful. My research on scorpionflies discussed in this essay was supported by grants from the National Science Foundation. I also acknowledge support from NSF grant BSR-8515377, the H.F. Guggenheim Foundation, and

Paul Risser, Vice President of Research, University of New Mexico.

LITERATURE CITED

Alexander, R.D. 1979. *Darwinism and Human Affairs.* Seattle, Washington: University of Washington Press.

____. 1987. *The Biology of Moral Systems.* New York: Aldine de Gruyter.

Altenbach, J.S. 1979. *Locomotion Morphololgy of the Vampire Bat, Desmodes rotundus. Special Publications of the American Society of Mammalogists* Number 6.

Arnold, S.J. 1983. Morphology, performance and fitness. *American Zoologist* 23, 347-361.

Arnold, S.J. & Wade, M.J. 1984. On the measurement of natural and sexual selection: Theory. *Evolution* 38, 709-719.

Austad, S.N. 1983. A game theoretical interpretation of male combat in the bowl and doily spider, *Frontinella pyramitela. Animal Behaviour* 31, 59-73.

Bell, G. 1984. Evolutionary and nonevolutionary theories of senescence. *American Naturalist* 124, 600-603.

Betzig, L. 1989. Rethinking human ethology: A response to some recent critiques. *Ethology and Sociobiology* 10, 315-324.

Bradbury, J.W. & Andersson, M.B. (eds.) 1987. . *Sexual Selection: Testing the Alternatives.* New York: John Wiley and Sons.

Burian, R.M. 1983. Adaptation. In: *Dimensions of Darwinism: Themes and Counterthemes in 20th Century Evolutionary Theory* (ed. by M. Grene), pp. 287-314. Cambridge: Cambridge University Press.

Charnov, E.L. 1976. Optimal foraging: the marginal value theorem. *Theoretical Population Biology* 9, 129-136.

____. 1982. *The Theory of Sex Allocation.* Princeton, New Jersey: Princeton University Press.

Clutton-Brock, T.H. (ed.) 1988. *Reproductive Success: Studies of Individual Variation in Contrasting Breeding Systems.* Chicago, Illinois: University of Chicago Press.

Cosmides, L. & Tooby, J. 1989. Evolutionary psychology and the generation of culture, Part II: Case study: A computational theory of social exchange. *Ethology and Sociobiology* 10, 51-98.

Crespi, B.J. & Bookstein, F.L. 1989. A path-analytic model for the measurement of selection on morphology. *Evolution* 43, 18-28.
Curio, E.B. 1973. Towards a methodology of teleonomy. *Experientia* 29, 1045-1058.
Daly, M. & Wilson, M. 1988. *Homicide*. New York: Aldine de Gruyter.
Darwin, C.R. 1859. *The Origin of Species*. London: Dent.
____. 1874. *The Descent of Man and Selection in Relation to Sex*. New York: Rand McNally.
Dawkins, R. 1982. *The Extended Phenotype*. Oxford: Oxford University Press.
____. 1986. *The Blind Watchmaker*. New York: W.W. Norton and Company.
Dingle, H. & Hegmann, J.P. (eds.) 1982. *Evolution and Genetics of Life History*. New York: Springer-Verlag.
Endler, J.A. 1986. *Natural Selection in the Wild*. Princeton, New Jersey: Princeton University Press.
Felt, E.P. 1895. The scorpionflies. In: *Tenth Report of the State Entomologist on the Injurious and Other Insects of the State of New Mexico*.
Fisher, R.A. 1958. *The Genetical Theory of Natural Selection*, 2nd Edition. New York: Dover.
Ghiselin, M.T. 1969. *The Triumph of the Darwinian Method*. Berkeley, California: University of California Press.
Gould, S.J. & Lewontin, R.C. 1979. The spandrels of San Marco and the panglossian paradigm: A critique of the adaptationist program. *Proceedings Royal Society London* B205, 581-598.
Grafen, A. 1984. Natural selection, kin selection and group selection. In: *Behavioral Ecology: An Evolutionary Approach*, 2nd Edition (ed. by J.R. Krebs and N.B. Davies), pp. 62-89. Sunderland, Massachusetts: Sinauer.
____. 1988. On the uses of data on lifetime reproduction. In: *Reproductive Success: Studies of Individual Variation in Contrasting Breeding Systems* (ed. by T.H. Clutton-Brock), pp. 454-471. Chicago, Illinois: University of Chicago Press.
Hamilton, W.D. 1966. The molding of senescence by natural selection. *Journal of Theoretical Biology*12, 12-45.
____. 1967. Extraordinary sex ratios. *Science* 156, 477-488.
Hamilton, W.D. & Zuk, M. 1982. Heritable true fitness and bright birds: A role for parasites? *Science* 218:384-387.

Hempel, C.G. 1966. *Philosophy of Natural Science.* Englewood Cliffs, New Jersey: Prentice Hall.

Howard, R.D. 1988. Reproductive success in two species of anuarans. In: *Reproductive Success: Studies of Individual Variation in Contrasting Breeding Systems* (ed. by T.H. Clutton-Brock), pp. 99-113. Chicago, Illinois: University of Chicago Press.

Lande, R. 1987. Genetic correlations between the sexes in the evolution of dimorphism and mating preferences. In: *Sexual Selection: Testing the Alternatives* (ed. by J.W. Bradbury and M.B. Andersson), pp. 83-94. New York: John Wiley and Sons.

Lande, R. & Arnold, S.J. 1983. The measurement of selection on correlated characters. *Evolution* 37, 1210-1226.

Lewontin, R.C. 1978. Adaptation. *Scientific American* 239, 156-169.

MacArthur, R.H. & Pianka, E.R. 1966. On optimal use of a patchy environment. *American Naturalist* 100, 603-609.

Majerus, M.E.N., O'Donald, P. & Weir, J. 1982. Female mating preference is genetic. *Nature* 300, 521-523.

Maynard Smith, J. 1982. *Evolution and the Theory of Games.* Cambridge: Cambridge University Press.

Maynard Smith, J. & Price, G.R. 1973. The logic of animal conflict. *Nature* 246, 15-18.

Mayr, E. 1983. How to carry out the adaptationist program? *American Naturalist* 121, 324-334.

McMinn, H. 1990. Effects of the nematode parasite *Camallanus cotti* on sexual and non-sexual behaviors in the guppy (*Poecilia reticulata*). *American Zoologist.*

McVery, M.E. 1988. 1988. The opportunity for sexual selection in a territorial dragonfly, *Erythemis simplicicollis*. In: *Reproductive Success: Studies of Individual Variation in Contrasting Breeding Systems* (ed. by T.H. Clutton-Brock), pp. 44-58. Chicago, Illinois: University of Chicago Press.

Mitchell-Olds, T. & Shaw, R.G. 1987. Regression analysis of natural selection: Statistical inference and biological interpretation. *Evolution* 41, 1149-1161.

Pagel, M.D. & Harvey, P.H. 1988. Recent developments in the analysis of comparative data. *The Quarterly Review of Biology* 63, 413-440.

Parker, G.A. 1984. Evolutionarily stable strategies. In: *Behavioural Ecology: An Evolutionary Approach* (ed. by J.R. Krebs & N.B. Davies), pp. 30-61. Oxford: Blackwell.

Pittendrigh, C.S. 1958. Adaptation, natural selection, and behavior. In: *Behaviour and Evolution* (ed. by A. Roe & G.G. Simpson), pp. 390-416. New Haven, Connecticut: Yale University Press.

Read, A.F. 1987. Comparative evidence supports the Hamilton and Zuk hypothesis on parasites and sexual selection. *Nature* 327, 68-70.

Ridley, M. 1983. *The Explanation of Organic Diversity*. Oxford: Clarendon Press.

Rose, M. & B. Charlesworth. 1980. A test of evolutionary theories of senescence. *Nature* 287, 141-142.

Sherman, P.W. 1988. The levels of analysis. *Animal Behaviour* 36, 616-618.

Sommerhoff, G. 1950. *Analytical Biology*. Oxford: Oxford University Press.

Stephens, D.W. & Krebs, J.R. 1986. *Foraging Theory*. Princeton, New Jersey: Princeton University Press.

Sutherland, W.J. 1985. Chance can produce a sex difference in variance in success and explain Bateman's data. *Animal Behaviour* 33, 1349-1352.

Symons, D. 1987. If we're all Darwinians, what's the fuss about? In: *Sociobiology and Psychology: Ideas, Issues and Applications* (ed. by C. Crawford, M. Smith & D. Krebs), pp. 121-146. Hillsdale, New Jersey: Lawrence Erlbaum Associates.

_____. 1989. A critique of Darwinian anthropology. *Ethnology and Sociobiology* 10, 131-144.

_____. In press. On the use and misuse of Darwinism in the study of human behavior. In: *The Adapted Mind: Evolutionary Psychology and the Generation of Culture* (ed. by J. Barkow, L. Cosmides, & J. Tooby). Cambridge: Cambridge University Press.

Thornhill, R. 1974. *Evolutionary Ecology of Mecoptera* (Insecta). Ph.D. Dissertation, University of Michigan, Ann Arbor, MI.

_____. 1979. Male and female sexual selection and the evolution of strategies in insects. In: *Sexual Selection and Reproductive Competition in Insects* (ed. by M.S. Blum & N.A. Blum), pp. 81-121. New York: Academic Press.

_____. 1980. Rape in *Panorpa* Scorpionflies and a general rape hypothesis. *Animal Behaviour* 18, 52-59.

_____. 1984a. Alternative female choice tactics in the scorpionfly *Hylobittacus apicalis* (Mecoptera) and their implications. *American Zoologists* 24, 367-383.

_____. 1984b. Alternative hypotheses for traits believed to have evolved by sperm competition. In: *Sperm Competition and the Evolution of Animal Mating Systems* (ed. by R.L. Smith), pp. 151-178. New York: Academic Press.

_____. 1984c. Scientific methodology in entomology. *Florida Entomologist* 67, 74-79.

_____. 1986. Relative parental contribution of the sexes to their offspring and the operation of sexual selection. In: *Evolution of Animal Behavior* (ed. by M.H. Nitecki & J.A. Kitchell), pp. 113-136. Oxford: Oxford University Press.

_____. 1987. The relative importance of intra-and interspecific competition in scorpionfly mating systems. *American Naturalist* 130, 711-729.

Thornhill, R. & Alcock, J. 1983. *The Evolution of Insect Mating Systems*. Cambridge, Massachusetts: Harvard University Press.

Trivers, R.L. 1988. Sex differences in rates of recombination and sexual on. In: *The Evolution of Sex* (ed. by R.E. Michod & B.R. Levin), pp. 270-286. Sunderland, Massachusetts: Sinauer.

Wade, M.S. 1976. Group selection among laboratory populations of *Tribolium*. *Proceedings of National Academy of Sciences* 73, 4604-4607.

_____. 1987. Measuring sexual selection. In: *Sexual Selection: Testing the Alternatives* (ed. by J.W. Bradbury & M.B. Andersson), pp. 197-207. New York: John Wiley and Sons.

Wade, M.J. & Arnold, S.J. 1980. The intensity of sexual selection in relation relation to male sexual behaviour, female choice and sperm precedence. *Animal Behaviour* 28, 446-461.

Ward, P.I. 1988. Sexual dichromatism and parasitism in British and Irish freshwater fish. *Animal Behaviour* 36, 1210-1215.

Williams, G.C. 1957. Pleiotropy, natural selection, and the evolution of senescence. Evolution 11, 398-411.

_____. 1966. *Adaptation and Natural Selection*. Princeton, New Jersey: Princeton University Press.

_____. 1985. A defense of reductionism in evolutionary biology. In: *Oxford Surveys in Evolutionary Biology*, Volume 2 (ed. by R. Dawkins and M. Ridley), pp. 1-27. Oxford: Oxford University Press.

Zuk, M. in press. Parasites & bright birds: New data and a new prediction. In: *Ecology, Evolution and Behaviour in Avian*

Parasite Interactions (ed. by J. Loy & M. Zuk). Oxford University Press.

Zuk, M., Thornhill, R., Ligon, J.D. & Johnson, K. 1990. Parasites and mate choice in red jungle fowl. *American Zoologist.*

3. The Units of Behavior in Evolutionary Explanations

Sandra D. Mitchell

Sociobiology is that branch of evolutionary biology which aims at providing biological explanations of social behavior. Sociobiology invokes no new general theories. Rather, it is characterized by its special domain. Given the assumption that natural selection has been the most significant force operating in evolutionary history, the explanation of the presence of a given behavior in a population is most often couched in terms of its adaptive significance. That is, given recent developments in evolutionary biology, the explanation of a behavior can employ any of a variety of analyses (including game-theoretic models, optimality models, kin selection models and reciprocal altruism models) to show how a given behavior, in a particular environment, affects the reproductive success of individuals who display that behavior. In this regard, sociobiological investigations of adaptive behaviors require the same evidence as other evolutionary inquiries, including measures of the consequences of trait possession on relative reproductive success, the genetic basis of the behavior, the historically available alternatives and the level at which selection is operating.

Though detailed evidence for all the parts of a justification of adaptive significance is difficult for any evolutionary explanation, sociobiological explanations are subject to further, domain specific, complications. While behavior is unquestionably part of an organism's phenotype (or gene's "extended phenotype") (Dawkins 1982), I will argue that special concerns regarding the target of selection in sociobiological explanations, i.e. individuating evolutionarily significant behaviors, are problematic. In this paper I will consider two such problems; the difficulties of individuating evolutionarily significant behaviors, and the collateral problem of recognizing similarity of behaviors across species. After a general discussion of these issues, I will turn to some recent studies of the adaptive significance of rape for illustration.

To claim that this or that behavior is an adaptation, rather than an aberration, or present just by chance, is to invoke a particular causal history. Here "adaptation" refers to a result of the historical process of evolution by natural selection. This usage of the term is common, although there are instances where "adaptation" is taken to refer generally to the "fit" of organism to its environment whatever the causal process that generated it. According to the second sense the chameleon "adapts" to the change of color of the background by the chemical process leading to a change in skin color, as well as peppered moths adapting to environmental pollution by changes over time in frequencies of melanism in the population. In this paper I will follow the first interpretation (Williams 1966; Vrba & Gould 1982; Burian 1983; Brandon 1985a; Mitchell 1987a) and embrace the historical connotation. Identifying a behavior as an adaptation then can be taken to offer an answer to the question "Why is this behavior, rather than another, present in the population?" Correctly identifying a behavior as an adaptation entails that this behavior, rather than some historically available alternatives, has evolved by means of natural selection because of its consequences on reproductive success in a specified environment. Of course, cases other than fixation of a single trait in a population can result. For example, frequency dependent selection issues in a population maintaining a variety of traits which are adaptive at specific frequencies.

B_1 is an adaptation entails:

B_1 is present in a population because, relative to historically available alternatives, B_2, B_3.....B_n, in environment E, B_1 yielded, on average, greater net inclusive fitness than B_2, B_3.....B_n.

We can separate the required evidence into three parts:

1. showing differential reproductive success results from having or not having B_1;
2. showing the proximate mechanisms of the behavior/environment interactions on reproductive success;
3. showing differential genetic transmission of B_1 results (and hence leads to differential expression of the phenotype).

Meeting conditions 1 and 3 ensures that B_1 will increase in frequency in the population over time, i.e. it will evolve.

Condition 2 is required in order to distinguish selection *for* B_1 from mere selection *of* B_1. This extremely useful distinction was drawn by Elliott Sober (1984) in order to clarify the nature of the causal process generating a particular trait. It allows us to distinguish mere evolution of a trait from evolution by natural selection for the trait. To be an adaptation, the behavior must be a direct result of evolution by natural selection. It must be B_1's relation to environmental conditions that results in relatively higher reproductive success, rather than its being associated with reproductive success by means of either an indirect selective process, such as chance (by drift) or by means of an indirect selective process (B_1's being genetically linked to another trait and then increasing in frequency when the linked trait is directly selected for its consequences in that environment). In short, the explanandum behavior must result in the alleged consequence on reproductive success, and that consequence must be directly causally relevant to the presence of that very behavior.

The difficulties in directly justifying an adaptation explanation are legion (Endler 1986). Indirect arguments for adaptive significance are also offered. Two types of indirect argument, comparative analysis and the bypassing of proximate causal mechanisms, are common in extending sociobiological explanations from the nonhuman to the human realm. Comparative analysis or analogical arguments are proposed to allow inference about the evolutionary significance of a trait from evidence gleaned from multiple populations or species. Sometimes the adaptive significance of a trait is inferred from the correlation between repeated instances of it and a specific ecological condition. Given the correlation it is inferred that the trait has evolved as an adaptation in response to that ecological condition. This inference assumes that similar selection pressures produce similar responses, divergent pressures produce divergent responses. (See Bock 1977 for a detailed account of the assumptions of this type of argument.) On other occasions, direct evidence of the adaptedness of a trait to certain environmental conditions obtained for one species is generalized to other species sharing the same trait and conditions. All such comparative arguments are based on the presumption of similarity of traits, ecological conditions, and selection pressures.

A second type of indirect argument is found almost exclusively in human sociobiology. It allows evidence of the reproductive consequences of alternative behaviors to justify ascription of adaptive significance by presuming that there must be some genetic basis for all phenotypes and hence whatever developmental sequence or environmental trigger directly causes the phenotype can be ignored. That is, the "ultimate" causes for any trait are based in the genetic substrate so "ultimate" evolutionary explanations can ignore the proximate mechanisms (Durham 1979; Irons 1979).

Clarifications of the evidence required "ideally" for justifying a claim that a trait is an adaptation have been developed, in part, in the context of distinguishing between different *levels* at which selection may operate. For example, describing the benefits as accruing at the group level, and the differential transmission of the traits by means of differential group propagation gives grounds for claiming the trait is a group adaptation and hence that selection has operated at the group level. This paper is not concerned with questions of the level of selection but rather with the *target* of selection. What behaviors are candidates for explanation by appeal to the process of evolution by natural selection (at whatever level)? In the case of direct experimental evidence, one may ask what counts as a behavioral "unit" which could be an adaptation. For indirect inference one must consider what counts as the "same" behavior in different species. I will consider these two questions in turn.

INDIVIDUATING BEHAVIORS

The "Adaptationist Program" has been criticized for accepting evolutionary explanations that presume every observable trait is adaptive and then conjure a story which justifies that assumption. One objection to this strategy concerns the "atomization" of traits. By assuming every conceptually distinct trait is an adaptation, the argument goes, we have made errors in identifying the actual objects of biological processes. Gould & Lewontin (1979) have argued that what appears to us as an individual trait may not always appear so to the forces of evolution. We intuitively begin by suggesting that what is perceptually distinct to us (like the height of an individual, or its color) is a trait that has evolved by means of natural selection. But this may fail to explain the presence of the trait, because what we have identified is not an adaptable unit. One example they offer is the change in accepted explanation for the shape of the human chin. While we can designate a portion of the anatomy of the face as "the chin" and can see variety in this feature within a population or over evolutionary time, it does not operate as an integral whole in the process of evolution by natural selection. What we see as "the chin", the process of evolution "sees" as the necessary consequence of two distinct growth fields. Gould & Lewontin claim that the chin is a developmental artifact of evolution operating on other discrete traits and not itself the object of the joint process of selection and evolution. Since adaptations result only when both processes operate on the same object, the chin fails to be an adaptation.

This example challenges the strategy of producing adaptation explanations for ignoring developmental constraints. We cannot tell an adaptation story about the chin, it is claimed, because selection cannot weld together what development has torn asunder. (But see Gould 1987; and Alcock 1987 for a disagreement about the significance of this type of claim.) I would like to suggest that adaptation explanations of social behavior may suffer a related hazard. In this context adaptive status is conferred onto individual behaviors which may, in fact, not be individually transmitted. An isolated behavior, like rape, may not be the correct subject of an adaptation story if it is an integral part of a complex behavioral strategy, or the outcome of a learning process. While isolated behaviors may be shown to have the requisite differential consequences for reproductive success, the genetic transmission of the behaviors may take a more complex route.

The possible relationships between genetic replicators and individual behaviors increases in number and complexity when we consider the role of learning in generating specific human behaviors. For the purposes of this discussion I am not concerned with the complexities occurring in the genome, i.e. whether a single allelic pair, multiple alleles closely aligned on the chromosome or a more complex interplay among disparate sections of the chromosome control phenotypic expression of a given trait. Rather it is the complexities at the phenotypic level and the path from whatever the relevant replicating structure in the genome that is central. Thus "gene" or "genetic replicator" is used in the sense employed by Dawkins (1982) and Hull (1981). Direct evidence of the genetic basis of many behavioral traits has been difficult to obtain. Given the variety of pathways from genetic replicators to behaviors, evidence of reproductive consequences alone will not be sufficient to endorse claims of adaptedness.

In order to evaluate the legitimacy of such explanations it is, thus, necessary to explicate the variety of possible causal pathways connecting genetic replicators and social behaviors. If phenotypic variation is the direct object of natural selection, one must understand the underlying relationship between the phenotypic expression and genetic replicators to argue that any such phenotypic trait is, or can be, an adaptation.

One-one Relationship: g <--> b_1 and g' <--> b_2

If it is plausible to assume that specific behaviors are genetically determined directly with little environmentally induced variability, then, given a history of genetic variation, the presence of a behavior in a population can be unproblematically explained by its effect of maximizing inclusive fitness. Obviously, traits are neither completely genetically determined nor completely environmentally determined. Everything always has genetic and environmental components. The question is rather when should we appeal to a genetic component to explain the presence of the trait, and when to an environmental component. If specific behavior b_1 is directly tied to a replicator, g, and b_2 tied to g' in a one-one relationship, then it is clear that the differential consequences of having b_1 relative to b_2 on an organism's reproductive success will cause one, say b_1, to be present in a population via evolution by natural selection. (See Figure 1, pathway I.) In this case, the explanation of a behavior can be identified by its consequence on reproductive success, and hence the adaptation claim is justifiable.

A one-one relationship is found in traits like wing color in peppered moths. Any trait will have a range of expression depending on the range of environments experienced during development, i.e. the norm of reaction of the trait. Phenotypic expressions are the result of both the genetic coding and the environment of development and expression. The same genotype developing in two different environments can have very different corresponding phenotypes. (See Ricklefs' 1973 discussion of the arrowleaf plant for an illustration.) Once the organism is developed, the trait will no longer vary substantially with variation in the environment. (See Figure 1, environmental input between g and b_1.) This category, however, cannot be applied to

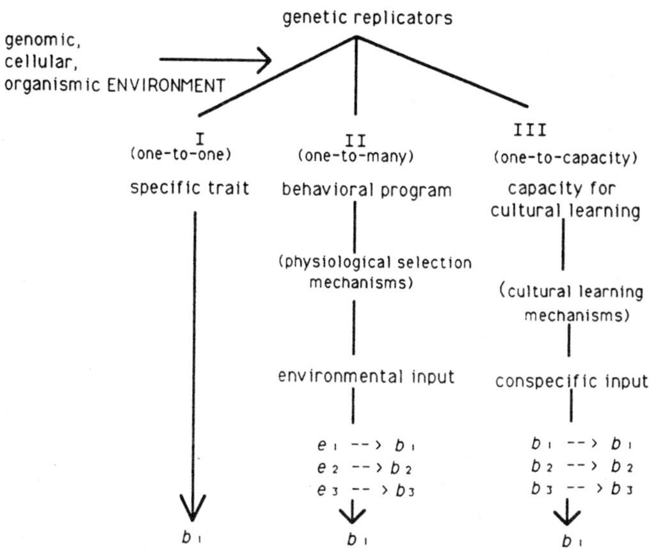

Figure 1. Causal paths from genes to behaviors.

social behaviors. It is in the very nature of what it is to be a social behavior, rather than a fixed trait, that the action involves a relationship with at least one other individual and is in part environmentally induced. The behavior will vary at least with respect to the presence or absence of relevant environmental input after development. This seems a plausible necessary condition even given our lack of detailed information of exact causal pathways from genetic replicator to behavior."Our ignorance of the pathways from genes to morphology is great. Our ignorance of the pathways from genes to behavior, pathways which surely vary with differences in the achieved morphology, is even greater." (Burian 1981-2: 54-5)

One-many Relationship: $g \dashrightarrow b_1$ or b_2 or b_3, and $g' \dashrightarrow b'_1$ or b'_2 or b'_3.

Suppose that specific behaviors are not tied directly to specific genetic counterparts, but are rather the result of environmental inputs via a genetically determined proximate mechanism. (See Figure 1, pathway II.) Thus behaviors are facultative, rather than obligate, or are governed by a "closed program." Here a particular *g* codes for a range of possible behaviors. Which behavior is expressed requires additional environmental input. Such environmental information is then mediated through proximate physiological mechanisms to generate a specific behavioral output.

$$\begin{aligned} g \text{ --> } & \text{if } e_1 \text{ --> } b_1 \\ & \text{if } e_2 \text{ --> } b_2 \\ & \text{if } e_3 \text{ --> } b_3 \\ g' \text{ --> } & \text{if } e_1 \text{ --> } b'_1 \\ & \text{if } e_2 \text{ --> } b'_2 \\ & \text{if } e_3 \text{ --> } b'_3 \end{aligned}$$

A specific behavior is expressed only by having both the genetic capacity that codes for the environment/behavior pair and having the appropriate environmental input. Thus behavioral variation within a population may be the result of either the variation in experiential histories of different individuals, i.e. differences in e, or may be due to similar environmental experiences and differences in the genetic replicator. If there is (or was) genetic variation (one of the prerequisites for evolution by natural selection), then the adaptive significance of a strategy including a set of behaviors can be identified with that set's consequence on reproductive success, and contributes to the justification of identifying the complete strategy as adaptive.

Since each individual acquires a behavior via the appropriate gene/environment conditions, the only means of transmission of traits across generations is the genetic pathways through differential replication. Thus evolution by natural selection is the appropriate causal history. Since that process takes consequences on reproductive success to be causally relevant, those consequences are explanatorily relevant.

The behavioral response in this case is "hard wired" by the genetic program - what Mayr calls a "closed program" in which "the program is contained completely in the fertilized zygote." (Mayr 1974: 652) Evidence for the genetic component of the behaviors in category II can be obtained from studies involving artificial selection. For example, it was noticed that there are two mating strategies adopted by male field crickets (*Gryllus integer*). A male will either call frequently or infrequently. A breeding program was developed to test for the genetic component to these alternate behaviors (there might clearly be environmental components having to do with the density of crickets in the area). Intense artificial selection effected a change in the frequency of the calling trait very quickly and hence provided evidence of a genetic component. (This case was reported in Trivers 1985: 95-98.)

Environmental Learning Plus Cultural Transmission: g --> (e_1 --> b_1) or (e_2 --> b_2) or (e_3 --> b_3) or (b_1 -->b_1) or (b_2 --> b_2) or (b_3 -->b_3).

Consider yet another way in which a specific behavior can be acquired. This is what some have called an open program, cultural learning or cultural transmission. Here the genome determines a capacity for learning from the environment or conspecifics. One manner in which learning occurs, allows an individual to adopt behavior b_1 by imitation of another instance of b_1 expressed by a conspecific. There clearly are genetic constraints even on such an open behavioral program. "An open program is by no means a *tabula rasa;* certain types of information are more easily incorporated than others." (Mayr 1974: 652) Here the informational input (a subset of general environmental input, namely that which is specifically from the behavior of another conspecific) is mediated through some proximate learning mechanism that selects a behavior as preferential to others based on some specified selection criteria (See Figure 1, Pathway III.). The specification of how this mechanism operates is the domain of learning theory. Some suggested proximate selection criteria include avoidance of pain and maximization of "satisfaction."

As outlined above, for a trait to be an adaptation two processes must occur and be appropriately connected. The first is that interaction of variants in a given environment resulted in differential reproduction. Whether it be through an individual trait bearer's own reproduction or that of genetic relatives, to be an adaptation the trait must have such an effect on reproductive success. The second is that the trait be differentially transmitted, which will occur if the trait is genetically produced and there is differential reproduction. That effect on reproductive success must cause the differential transmission of genetic replicators. And the differential replication of genes must in turn be responsible for the consequent frequencies of the behaviors.

It is obvious that different behaviors can have varying effects on the reproductive success of an individual and his/her genetic relatives. Abstaining from reproduction and not aiding in the survival and reproduction of kin is a behavioral trait that will fare ill on this test compared to having lots of offspring and helping kin. But the celibate hermit's behavior would be maladaptive in the biological sense only if the genetic replicators that get transmitted differentially as a result of differential reproductive success are responsible for the behavioral trait in question.

In the case of cultural learning, the step of selection for increased reproductive success may be severed from the step of transmission necessary for the evolution of the selected trait. No matter how greatly the trait enhances genetic reproductive success, that factor may have no role in the presence of the trait in future generations - transmission follows a different path. Indeed, not only is increased fitness not sufficient for explaining the presence of the trait, it may not be necessary. (For a detailed argument, see Boyd & Richerson 1985; Brandon 1985b; Mitchell 1987b.)

When we observe or presume changes in behavioral traits, the inclination to treat those changes as adaptively significant requires strong assumptions about both reproductive consequences and the pathway from genetic replicators to behavior. In sum, for an individual behavior to be an adaptation it must be the direct cause of differential reproductive success and it must be directly transmitted as a result. In the case of complex strategies (the one-many relationship), an individual component behavior is not the unit of adaptation. Since adaptations involve both selection and evolution, the correct unit is the complete strategy, for it is the unit that gets transmitted as an integral whole. In the case of a learned behavior, it is the proximate learning mechanisms themselves that have been the object of evolution, and not any one or complex set of behaviors. Having identified the criteria for a unit of adaptation, one can then garner direct evidence that a given behavior (or behavioral strategy) is in fact an adaptation. Artificial selection experiments may be designed to justify the genetic underpinnings of a behavior and field and lab studies used for determining the effects on reproductive success. Appropriate alternatives (either individual behaviors, strategies, or proximate cultural learning mechanisms) may then be proposed and discerned.

CLASSIFICATIONS OF SIMILARITY

Up to now I have been concerned with what counts as an individual behavioral trait. Once the unit of adaptive behavior has been clarified, a derivative problem arises in comparing behaviors in one population or species with those found in another. This issue becomes especially relevant in the employment of the comparative method. The new question is what are the criteria for similarity across populations or species or even taxa, that justify ascribing the same name to two behaviors? That is, what are the criteria for similarity which allow the same evolutionary explanation to be inferred. The problems surrounding the link between behavior and genetic determinant might be hidden if the descriptions of significant social behaviors are cavalierly attached to correlative genetic replicators. For behaviors to be explained by their adaptive significance, the very same item which results in relatively higher reproductive success must be present because of that consequence. Sociobiological explanations often leave this presupposition ungrounded.

Burian argues that one of the problems with establishing the genetic determinism of social behaviors arises, in part, from the use of two different descriptive paradigms or "units of behavior." (Burian 1981-2: 53) We describe social behaviors from the perspective of their human significance using terms like "aggressiveness" or "rape." But these do not necessarily correspond directly to the genetic units of behavior, namely particular genetic replicators or gene complexes that both operate as coherent wholes in transmission and are responsible for the development of the behavior. Since adaptation explanations require the operation of both selection of phenotypes and transmission of corresponding replicators, there needs to be an account of how socially significant behaviors are mapped onto biologically significant units.

The blurring of descriptive categories is most likely to occur in indirect sociobiological arguments. The "comparative method" is often employed in order to justify sociobiological explanations of human behavior. That is, the "same" or similar behavior is studied in a variety of species who share certain environmental or structural similarities. Evidence is obtained for explaining the behavior's presence in one (or some) species and, by analogy, is extended to account for the presence of the "same" behavior in humans. (Kitcher 1985: 184)

The Case of Rape

An example of the illegitimate grouping together of disparate behaviors under one descriptive category is found in Thornhill's work on rape in scorpionflies and humans (Thornhill 1980, 1984; Thornhill & Thornhill 1983). Thornhill's initial study of "rape" in scorpionflies suggested a generalized hypothesis about rape as an adaptive copulation strategy in any population where females exhibit choice of mates and males can secure material assets. He observed "rape" in the scorpionflies, and tested to determine the reproductive consequences of the behavior as well as describing the associated ecological conditions. On the basis of direct investigation of the different copulation behaviors of scorpionflies (including copulation with and without the presentation by the male of a "nuptial gift" of a dead insect or salivary mass to the female), combined with information gleaned from studies of territorial fish and mallard ducks, Thornhill formulated a general hypothesis about the conditions under which "rape" as a copulation strategy would evolve by natural selection in any population. Humans were explicitly included in the scope of the hypothesis. He later (Thornhill & Thornhill 1983) was involved in a more direct consideration of the evolutionary significance of human rape; that is, he tested the predictions of the hypothesis he developed from his scorpionfly studies. To endorse a general hypothesis and be compelled to further test applications of it for humans presumes that "rape" in scorpionflies and "rape" in humans is a similar behavioral strategy, that the necessary ecological conditions are shared, that the causal history which generated "rape" in each case would be the same and hence has the same adaptive significance for both.

Is it uncontroversial in the case of the flies, that their behavior is genetically controlled, and hence explicable by consequences on reproductive success? The "unit" of behavior question is not entirely straightforward even in this case. No individual male adopts a single copulation behavior obligately. Rather, the evidence from Thornhill's own experiments is that any male will "rape" under the correct triggering environmental conditions. Similarly any will, under appropriate conditions, send out a long distance pheromone when either guarding a dead insect or after producing a salivary mass. Hence "rape" is not an isolated behavioral alternative subject to selection, but rather part of a complex strategy for copulation that includes a set of outcome behaviors that depend on environmental triggers. In short, it is a component of a conditional male reproductive strategy composed of behavioral alternatives. It might be better represented as the ordered sequence: 1. If possessing dead arthropod, emit long distance pheromone, then copulate. 2. If no dead arthropod, produce salivary mass, emit long distance pheromone, then copulate. 3. If no dead arthropod and unable to produce salivary mass, and female is present, secure female with physical force, then copulate. Understanding it thus it is clear that identifying "rape" as an adaptation is shorthand for claiming that the complex conditional strategy which includes "rape" is adaptive. One must then make a case for variant copulation strategies being present in the evolutionary history of the scorpionflies such that some included the "rape" component and others did not. It should be pointed out that in more recent writings (Thornhill & Thornhill 1983, 1987) the conditional, facultative nature of copulation strategies is explicitly acknowledged. However, the evolutionary hypothesis is still framed in terms of *rape* being adaptive or maladaptive.

Clarifying what are alternative strategies subject to evolutionary explanation is important in the context of game-theoretic analysis as well. Those studies focus on explaining the maintenance of behavioral variation in a population by means of frequency dependent or disruptive selection processes. Austad has pointed out that confusion resulted from the absence of common terminology for describing behavioral alternatives. The use of both "tactic" and "strategy" indiscriminately in referring to behavioral components of complex strategies and obligate behaviors as well as the complex strategy as a whole produced the ambiguity (Austad 1984). Clearly if the unit of behavior is ambiguous, identifying alternative behaviors will be hopeless.

Suppose the evidence is convincing that copulation strategies which include "rape" are adaptive for the scorpionflies. Can we then infer that human rape is similarly explained? For an analogical argument to support an explanation of human behavior as adaptive, the properties appealed to as shared must be relevantly similar. Serious doubts can be raised to the successful grouping together of human behavior and fly behavior as "rape." If this similarity fails, then the evidence that the behavior in the one species is adaptive lends little credence to the claim that it is so for the other species as well.

Let us look more closely then, at the identification and explanation of "rape" in scorpionflies in order to see if evolutionary explanations of its adaptive nature are justified there, and if so, if they can be extended to explain rape in humans as well. Thornhill describes the observable sequence of events that constitute "rape" in scorpionflies.

> A rape attempt involves a male without a nuptial offering (i.e. dead insect or salivary mass) rushing toward a passing female and lashing out his mobile abdomen at her. On the end of the abdomen is a large, muscular genital bulb with a terminal pair of genital claspers. If the male successfully grasps a leg or wing of the female with his genital claspers, he slowly attempts to reposition the female. He then secures the anterior edge of the female's right forewing in the notal organ...Females flee from males without nuptial gifts. If grasped by such a male's genital claspers, females fight vigorously to escape. When the female's wings are secured, the male attempts to grasp the genitalia of the female with his genital claspers. The female attempts to keep her abdominal tip away from the male's probing claspers. The male retains hold of the female's wing with the notal organ during copulation, which may last a few hours in some species. (Thornhill 1980: 53)

Is what is described an instance of "rape?" In order to demonstrate that it is Thornhill suggests two criteria which must be met: "...it is necessary to (1) clearly distinguish female coyness from rape and (2) show that males that rape enhance their own fitness." (Thornhill 1980: 52) What is the motivation for these criteria? The descriptive content of the term "rape" must be derived from its use in human social contexts. For humans, purely behavioral information is insufficient to determine that a social interaction counts as rape (throughout rape will mean only heterosexual rape). There is an essential intentional component. For it to be rape two psychological conditions must be true, the female must be unwilling to engage in sexual intercourse, and the male must be willing. No behavioral expression is either necessary or sufficient to characterize this behavior.

This definition is unassailable if you consider a set of behavioral observations and ask yourself if any count as cases of rape. Consider a case of copulation where there is physical struggle between male and female. On the face of it, this can be either consenting sadist-masochistic behavior or it could truly be rape. What makes the difference is the intentional attitudes of the participants - not the behavioral counterparts. What if there is no physical struggle associated with intercourse? That behavioral set does not guarantee that what is going on is not rape. Fear of physical harm induced by threats can easily account for cases when the female is not consenting and yet not physically struggling. Again, what makes it genuine rape has to do with intentional attitudes. Since all such cases are male induced actions, the assumption of male willingness is unproblematic.

It is crucial to separate ontological from epistemological or evaluative judgments. The legal definition of rape involves assigning culpability and degree of punishment appropriate to a given case. Here the kind of behavioral evidence I have claimed to be inessential in defining rape may well come into play. But this is not a question of whether or not the action *is* rape, but whether or not the parties involved had good evidence for knowing that. For purposes of evaluation, one must first be able to identify the behavior.

How do the criteria for human rape correspond to the criteria used in Thornhill's study? In order to apply a concept like "rape" which is essentially intentional, not just correlated with intentions, an analog to unwillingness must be found. Thornhill's first criterion, distinction from female coyness, presumably plays that role. But what is packed into this distinction? Female coyness is taken to be a way for the female to exercise a discriminating role in interactions. By not engaging immediately in copulation behavior, the female may be able to elicit information about the male's fitness. Thus, in the case of the scorpionfly behavior Thornhill described above, it is plausible that the female by struggling to free herself from the grasp of the male, is in fact determining if he is strong enough to hold her, and hence likely to have those features which make him well adapted to a hazardous environment. Thornhill points out that 65% of adult mortality is due to predation by web-building spiders (Thornhill 1980: 54.). If the female is using the struggle to evaluate male fitness, then the behavior described is not "rape." What Thornhill argues, however, is that given his studies which show an ordered preference of females for males with large, rather than small dead arthropods as nuptial gifts, and dead arthropods rather than salivary masses, that there is no reproductive advantage for the female to copulate with a male who fails to present a nuptial gift (Thornhill 1984: 91). To be fair, Thornhill has refined his hypotheses regarding scorpionfly preferences, testing the contributions of body size, prey size, and frequency of males on female choice (see Thornhill 1984 for a list of his relevant studies). By receiving a gift, the female acquires nourishment which otherwise she would have to obtain by means of risky foraging in a hostile environment. Both the fact that a male offering a gift displays his ability to acquire food in that environment (producing a salivary mass is only possible after a male has recently fed) and the fact that she directly benefits materially from the food, make copulation with a "rapist" less beneficial. So to defend the view that the behavior is "rape" Thornhill must presuppose that to distinguish it from coyness the behavior must be clearly *not* in the reproductive interests of the female. Thus the analog to unwillingness is reproductive disadvantage.

This interpretation is consistent with the other half of the set of criteria. A male's fitness is enhanced, so Thornhill argues, by inseminating a female without having to put himself at risk foraging to acquire a dead arthropod to present or as a means of producing a salivary mass. What makes the incidence of "rape" in the flies so rare is the lower frequency of successful insemination for that behavior. Thus for a behavior to be "rape" it has to be, at the same time, in the reproductive interests of the male and against the reproductive interests of the female.

There are two criticisms to be raised to this analysis of "rape" in scorpionflies. The first is that it is not clear that the two criteria that Thornhill has set out are met. That is, it is not clear that the behavior is really "rape." The behavior of the female in struggling to free herself from the grasp of the male has not been sufficiently distinguished from coyness. Her behavior may elicit just the sort of fitness information she requires to make copulation in her reproductive interests. Furthermore, given the aggressive nature of the interactions with conspecifics, heterospecifics, and predators in that environment, the female's behavior may have nothing whatsoever to do with her willingness or unwillingness to copulate with a given male. Rather than struggle indicating an instinct to avoid copulation, it may be part of a different behavioral set. She may be avoiding being grasped in general, a behavior much to her advantage. Supporting this view are the following facts: there is no pheromone release from the "rapist" male, pheromones are used for species recognition, there is aggressive behavior over food between conspecific males, heterospecific pairs of any sex and with predatory spiders (Thornhill 1984: 81-83). Given this, it is plausible that the female's struggle with the "rapist" has nothing to do with copulation at all.

The second criticism is more global. If, in fact, the criteria are met, and the behavior is thus classified as "rape", by so doing Thornhill can not ask the further question, "Is 'rape' adaptively significant?" That is because to be rape in the first place, by definition entails that the behavior is in the reproductive interests of the male and contrary to the reproductive interests of the female. A more neutral classification of the behavior would allow the question of adaptive significance to be raised, tested and disputed.

With respect to the extension of the adaptive significance of "rape" in scorpionflies to rape in humans, two points must be made. The first has to do with the analog of unwillingness. In the case of humans, unwillingness refers to some proximate psychological mechanism that, while itself having an evolutionary history, can generate behaviors that are not directly subject to the process of evolution by natural selection. That means that, for humans, unwillingness is not directly correlated with consequences of reproductive failure. This is just the point of the distinction drawn above between behaviors that are part of a closed behavioral program and behaviors that are generated by cultural learning mechanisms. The relevant causal processes for these two categories are distinguishable. For that reason, the adaptive significance for what we may describe as "rape" in one case cannot be generalized to cover a behavior with a different, and not necessarily complimentary, causal history in the other case.

This slide between components of closed and open behavioral programs is made by those who believe that because a cultural learning program has itself evolved by means of natural selection, particular behaviors proximately generated by it must necessarily be in the reproductive interests of the actors. Thornhill and Thornhill (1983: 139) while recognizing the different causal routes involved in the evolution of human behavior - i.e. evolution via natural selection and learning via cultural selection models, nevertheless insist that "Both...routes are expected to result in behavior that promotes inclusive fitness (number of descendent and nondescendent kin) of individuals" (see also Durham 1979; Irons 1979; Shields & Shields 1983). However, Boyd & Richerson have convincingly argued that there is a consistent evolutionary account of how a cultural learning model can arise by natural selection and yet generate behaviors which do not promote the genetic fitness interests of the actors who express them (Boyd & Richerson 1985; Brandon 1985b; Mitchell 1987b). Once that inexorable relationship between behaviors generated by a proximate learning mechanism and consequences which increase the reproductive fitness of actors is given up, then all the statistical evidence that Thornhill and Thornhill provide on the reproductive consequences for rapists and raped women are beside the point.

Sociobiologists attempt to explain human behavior as biologically adaptive. Why a social behavior is present in a specific human society is answered by appeal to the consequences of that behavior for maximizing inclusive fitness. To be adequate to the task, such explanations require that there be an appropriate connection between the causal background processes. While it may be shown that a behavior confers a relative advantage for reproductive success onto the actor, for the behavior to be an adaptation, that advantage must be the cause of the behavior being present in the population. Such is the case when a trait is genetically determined since the advantage in reproductive success directly corresponds to an increase in genetic replication. However, when the direct connection between behavior and genetic replicators is severed, then the causal significance of the consequence of a behavior on reproductive success becomes questionable.

Whereas some types of behavioral variability are indeed the result of evolution by natural selection, namely those that are generated and transmitted by genes, other types of behavioral variability have a different etiology. The consequences on inclusive fitness of behaviors of the first type of variability can indeed explain why such behaviors are present in a population, even though there is no direct link between genetic replicator and specific behavior. The consequences on inclusive fitness will not, however, explain why a behavior is present that is learned and transmitted by imitation. Neither direct nor indirect arguments for the adaptive significance will help explain behaviors which are the result of a causal history independent of the combined processes of natural selection and evolution. (Thornhill and Thornhill's most recent collaboration (1989) shifts the focus of explanation of human rape to the psychological mechanisms generating male rape behavior and away from the behavior itself.)

In launching an investigation into the adaptive significance of a trait, it is advantageous to clearly identify the target of evolution by natural selection. Making explicit the assumptions concerning the relationship of the target to the genetic substrate, the specification of phenotypic alternatives, and the environmental conditions and reproductive consequences of the alternatives will aid in avoiding the comparison of incommensurable traits across species. Inferences of the same adaptive significance for similar targets of selection involves justification that the same causal forces are at work. Ultimately it is the causal history of the trait that determines whether it is an adaptation and the nature of its adaptiveness.

ACKNOWLEDGEMENTS

I wish to thank the following people for helpful discussions of the topics in this paper: James Bogen, Michael Dietrich, Jerry Downhower, Philip Kitcher, Peter Machamer, and Rob Page. Thanks also to Randy Thornhill for sending me his recent manuscripts.

LITERATURE CITED

Alcock, J. 1987. Ardent adaptationism. *Natural History* 96, 4.

Austad, S.N. 1984. A Classification of Alternative Reproductive Behaviors and Methods for Fieldtesting ESS Models. *American Zoologist* 24, 309-19.

Bock, W.J. 1977. Adaptation and the comparative method. In: *Major Patterns in Vertebrate Evolution* (ed. by M.K. Hecht, P.C. Goody & B.M. Hecht), pp. 57-81. New York: Plenum Press.

Boyd, R. & Richerson, P.J. 1985. *Culture and the Evolutionary Process.* Chicago, Illinois: University of Chicago Press.

Brandon, R.N. 1985a. Adaptation explanations: Are adaptations good for interactors or replicators? In: *The New Biology and the New Philosophy of Science* (ed. by D. Depew & B. Webber), pp. 81-96. Cambridge, Massachusetts: MIT Press.

_____. 1985b. Phenotypic plasticity, cultural transmission and human sociobiology. In: *Sociobiology and Epistemology* (ed. by J.H. Fetzer), pp. 57-74. Dordrecht: Reidel.

Burian, R.M. 1981-2. Human sociobiology and genetic determinism. *The Philosophical Forum* 13, 43-66.

Burian, R. 1983. "Adaptation." In *Dimensions of Darwinism: Themes & Counterthemes in Twentieth-century Evolutionary Theory* (ed. by M. Grene), pp. 287-314. Cambridge: Cambridge University Press.

Dawkins, R. 1982. *The Extended Phenotype: The Gene as the Unit of Selection.* Oxford: Oxford University Press.

Durham, W.H. 1979. Towards a coevolution theory of human biology and culture. In: *Evolutionary Biology and Human Social Behavior: An Anthropological Perspective.* (ed. by N.A. Chagnon & W. Irons), pp. 39-59. North Scituate, Massachusetts: Duxbury Press.

Endler, J.A. 1986. *Natural Selection in the Wild.* Princeton, New Jersey: Princeton University Press.

Gould, S.J. 1987a. Freudian Slip. *Natural History* 96, 14-21.

_____. 1987b. Stephen Jay Gould replies. *Natural History* 96, 4-6.

Gould, S.J. & Lewontin, R.C. 1979. The spandrels of San Marco and the Panglossian paradigm: A critique of the adaptationist programme. *Proceedings of the Royal Society of London* 8, 205-58.

Gould, S.J. & Vrba, E.S. 1982. Exaptation - a missing term in the science of form. *Paleobiology* 8, 4-15.

Hull, D.L. 1981. Units of evolution: A metaphysical essay. In: *The Philosophy of Evolution* (ed. by U. L. Jensen & R. Harre), pp. 2-44. Brighton: Harvester Press.

Irons, W. 1979. Cultural and biological success. In: *Evolutionary Biology and Human Social Behavior: An Anthropological Perspective.* (ed. by N.A. Chagnon & W. Irons), pp. 257-72. North Scituate, Massachusetts: Duxbury Press.

Kitcher, P. 1985. *Vaulting Ambition.* Cambridge, Massachusetts: MIT Press.

Mayr, E. 1974. Behavior programs and evolutionary strategies *American Scientist* 62, 650-9.

Mitchell, S.D. 1987a. Competing units of selection?: A case of symbiosis. *Philosophy of Science* 54, 351-67.

_____. 1987b. Can evolution adapt to cultural selection? *PSA 1986, Volume 2,* (ed. by A. Fine & P. Machamer), pp. 87-96. East Lansing, Michigan: Philosophy of Science Association

Ricklefs, R. 1973. *The Economy of Nature,* 2nd Edition. New York: Chiron Press.

Shields, W.M. & Shields, L.M. 1983. Forcible rape: An evolutionary perspective. *Ethology and Sociobiology* 4, 115-136.

Sober, E. 1984. *The Nature of Selection.* Cambridge: MIT Press.

Thornhill, R. 1980. Rape in panorpa scorpionflies and a general rape hypothesis. *Animal Behavior* 28, 52-59.

_____. 1984. Scientific methodology in entomology. *Florida Entomologist* 67(1) 74-96.

Thornhill, R. & Thornhill, N.W. 1983. Human rape: An evolutionary analysis. *Ethology and Sociobiology* 4, 137-173.

_____. 1987. Human rape: The strengths of the evolutionary perspective. In: *Sociobiology and Psychology: Ideas, Issues and Applications* (ed. by C. Crawford, M. Smith, & D. Krebs), pp. 269-292. Hillsdale, New Jersey: Lawrence Erlbaum Associates.

_____. 1989. The evolutionary psychology of rape. Unpublished manuscript.

Trivers, R.L. 1985. *Social Evolution.* Menlo Park, California: Benjamin/Cummings Publishing Company, Inc.

Williams, G.C. 1966. *Adaptation and Natural Selection.* Princeton, New Jersey: Princeton University Press.

4. Inference in Social Evolution Theory: A Case Study

John A. Byers and Marc Bekoff

SOCIAL EVOLUTION THEORY AND THE FALLACY OF AFFIRMING THE CONSEQUENT

The application of evolutionary theory to the study of social behavior often involves deductive and inductive reasoning. Here we will be concerned with a logical fallacy that often plagues deductive arguments, specifically, the *fallacy of affirming the consequent*, and show how it has been employed in the explanation of an important problem in evolutionary behavioral biology, namely, differential parental behavior toward offspring of different sexes. This fallacy is also frequently encountered in studies of human sociology (Pawson 1989).

The fallacious argument has the following logical form: (1) "if *p* is true then so is *q*" ; (2) evidence indicates that *q* is true; therefore (3) *p* is true. The conclusion (3) may be false even if the premises (1 and 2) are true. In a deductively valid argument, the conclusion must be true if the premises are true. A related valid argument form is called *modus tollens* (see Hempel 1966 for a general discussion). Thus: (1) if *p* , then *q*; (2) it is not the case that *q*; therefore (3) it is not the case that *p*. In this case, if the premises are true, then so is the conclusion.

Scientific inquiry also employs inductive reasoning, which produces general conclusions from premises about specific cases and in which the premises imply the conclusion with various degrees of certainty (see below). Comparisons between the views of Karl Popper, who claimed that all knowledge is fallible and who favored deduction over induction, and Rudolf Carnap, who thought that knowledge had foundations and stressed the importance of induction, are reviewed by Hacking (1983).

Although differences in opinion about the importance of induction have been, and will probably always be, put forth, logically invalid deductive arguments have produced a lot of confusion in efforts to interpret and explain observed patterns of behavior. This is so in research on the evolution and adaptive significance of differential parental care provided to male and female offspring, the focus of our paper. (It should be mentioned that scientists are not necessarily unable to assess conditional inferences; rather, biases in how data are viewed frequently color the way in which data are interpreted [Tweney & Yachanin 1985; Bekoff & Jamieson 1990].)

ADAPTIVE STORIES AND SOCIOBIOLOGY

It is easy to understand, and even sympathize with, the rush into sociobiology, especially for anyone who was required as a graduate student in animal behavior to read tracts such as Hinde's (1970) encyclopedic monograph on ethology and comparative psychology. Hinde's book was the clearest and most thoughtful analysis then available, but still, it was tough going. Animal behavior seemed often mired in semantic wrangling (the nature-nurture issue), or in tortuous reasoning (motivational models, reinforcement schedule theory). Although valuable progress had been made on rules of inference linking observed behavior to central nervous system events (Marler & Hamilton 1966), progress had slowed, and the field seemed to be stagnating (Barlow 1989). Enter the fresh wind of evolutionary theory, and presto! - a huge body of new, *testable hypotheses* was revealed. Even better, the hypotheses seemed walled-off from the messy concerns of ethology and comparative psychology. It wasn't necessary to worry about the mechanisms that produced behavior or development; one needed simply to ascertain whether or not animals acted in accordance with evolutionary prediction. To wit, Barash (1982: 45) wrote: "Although complex motivational processes cannot be ruled out in most cases, they are certainly not needed for evolutionary analysis based on the central principle of sociobiology, which states that individuals will tend to behave in a manner that maximizes their inclusive fitness..."

The sociobiological revolution was valuable in at least two ways. First, it offered theoretical rejuvenation to ethology (Barlow 1989; for a specific example, see the Caryl-Hinde exchanges [Caryl 1979, 1982; Hinde 1981]). Second, its empirical findings showed us that behavior of individual animals often is much more finely tuned to specific circumstances than previously would have been believed (see Byers & Byers 1983; Lendrem 1983, for examples).

Nonetheless, the strictly functional (adaptationist) approach has pitfalls, premier among which is *facile adaptationism* or "vulgar Darwinizing" (Kitcher 1987), which comes in two varieties. The first, mentioned as a problem by Wilson (1975: 27), and later roasted and carved by Gould and Lewontin (1979), has been characterized as "adaptive storytelling" (see also Caplan 1984). Observations of behavior are made, then adaptive explanations - "just-so stories" - are provided. Reasoning of this sort has diminished in frequency in the behavior literature and catches few victims nowadays. However, the second, more subtle, version of adaptive story-telling continues to take its toll. Here, prediction from theory about optimal behavior is made. Then, animals are observed in order to ascertain if their behavior matches the theoretically defined optimum. If there is a good match, the behavior is interpreted as an adaptation, *a design feature modified by natural selection in response to the selection pressures outlined in the predictive theory.*

The problem with this sort of reasoning is that the observed conformity to optimality may have been produced either by selection pressures, or other processes different from those on which the prediction was based (Gould & Lewontin 1979). The fallacy of affirming the consequent is a common error in the evolutionary interpretation of behavior because, more often than not, several plausible theoretical formulations predict the same behavioral result. Also, for many social traits, such formulations often are complementary, rather than competitive (Wittenberger 1981).

When initial findings support a hypothesis, many scientists routinely begin stepwise elimination of alternative explanations for the results. In this process argument, observation, and experimentation are used to show that the probabilities of alternate explanations being correct are low. Thus the probability that the original hypothesis is true becomes increasingly more certain. Such reasoning is known as Bayesian confirmation (Franklin & Howson 1988; Salmon & Fogelin 1988; Howson 1989; Howson & Urbach 1989). Regrettably, scientists who test social evolution theory do not routinely practice Bayesian confirmation, despite the existence of abundant alternative explanations for most observations in this area of inquiry.

BAYESIAN CONFIRMATION IN THE STUDY OF ADAPTATION

Dunbar (1982: 22) noted that "Distinguishing between possible explanations for an observed phenomenon will invariably require us to understand in detail the biological mechanisms that brought the phenomenon about." We agree. It seems that in most attempts to demonstrate an adaptation's function, it will be *necessary*, but not *sufficient*, to show conformity to predicted structure. Sufficient evidence will involve information that excludes other, competing explanations of the observed behavior (see Thornhill this volume). Often, sufficient evidence will involve data on the development of behavior. In the following section, we offer a detailed treatment to support this contention. We illustrate an example of the fallacy of affirming the consequent and show how data from studies of behavioral development are necessary to test the supposition of adaptation.

MATERNAL ADAPTATION TO INVEST MORE IN INDIVIDUAL SONS: HAVE ALTERNATIVE EXPLANATIONS BEEN ELIMINATED?

Maynard-Smith (1980) considered the following situation: (1) Sex ratio at birth is fixed at unity, (2) parents can adjust physiology and behavior according to sex of individual offspring, (3) sex-specific probabilities of offspring survival are equal, and (4) in response to greater than average parental investment, offspring of one sex gain in fitness more than offspring of the other sex. He showed that given these conditions, selection for differential parental investment, according to offspring sex, will exist and favor parents that invest more in individual offspring of the sex with the greater investment-specific elevation of fitness.

Polygynous mammals are considered to fulfill the assumptions of Maynard Smith's model (Clutton-Brock et al. 1981). In strongly polygynous mammals, birth sex ratios usually are close to unity (although emerging data reveal that the extent of sex-ratio adjustment remains unknown [Clutton-Brock & Iason 1987]), mothers, with few exceptions, learn individual identities of their offspring, and male success in intrasexual competition often is correlated with either body size or early dominance status, characters that conceivably could vary with maternal investment. Because polygynous mammals have these characteristics, it is predicted (Maynard-Smith 1980; Clutton-Brock et al. 1981) that mothers in such species should invest more heavily in individual male than in individual female offspring, because males are presumed to have the greater investment-specific elevation of fitness.

Before discussing these issues in detail, we need to consider briefly the use of the words *investment, effort, and cost.* The term "investment" traditionally implies an increase in offsprings' chances of surviving and a decrease in reproductive value (expected future reproduction) of care-givers (Trivers 1972). Any investment involves reproductive effort, usually measured as energy cost in joules (Pianka & Parker 1975), but sometimes also measured as elevated risk of injury that accompanies reproductive behavior (Maher & Byers 1987). Although an increase in reproductive effort implies an increase in investment, and vice versa, the mathematical function that relates effort to investment is unknown in most species. Therefore, differences in the reproductive effort required to raise male versus female offspring do not necessarily imply differences in investment (LeBoeuf et al. 1989; Clark et al. 1990). However, most authors we cite have used the term "investment" as if it were synonymous with effort; to avoid cumbersome prose we shall use the term "investment" throughout, although in most instances it is really *effort* that has been measured. Indeed, demonstrating enhanced survival of offspring *and* reduced reproductive output of care-givers is very difficult, especially under field conditions.

With few exceptions, data published to address predictions about differential investment in sons, or claimed in secondary citation to address this prediction, have focused on the fact that sons are energetically more expensive to produce than are daughters. Thus, evidence for differential maternal investment has used data on: (1) sex differences in birth weights; (2) sex differences in growth rates, or weights at weaning; (3) sex differences in various estimates of milk intake; and (4) differences in future fecundity, or estrus date, of females that have raised male-versus female offspring. By using this information, maternal investment skewed in favor of sons has been shown in Mountain sheep, *Ovis canadensis* (Hogg et al. 1990), Red deer, *Cervus elaphus* (Clutton-Brock et al. 1981; 1982), feral horses, *Equus caballus* (Duncan et al. 1984; Berger 1986), African elephants, *Loxodonta africana* (Lee & Moss 1986), several species of Phocid seals (Reiter et al. 1978; Kovacs & Lavigne 1986a; Anderson & Fedak 1987), Old world rabbits, *Oryctolagus cuniculus* L. (Boyd 1985), red-necked wallabies, *Macropus rufogriesus* (Johnson, 1986), coypus, *Myocastor coypus* (Gosling et al. 1984), and Mongolian gerbils,*Meriones unguiculatus* (Clark et al. 1990).

The question is, based on these studies and on many other demonstrations of greater birth weights, growth rates, and food requirements of male versus female young in mammals (Clutton-

Brock et al. 1981), can we infer that Maynard-Smith's (1980) theory of sexual investment is confirmed in the Bayesian sense? We think not.

The main problem is that in order to confirm Maynard-Smith's theory it is necessary but not sufficient to show that males are the more costly sex to produce. There are at least two competing explanations for differential investment. The first is that mothers control level of investment and that selection has operated on mothers to make greater investment higher in sons than daughters; this is Maynard-Smith's hypothesis. The second explanation is that offspring control investment, either through competition with each other or with the mother. Maynard-Smith mentioned this problem, with respect to the specific issue, raised by Clutton-Brock et al. (1982), of whether sex differences in offspring dispersal affect sex-specific investment. However, we assert that the problem is general, and applicable to all aspects of parental investment, including placental transfer and delivery of milk.

The second possibility, that offspring control investment, even to the possible detriment of mothers, has received little attention in the literature on differential investment. Kovacs & Lavigne (1986b: 1942) noted that differential investment "...may not be dependent on conscious prejudice by the parent..." and Lee & Moss (1986: 360) suggested that "...mothers may be constrained in their ability to choose how much to invest in calves of either sex." Otherwise, authors seem to have assumed that differential investment proves that selection has operated on mothers to invest differentially (Clutton-Brock et al. 1981).

A related argument, that identification of who controls investment (mother or offspring) is unimportant, was stated most clearly by Gosling et al. (1984: 293) when discussing their results on coypus:

> ...in terms of parental investment theory it is immaterial whether the mother appears to give resources to her offspring or the young appear to take the resources (as in the case of coypus). The critical aspect of the transaction is the transfer of particular amounts of resources from the mother to each offspring (i.e. the cost of each offspring to its mother).

We believe, contrary to Gosling et al. (1984), that the issue of control of investment is vital, because the target of selection (mothers or sons) is at issue. For example, does differential investment exist because mothers with traits that support excess investment in sons produce increased numbers of granddaughters (with the traits) through their successful sons (Maynard-Smith's hypothesis), or does differential investment exist solely because of selection on males to demand more from mothers during the period of maternal investment?

If the first case is true, maternal actions and physiology would have been modified by selection, to accommodate elevated demands by sons. If the second is true, maternal actions and physiology would *not* have been modified by selection. It is juvenile males that extract extra investment. If this second case is true: either it is an evolutionarily unstable moment, when sons extract more investment than is optimal for mothers (sons are the current winners of parent-offspring conflict [Trivers, 1974]), or it is part of an evolutionary equilibrium in which differential investment, optimal for mothers, has been achieved without modification of maternal traits. Although equilibrium presumably is what is implied by Gosling et al. (1984), either condition is different from the evolutionary process discussed by Maynard-Smith (1980) in which a rare parental allele coding for differential investment increases in frequency. As Maynard-Smith (1980: 250) noted: "Throughout the analysis, I have supposed that the level of investment is determined by the mother. Identical conclusions follow if it is determined by the father."

Now, it seems clear that to interpret differential investment as confirmation of the evolutionary process Maynard-Smith discussed, one must examine evidence that, during postnatal ages when differential investment is observed, mothers control level of investment. Otherwise, the alternate explanation of differential demands by sons is equally, if not more, plausible. Therefore, data that show only an elevated transfer of energy to sons provide incomplete support for the Maynard-Smith hypothesis; they do not demonstrate maternal adaptation. There lies the problem: *most work on this issue has focused on elevated cost of sons.* Few data have been published on specific maternal actions that support extra investment in sons. Thus, we conclude that existing evidence at best provides provisional support for Maynard-Smith's hypothesis.

Implicit in the above quotation from Gosling et al. (1984) was the following argument. When differential investment is found, it doesn't matter whether or not we find evidence that mothers control investment. This is so, because if it were disadvantageous for mothers to invest more in sons than in daughters, then selection would have operated on mothers to deny the extra investment to sons. Therefore, when sons manage to take more from mothers than daughters, we can assume that the situation is to the mother's advantage, or, at worst, selectively neutral to the mother. Such an argument clearly is circular; it requires that we *assume* maternal adaptation to prove maternal adaptation. It is fundamentally antievolutionary to insist that all animal traits represent optimal, or even currently stable, states (see Dupré 1987).

What would Show that Maynard-Smith's Theory of Sexual Investment is Correct?

Given the great current interest in empirically testing various theories of social evolution (see Dupré 1987; Krebs & Davies 1987; see also Blaustein & Porter, Koenig & Mumme, and Lima this volume), it is important to ask what types of data are needed to evaluate fully the Maynard-Smith hypothesis? Consider first different sorts of data on the relative costs of sons and daughters. Confirmation of differential investment by at least one of these measures seems *necessary* to prove Maynard-Smith's hypothesis, but is any measure alone also *sufficient*? Does any measure of cost also indicate whether mother or offspring controls the interaction? Information on birth weight or gestation length can show that males cost more *in utero,* but it is not obvious that mothers have complete control of placental exchange or length of gestation (Renfree 1982; Guyton 1987; Ingermann 1987). Data on postnatal growth rates are, at best, a measure of cost (energy flow); they do not indicate whether mothers or offspring control cost. Data on fecundity, or timing of estrus, of mothers that raised male versus female offspring in the previous year (Clark et al. 1990) are good measures of the relative costs of sons and daughters, but, like growth rate data, they say nothing about who controls the exchange of energy. Also, sex differences in growth rates do not necessarily imply differences in maternal investment; they might be due to differences in activity budgets of male and female offspring (Byers & Moodie 1990). Finally, we are left with estimates of milk transfer, and associated behavior of mothers and offspring. Here, we believe, data might be found to provide sufficient support of Maynard-Smith's hypothesis.

Because mothers clearly have the possibility, at least, of controlling nursing bout frequency and duration, and because such exchanges between mothers and offspring can be observed, evidence for specific modification of maternal behavior to support increased investment in sons could be obtained. For example, if it could be shown 1) that sons cost more to produce than daughters (by any of the post-natal cost [energy transfer] measures listed above), and 2) that sons suckled at a higher rate than daughters, and 3) that mothers always controlled nursing bout onset and duration during the period of differential investment, then it would seem plausible that maternal behavior had been modified by the kind of selection Maynard-Smith envisioned. Similarly, it might be shown that sons suckle at higher rates because mothers reject their suckle attempts less frequently than those of daughters, or because mothers wean (rejection of all suckling attempts) daughters at an earlier age.

We are not arguing that suckling rate data are the best (or even necessarily a reliable) measure of the relative costs of sons and daughters; suckling rate may (Carl & Robbins 1988) or may not be (Mendl & Paul 1989) an accurate measure of energy intake. Our argument is that, when sons are shown to be more costly, suckling and related interactions are most likely to reveal the extent to which mothers control energy transfer to offspring, and whether the control exercised seems male-biased. Probably, the exact kinds of data required to demonstrate maternal control of milk transfer will be diverse, and often species-specific.

At present, these data are far more rare than the many demonstrations that males require more energy (or are more costly) to produce. Reiter et al. (1978) showed that male elephant seal (*Mirounga angustirostris*) pups are weaned one day later than are female pups, and that weaning is controlled by mothers; because milk is the sole source of nutrition during the period of postnatal parental investment in pinnipeds, weaning date was a good index of parental cost. However, LeBoeuf et al. (1989) showed that subsequent fecundity of elephant seal mothers that raised sons was no less than fecundity of mothers that raised daughters. As we noted earlier, elevated costs of one sex need not imply elevated investment in that sex. Gauthier & Barrette (1985) showed that male and female white-tailed deer and fallow deer suckle at equal rates, and have equal frequencies of rejected suckle attempts. Byers & Moodie (1990) showed that male and female pronghorn suckle at equal rates (at some ages, females at higher rates), that mothers cut off suckling bouts of male and female fawns at equal frequencies at all ages, and that, at later ages when male and female suckling rates are equal, female fawns have higher rates of rejected suckle attempts than do male fawns.

To summarize, it is too soon to say whether the increased investment in sons that has been found in several polygynous mammals has resulted from selection on mothers to accommodate elevated demands by sons, or from selection solely on sons, with no modification of maternal traits. Our major argument has centered on the idea that demonstration of an elevated cost of male offspring, while *necessary* to support Maynard-Smith's hypothesis, is not alone *sufficient*. The issue of who controls investment is crucial, and further empirical study is needed. The problem cannot be dismissed by assuming maternal adaptation.

ACKNOWLEDGEMENTS

We thank Dale Jamieson, Bennett Galef, Sandra D. Mitchell, Allen Moore, Thomas Daniels, Randy Thornhill, and Margaret Dussault for comments on an earlier draft of this paper. Professor Mitchell suggested that we adopt a Bayesian approach and provided useful references. A previous version of this manuscript was also circulated for review and perhaps a bit too liberally paraphrased without the authors' consent. Oh well!

LITERATURE CITED

Anderson, S.S. & Fedak, M.A. 1987. Grey seal, *Halichoerus grypus*, energetics: Females invest more in male offspring. *Journal of Zoology* 211, 667-679.

Barash, D.P. 1982. *Sociobiology and behavior,* 2nd Edition. New York: Elsevier.

Barlow, G.W. 1989. Has sociobiology killed ethology or revitalized it? *Perspectives in Ethology* 8, 1-45.

Bekoff, M. & Jamieson, D. 1990. Reflective ethology, applied philosophy, and the moral status of animals. *Perspectives in Ethology* 9, in press.

Berger, J. 1986. *Wild Horses of the Great Basin: Social Competition and Population Size.* Chicago, Illinois: University of Chicago Press.

Boyd, I.L. 1985. Investment in growth by pregnant wild rabbits in relation to litter size and sex of the offspring. *Journal of Animal Ecology* 54, 137-147.

Byers, J.A. & Byers, K.Z. 1983. Do pronghorn mothers reveal the locations of their hidden fawns? *Behavioral Ecology and Sociobiology* 13, 147-156.

Byers, J.A. & Moodie, J.D. 1990. Sex-specific maternal investment in pronghorn, and the question of a limit on differential provisioning in ungulates. *Behavioral Ecology and Sociobiology* 26, 157-164.

Caplan, A.L. 1984. Sociobiology as a strategy in science. *The Monist* 67, 143-160.

Carl, G.C. & Robbins, C.T. 1988. The energetic cost of predator avoidance in neonatal ungulates: Hiding versus following. *Canadian Journal of Zoology* 66, 239-246.

Caryl, P.G. 1979. Communication by agonistic displays: What can games theory contribute to ethology? *Behaviour* 68, 136-169.

_____. 1982. Animal signals: A reply to Hinde. *Animal Behaviour* 30, 240-244.

Clark, M.M., Bone, S., & Galef, B.G. 1990. Evidence of sex-biased postnatal maternal investment by Mongolian gerbils. *Animal Behaviour* 39, 735-744.

Clutton-Brock, T.H. & Iason, G.R. 1987. Sex ratio variation in mammals. *Quarterly Review of Biology* 61, 339-374.

Clutton-Brock, T.H., Albon, S.D. & Guinness, F.E. 1981. Parental investment in male and female offspring in polygynous mammals. *Nature* 289, 487-489.

Clutton-Brock, T.H., Guinness, F.E. & Albon, S.D. 1982. *Red Deer: Behavior and Ecology of Two Sexes.* Chicago, Illinois: University of Chicago Press.

Dunbar, R.I.M. 1982. Adaptation, fitness and the evolutionary tautology. In: *Current Problems in Sociobiology* (ed. by King's College Sociobiology Group), pp. 2-28. Cambridge: Cambridge University Press

Duncan, P., Harvey, P.H. & Wells, S.M. 1984. On lactation and associated behaviour in a natural herd of horses. *Animal Behaviour* 32, 255-263.

Dupré, J. (ed.) 1987. *The Latest on the Best: Essays on Evolution and Optimality.* Cambridge, Massachusetts: MIT Press.

Franklin, A. & Howson, C. 1988. It probably is a valid experimental result: A Bayesian approach to the epistemology of experiment. *Studies in the History and Philosophy of Science* 19, 419-427.

Gauthier, D. & Barrette, C. 1985. Suckling and weaning in captive white-tailed and fallow deer. *Behaviour* 94, 128-149.

Gosling, L.M., Baker, S.J. & Wright, K.M.H. 1984. Differential investment by female coypus (*Myocastor coypus*) during lactation. In: *Physiological Strategies in Lactation* (ed. by M. Vernon, R.G. Peaker & C. H. Knight), pp. 273-300. London: Academic Press.

Gould, S.J. & Lewontin, R.C. 1979. The spandrels of San Marco and the panglossian paradigm: A critique of the adaptationist programme. *Proceedings of the Royal Society of London, B* 205, 581-598.

Guyton, A.C. 1987. *Human Physiology and Mechanisms of Disease.* Philadelphia, Pennsylvania: W.B. Saunders.

Hacking, I. 1983. *Representing and Intervening: Introductory Topics in the Philosophy of Natural Science.* New York: Cambridge University Press.

Hempel, C.G. 1966. *Philosophy of Natural Science.* Englewood Cliffs, New Jersey: Prentice-Hall.

Hinde, R.A. 1970. *Animal Behaviour. A Synthesis of Ethology and Comparative Ethology.* Second Edition. New York: McGraw-Hill.

_____. 1981. Animal signals: Ethological and games theory approaches are not incompatible. *Animal Behaviour* 29, 535-542.

Hogg, J.T., Hass, C.C. & Jenni, D A. 1990. Sex-biased maternal investment in Rocky Mountain bighorn sheep. *Canadian Journal of Zoology* , in press.

Howson, C. 1989. Accommodation, prediction and Bayesian confirmation theory. *Philosophy of Science Association* 2, 381-392.

Howson, C. & Urbach, P. 1989. *Scientific Reasoning: The Bayesian Approach.* LaSalle, Illinois: Open Court.

Ingermann, R. 1987. Control of placental glucose transfer. *Placenta* 8, 557-571.

Johnson, C.N. 1986. Philopatry, reproductive success of females, and maternal investment in the red-necked wallaby. *Behavioral Ecology and Sociobiology* 19, 143-150.

Kitcher, P. 1987. Why not the best? In: *The Latest on the Best: Essays on Evolution and Optimality* (ed. by J. Dupré), pp. 77-102. Cambridge, Massachusetts: MIT Press.

Kovacs, K.M. & Lavigne, D.M. 1986a. Maternal investment and neonatal growth in phocid seals. *Journal of Animal Ecology* 55, 1035-1051.

_____. 1986b. Growth of gray seal (*Halichoerus grypus*) neonates: Differential maternal investment in the sexes. *Canadian Journal of Zoology* 64, 1937-1943.

Krebs, J.R., and Davies, N.B. 1987. *An Introduction to Behavioural Ecology.* Sunderland, Massachusetts: Sinauer Associates.

LeBoeuf, B.J., Condit, R. & Reiter, J. 1989. Parental investment and the secondary sex ratio in northern elephant seals. *Behavioral Ecology and Sociobiology* 25, 109-117.

Lee, P.C. & Moss, C.J. 1986. Early maternal investment in male and female African elephant calves. *Behavioral Ecology and Sociobiology* 18, 353-361.

Lendrem, D.W. 1983. Sleeping and vigilance in birds. I. Field observations of the mallard (*Anas platyrhynchos*). *Animal Behaviour* 31, 532-538.

Maher, C.R. & Byers, J.A. 1987. Age-related changes in reproductive effort of male bison. *Behavioral Ecology and Sociobiology* 21, 91-96.

Marler, P. & Hamilton, W.J. III. 1966. *Mechanisms of Animal Behavior.* New York: John Wiley.

Maynard-Smith, J. 1980. A new theory of sexual investment. *Behavioral Ecology and Sociobiology* 7, 247-251.

Mendl, M., & Paul, E.S. 1989. Observation of nursing and suckling behaviour as an indicator of milk transfer and parental investment. *Animal Behaviour* 37, 513-515.

Pawson, R. 1989. *A Measure for Measures: A Manifesto for Empirical Sociology.* New York: Routledge.

Pianka, E.R. & Parker, W. S. 1975. Age-specific reproductive tactics. *American Naturalist* 109, 453-464.

Reiter, J., Stinson, N.L. & LeBoeuf, B.J. 1978. Northern elephant seal development: The transition from weaning to nutritional independence. *Behavioral Ecology and Sociobiology* 3, 337-367.

Renfree, M.B. 1982. Implantation and placentation. In: *Reproduction in Mammals Book 2: Embryonic and Fetal Development* (ed. by C. R. Austin & R. V. Short), pp. 26-69. Cambridge: Cambridge University Press.

Salmon, M.H. & Fogelin, R.J. 1988. *Introduction to Logic and Critical Thinking. Second Edition.* New York: Harcourt Brace Jovanovich, Publishers.

Trivers, R.L. 1972. Parental investment and sexual selection. In: *Sexual Selection and the Descent of Man* (ed. by B. Campbell), pp. 139-179. Chicago, Illinois: Aldine.

_____. 1974. Parent-offspring conflict. *American Zoologist* 14, 249-264.

Tweney, R.D., and Yachanin, S.A. 1985. Can scientists rationally assess conditional inferences. *Social Studies of Science* 15, 155-173.

Wilson, E.O. 1975. *Sociobiology: The New Synthesis.* Cambridge, Massachusetts: Harvard University Press.

Wittenberger, J.F. 1981. *Animal Social Behavior.* Boston, Massachusetts: Duxbury Press.

5. Natural History and the Superorganic in Studies of Tool Behavior

Thomas Wynn

INTRODUCTION

Like humans, chimpanzees make and use tools. This has been generally appreciated since the work of Jane Goodall in the 1960s (van Lawick-Goodall 1970). However, despite this common ground, the implications of nonhuman tool behavior for human evolution have only rarely been pursued in more than a superficial way. Indeed, for many students of human evolution, tool behavior remains a great evolutionary divide separating man from animal. "Thus tool modification by other primates does not erode the significance of human tool-making; it serves, if anything, to highlight how much further we have gone in that direction than any species" (Gowlett 1984: 175). At first consideration there is a dramatic difference between human technology and ape tool behavior; so, in a sense, Gowlett is justified in highlighting the distinction. In most domains of behavior, however, the comparative evidence has led naturally to hypotheses of evolutionary continuity. The comparative evidence for foraging has been used persuasively in discussions of early hominids, for example. This has rarely been the case for tool behavior. I believe that this curious lacuna can be traced to a fundamental mismatch in the theories and methods employed by natural scientists studying nonhuman tool behavior and the social scientists studying early hominid culture. In effect, the two study very different phenomena, make very different assumptions, and reach noncomparable conclusions.

In the following essay I will address this difference between the "natural historic" approach to tool behavior and the "superorganic" approach to technology. I will then discuss the problems that emerge at the point of convergence between the two perspectives - the emergence of human tool behavior. Finally, I will make some suggestions about how the continuity between nonhuman and human tool behavior can be reestablished and exploited for a greater understanding of early human evolution.

THE NATURAL HISTORY TRADITION

Most treatments of nonhuman tool behavior fall within the long-established tradition of natural history. Until fairly recently, the emphasis of such studies had been descriptive and anecdotal, with the tacit goal of documenting the range of tool behaviors encountered in natural populations (I will here omit discussion of tool behavior in captive settings). In the last thirty years or so, more systematic studies of the adaptive role of tool behavior in specific niches have also appeared. Beck (1980) has written a comprehensive review of this literature that both summarizes what is known from anecdotal and systematic reports and examines other general issues. Suffice it to say that tool behavior, including tool-using and tool-making, is known for both birds and mammals of various species. General phylogenetic trends are not obvious, however, and Beck argues that the diversity of tool behavior in the animal world is a result of convergence and parallelism, not phyletic continuity. In the following discussion I will focus on the tool behavior of nonhuman primates. Presumably it is within this narrower focus that one could hope to find phyletic continuity, in particular continuity between apes and humans. More importantly, I hope to identify the interpretive and methodological orientation of the literature in order to contrast it with studies of human technology.

Much of the literature on nonhuman primate tool behavior remains anecdotal. By anecdotal I do not mean heresay accounts, but sober descriptions of behavior observed in the field. However, these descriptions have minimal interpretive intent. For example, Hamilton et al. (1975) published a description of defensive stoning by chacma baboons (*Papio ursinus*) in South West Africa. The report includes a general description, tabulation and correlation of incidents, some measurement of projectiles, and an amusing description of specific episodes. "This frequently resulted in stones whipping over our ears. Usually we could dodge, but occasionally two or more individuals release at approximately the same time, complicating evasion" (p. 488). This account, and many others like it, operates to expand the range of known tool behaviors. It makes no attempt to generalize or raise theoretical issues. It is clearly within a long tradition of natural history that is based on cataloguing organisms and behaviors. This tradition supplies the data upon which interpretations can be based even if they are not interpretive. Occasionally such anecdotes have more general implications for understanding tool behavior because they expand the sample in unexpected directions.

Brewer & McGrew (1990) describe a single incident of chimpanzee (*Pan troglodytes*) tool behavior in which a rehabituated female (wild born) used a series of four tools to extract honey from the nest of a species of stingless bee. The first was a "stout tool" to break away a layer of bitumen. The second was shorter, thinner and "sharper pointed" and used with a different grip to widen the excavated hole. The third was a "bodkin" used forcefully to puncture the nest and the fourth was a flexible "dip stick" used to extract the honey. This brief report supplies sizes of the tools and duration of use and ends with the observation that "even if these were exceptional actions of an individual specialist, the skillful performance is impressive" (Brewer & McGrew 1990). Even though Brewer & McGrew do not press the interpretive implications of the incident, it has considerable significance. The anecdote is important because it is unique and documents something unexpected. Nevertheless, it is not itself interpretive and functions, as the first example, to expand our catalogue.

While many reports of tool behavior are anecdotal, a few are more systematic. They present comprehensive analyses of specific behaviors within a broader context, usually that of adaptive niche.

A fine example of this is Uehara's (1982) study of seasonal changes in termiting technique by chimpanzees in the Mahale Mountains of Tanzania. Uehara observed two groups of Mahale chimpanzees and noted that termites constituted a higher proportion of the diet in one group than the other. Moreover, the chimpanzees in one of the groups monitored the condition of termite mounds by taste in order to determine whether the mounds could be usefully exploited. Prior to the termites' seasonal swarming the chimpanzees used probes manufactured from plant stems to extract termites from the mounds, but later in the wet season they simply break open the mounds by hand and pluck the termites from the surface. The Mahale chimpanzees also manufacture a different kind of probe to "fish" for ants, a variety of foraging practiced year round. Uehara's study includes the usual descriptions of tool size, grip used, and so on. His study differs from an anecdotal account in terms of the number of observations, its comparative nature, and its attempt to integrate tool behavior into a more general picture of chimpanzee foraging.

Another example of an in-depth study of one variety of tool behavior has been made on chimpanzees in the Tai Forest of the Ivory Coast (Boesch & Boesch 1978, 1984). The Boeschs' observed chimpanzees using hammers to crack open several species of nut and also made systematic measurements of the hammers, the raw material, and the distances hammers were carried. They make the important documentation of selectivity - stone hammers for hard nuts like *Panda* and wooden hammers for *Coula*. The Boeschs' further argue that chimpanzees must employ some sort of cognitive map of their surroundings, which enables them to judge transport distances in a cost-benefit fashion. They emphasize what this tells us about chimpanzee cognition, rather than what this tells us about chimpanzee diet or seasonality.

Both Uehara's and the Boeschs' studies are more detailed and interpretive than anecdotal accounts. They provide extensive information and place their observations within an interpretive framework. Uehara's is more typical of systematic studies because his framework is a common one - the foraging aspects of adaptive niche. The cognitive interpretation of the Boesch study is much less common, though not unique (see for example Menzel 1978), but it also focuses on an aspect of the chimpanzees' current niche, their organization of space. Neither pursues the evolutionary implications of their results or, indeed, places their results in a general comparative context, beyond obligatory citations. Their intent, like that of anecdotal accounts, is to describe behavior and, unlike anecdotal accounts, place it in its adaptive context. Once again this is well within the natural historic tradition of description. Very few studies directly address the evolutionary implications in any depth.

Beck (1980) devotes a chapter of his book to the question of the evolution of tool behavior. Much of his discussion deals with the attempts to explain the emergence of tool behavior. He discusses, for example, object manipulation and agonistic behavior (specifically redirection of aggression) as possible sources and the role of "uncharacteristic niches" (that is, an animal morphologically ill-adapted to its current niche). Beck finds none particularly convincing. Indeed he finds it difficult to discuss tool behavior in evolutionary terms. "The evolution of a novel tool phenotype is disjunctive: the behavior does not exist until the moment that a novel stimulus-response sequence incorporating an object is produced and reinforced. Such quantum changes are not easily accommodated within modern evolutionary theory. Perhaps only the general genetic determinants of learning, and not learned behavior patterns *per se*, can be said to evolve" (Beck 1980: 183-4). By redirecting the question of evolution to a question of competence Beck reabsorbs tool behavior into the general topic of behavioral evolution. This is in keeping with the tone of anecdotal and systematic descriptions like the examples I presented above: Tool behavior is an aspect of adaptive niche (especially foraging), of no greater importance than any other, and evolves as other behavior evolves.

Huffman & Quiatt (1986) reiterate many of Beck's points in their discussion of stone handling by provisioned Japanese macaques (*Macaca fuscata*). In stone handling, the macaques collect and manipulate stones and pebbles in various contexts but do not use them as tools. The topic of stone handling is of potential interest to the question of stone *use* and Huffman & Quiatt are explicit about how stone handling might lead to tool use. It would require "environmental opportunity." In this example, provisioning by humans has relaxed pressure on the niche, in a sense permitting stone handling. If pressure were relaxed long enough then certain adaptive advantages of stone handling might appear and their competence selected for. Thus tool behavior evolves like other behavior through environmental change and niche transformation. Perhaps more interesting than their espousal of classic evolutionary process is their explicit denial of an alternative. "It [stone handling] encourages us not to think of the invention of stone tools either as a serendipitous solution to problems set by nature or as the automatic endpoint of a natural chain of events: noninstrumental manipulation of stones ➜ use of stones as tools ➜ manufacture of stone tools" (Huffman & Quiatt 1986: 422). This is an explicit denial of an internal dynamic of change, or to put it another way, a denial of progress as a factor in the evolution of tool behavior. Once again, this is well within the current tradition of natural history and neo-Darwinian evolutionary theory.

It is not that studies of nonhuman tool behavior have avoided the question of evolution. Many refer to it in passing, including mention of relevance to hominid evolution, and, as we have seen, a few address the matter in some depth. What is missing is *technological* change. In the natural historic literature the evolution of tool behavior is invariably phrased in terms of environmental constraint and niche alteration. Competence for tool behavior evolves in classic Darwinian fashion, and environmental opportunity or restriction determines whether a particular example of tool behavior, whatever its origin, survives or disappears. Despite attention to biomechanics of grip, size of tools, sex differences in foraging, and so on, the tools themselves are not seen to change. I mean the latter in both senses. Nonhuman tools have not been observed to change but, more to the point, natural scientists seem not to expect them to change, except to disappear or reappear as niches change. And it is this omission that most clearly distinguishes natural historic approaches to tool behavior from those of the "superorganic" tradition.

THE SUPERORGANIC TRADITION

Social science approaches tool behavior from an entirely different perspective. Whereas natural science views tool behavior from the perspective of adaptive niche, social science views tool behavior from the perspective of social context. Emphasis is on the ramifications that tool behavior has within the social, economic, symbolic, and ethical lives of the participants. Tool behavior becomes technology, a quasi-autonomous element that enters into the complex web of social and cultural dynamics. Social theory varies considerably as to the role of technology in human culture and history. Indeed, in the last forty years the disciplines of history of technology and the philosophy of technology have gained their own practitioners and literature. It is unnecessary to review this literature (the journal *Technology and Culture* is devoted to these questions, for example). Instead I will focus on two major threads that have directly or indirectly influenced consideration of the origin of human tool behavior: (1) the idea that technology is an autonomous force and (2) the idea of progress.

One of the major themes in philosophical literature is the idea that technology is an autonomous entity whose nature has profound consequences for human society. Jacques Ellul, one of the leading figures in the field, goes so far as to argue that man is now embedded in a technological milieu (something he terms Technique) rather than a natural milieu (Ellul 1963). This milieu is internally consistent, like nature, and is self-determining in the sense that it is independent of human intervention. Social phenomena exist *within* this context, not as separate equivalent domains of human behavior. The technological milieu is value neutral, neither necessarily benign nor necessarily evil. Nevertheless, it is the essential context in which such matters as human freedom must be addressed. Ellul's is perhaps an extreme position, but he makes explicit an idea tacitly held by many scholars in the history and philosophy of technology: technology as an autonomous force. Not all philosophers agree, of course (Lewis Mumford 1966, for example). Moreover, few, if any, of the scholars who deal with the question of early human tools cite the philosophical literature. As a consequence, we need not pursue exegesis of Ellul or Mumford. However, the notion of technology as an autonomous, superorganic force appears in social theory as well, the most well-known being Marx, who argued that the economic base of a society determined the other characteristics of that society. "Material forces of production," including tools, techniques, and control of physical power, determined the "relations of production," which are the relationships between individuals and institutions (nature of authority, for example). These then determine features of the "superstructures," which include government, religion, philosophy and so on. Changes in technology force accommodation in the relations of production and, in turn, in the superstructures (Somerville 1967).

More recent materialist theories take a similar tack, though they differ considerably in detail. In anthropology the best recent examples are the cultural evolutionism of Leslie White (1959) and the cultural materialism of Marvin Harris (1979). These view cultural adaptation, especially the technologically productive domains of culture, as determining the major characteristics of other social and symbolic systems and as being the leading elements of change. In this guise, "superorganic" technology has had considerable influence on the question of early tool behavior because most archaeologists are explicitly or implicitly materialists and archaeologists are the scholars most directly involved.

A second thread in the social science literature is the idea of technological progress. This is an idea that is so pervasive in Western thought that it is difficult to find sober definitions. Skolimowski (1966: 375) sees it as "...the ability to produce more and more diversified objects with more and more interesting features, in a more and more efficient way;" this is a reasonable approximation of the common sense notion. The idea of progress is very much a nineteenth century idea (Staudenmeir 1985) and, indeed, Mumford (1961) blames the scholarly popularity of progress on anthropologists and archaeologists of the period who devoted considerable energy to constructing classifications and sequences of artifacts based largely on complexity. The most influential of these anthropologists was Pitt-Rivers.

Born Augustus Henry Lane Fox in 1827, Pitt-Rivers' life coincided with the Victorian era. As the younger son of a Yorkshire landed family, he pursued a military career during which he was a central figure in the testing of the new Enfield rifles (1850s). It is here that he developed an interest in the history of firearms that rapidly expanded into a general interest in the history of technology. He acquired an extensive collection of artifacts from around the world. In 1880 he came into a large inheritance, which required that he use the Pitt-Rivers surname and also allowed him the time and resources to direct extensive archaeological excavations in Great Britain. He died in 1900 (Thompson 1977). Pitt-Rivers primary interest was the description and explanation of the evolution of material culture. His evidence consisted of an extensive ethnographic collection of indigenous material culture from all over the world. At the core of his analysis was a classification based on a direct analogy to biological classification, it being Pitt-Rivers' contention, unusual for the time, that a study of culture could be as scientific as a study of the natural world (Thompson 1977). Whereas previous classifications of artifacts had been based largely on geographic or ethnic origin, Pitt-Rivers classified exclusively on the basis of "affinities" of form, disregarding context. In doing so he was able to establish a formal series for throwing sticks, bows, boats, and so on, each example in a series sharing affinities with adjacent examples. Pitt-Rivers then argued, again on biological analogy, that the formal series also represented gradual evolutionary sequences. He initially drew on Darwin but later on Herbert Spencer for his evolutionary mechanism. He was always careful to emphasize that artifacts do not reproduce, and attributed the increasing complexity of artifacts over a series to a "succession of ideas" tied to the increasing utility of the artifact for its task (Lane Fox/Pitt-Rivers 1875). He argued, based largely on Spencer, that this succession of ideas would lead to selection for appropriate neural structures. Pitt-Rivers had considerable general influence at the time because he corroborated the Victorian notion of progress and the advance of technology. He was influential in the field of prehistoric archaeology because his ideas of "gradual sequence" of change could be applied directly to archaeological remains. For example, Pitt-Rivers traced the evolution of Neolithic spear points out of early Paleolithic "side tools" (side-scrapers in modern terms) through stages of "oval tools" and "leaf shaped tools." Pitt-Rivers also took leading roles in the Ethnological Society and the International Congress of

Prehistoric Archaeology so that, even though he did not write extensively himself, his ideas received wide circulation (Thompson 1977). In part he simply formalized the "progress" thinking of the times and combined it with the surge of evolutionary enthusiasm engendered by Darwin. But the ultimate result of his work was the establishment of the idea that artifacts have genealogies. His repeated observation that tools do not reproduce faded into the background in favor of an emphasis on gradual sequences of form driven by the progressive force of increasing "utility."

Of the two threads - technology as an autonomous force and progress - the former is the more pervasive idea in current social science. Even theoretical approaches that are not explicitly materialist acknowledge the importance of technology. "To say that technology is humanized nature is to insist that is a fundamentally *social* phenomenon: it is a social construction of the nature around us and within us, and once achieved, it expressed an embedded social vision, and it engages us in what Marx would call a form of life" (Pfaffenberger 1988: 244). While Pfaffenberger explicitly argues against technological determinism, it is clear from this quotation that he sees technology as something with ramifications throughout cultural systems. This is more in line with current fashion in social theory. For the most part, modern social theory deals with technology as a monolithic entity and rarely considers its components. There is rarely any consideration of actual tool behavior, so that we are remarkably naive about how people actually use tools (Pfaffenberger 1988). There have been a few recent studies that take a cognitive approach to tool behavior in modern contexts (e.g. Gatewood 1985; Daugherty & Keller 1982; Lemonnier 1985) but these are uncommon and not yet well-developed. Social science appears interested in how technology drives, or constrains, or is in turn affected by cultural context but it is not very interested in how the tools themselves are used.

The idea of progress, on the other hand, has few active practitioners on any level. The only recent example outside of the historical and archaeological literature is Wendell Oswalt (1976), who has made a comparative study of food-getting technology that is very much in the tradition of Pitt-Rivers. Indeed, the progress idea has mainly persisted in social science as a rarely examined assumption. It has changed little in the last one hundred years and is, in fact, a kind of "survival" from the heyday of nineteenth century evolutionism. As a consequence it does not fit well into modern notions of evolutionary mechanism. Indeed, superorganic conceptions of technology do not easily incorporate the important features of modern synthetic theory. What is the source and nature of technological variation? How are tools selected? Although there have been relatively simplistic attempts to equate invention with mutation and utility with selection (as discussed in Staudenmeier 1985), none have been very persuasive. And the idea of a gradual change in gene frequency just does not seem to apply to tools. The natural historic and superorganic perspectives on tools do share the concept of evolution but do not share a concept of evolutionary mechanism. In the absence of a common understanding of the mechanism of change, they say little of relevance to one another.

Before addressing the issue of early hominid tools I would like to summarize the major differences between the natural historic and superorganic approaches to tool behavior.

The natural historic approach emphasizes descriptions of tool behavior. These accounts are often detailed and exhaustive, and focus mainly on questions of adaptive context. The natural historic approach does not often attend to the ramifications that tool behavior has outside of the specific context of use. It shows little interest in evolution and no interest in technical change. Evolution, if mentioned, is seen largely as a matter of niche transformation. The natural historic approach is well within the mainstream of ecological, neo-Darwinian ethology.

The superorganic approach emphasizes the systemic role of technology within cultural systems, especially its relationship with social behavior. It sees technology as an autonomous force and as progressive - one technical idea leading to a better technical idea as a matter of course. Despite this emphasis on progress, the superorganic approach is oddly uninterested in the details of tool behavior. Its perspective is evolutionary, but more in the tradition of nineteenth century evolutionism than in the tradition of the modern synthetic theory.

It is clear that the two perspectives have little in common. They ask different questions, attend to different evidence, and have very different intellectual histories. They share only the subject of tools. This mismatch would be only a curiosity in the history of ideas - after all they both successfully address concerns in their respective intellectual fields and there is no need that they be in any way reconciled - if it were not that they collide on the one crucial issue of the origins of human tool behavior.

THE PROBLEM OF EARLY HOMINID TOOLS

One could perhaps make the case that modern human technology does indeed act as an autonomous, superorganic force and that herein lies the major difference between humans and apes. But even if true, it remains axiomatic that this quality evolved from something less impressive. Documenting this evolution should be one of the central tasks of students of human evolution. Comparative evidence from blood chemistry and DNA indicate that humans and chimpanzees are very closely related and probably share a common ancestor within the last six to ten million years (Tattersall et al. 1988). By 2.5 million years ago we have an archaeological record of chipped stone tools and by 2.0 million years ago we have archaeological refuse of tool-use in the form of smashed bones from mammalian prey (Toth & Schick 1986). Even conservative evolutionary thinking should lead one to explore the close evolutionary relationship between humans and chimpanzees to reach an understanding of the archaeological remains. In other words, knowledge of both modern chimpanzee and modern human tool behavior should be brought to bear to understand these very early hominid tools. This has almost never been done. Because studies of nonhuman primate tools never address technical change and rarely address the ramifications that tool behavior has for behavior in general and because studies of hominid tools always include these perspectives, there remains a fundamental mismatch in the literature. When paleoanthropologists search the primate tool literature for useful insights, they find little of relevance to the questions they ask. Interpretations of early tools have been made almost exclusively from the perspective of the superorganic approach.

The idea of progress can be found, little changed from that of Pitt-Rivers, in the archaeological literature on early stone tools. Chavaillon (1976: 572), for example, proposes the following "developmental lineage" for the earliest stone tools:

> First stage - utilization of a pebble as a hammerstone, necessarily associated with fragments detached during percussion. Second stage - the hominids continued to use pebbles as hammers, but in addition used the scraps, sharp fragments obtained during the course of manufacture. Third stage - the hominids deliberately broke pebbles in order to obtain flakes, the pebble thereby becoming a core, with the flake as one of the

> aims of the operation...Fourth stage - ...the flake continued to be sought after, but cores which carried cutting edges could themselves be used as tools.

The "lineage" of ideas here is classic progress in the Pitt-Rivers sense of a succession of ideas tied to utility. There is also something oddly missing. Chavaillon's first stage describes a kind of tool behavior well-known for chimpanzees (McGrew [personal communication] has even documented the unintentional production of flakes detached from stone anvils.) Even assuming that Chavaillon was unaware of systematic studies of chimpanzee use of stone hammers, anecdotal accounts had appeared as early as 1843 (Struhsaker & Hunkeler 1972). He makes no reference to them precisely because, I believe, he did not consider that progress could apply to nonhuman tools. *Chimpanzee* hammers are frozen in time, but *hominid* hammers progressed. Chavaillon is not alone in his implicit advocacy of progress. Gowlett (1986), for example, traces the development of the handaxe (a standardized tool that first appeared about 1.5 million years ago) as a series of stages: 1) bifacially flaked cobble, 2) discoid (a cobble flaked bifacially around the entire circumference), 3) an elongated discoid, and 4) a rudimentary handaxe with bilateral symmetry. He, too, refers to a sequence of ideas and to utility. There is no *a priori* reason why one could not adhere to a notion of progress and extend the sequence to known nonhuman tools. But this is never done. The sequences begin with hominid tools. Interestingly the progress approach never considers the possibility that flaked stone tools could have been acquired and then abandoned (in response to a niche transformation, for example). Once acquired the technology seems to have led slowly but inexorably to modern tools. There is an implied unilinearity in tool evolution, a unilinearity not encountered in the literature on hominid fossils because this mostly anatomical literature is based on current ideas of evolutionary mechanism, not on nineteenth century evolutionism.

Considerations of technology as an autonomous force are less baldly stated. One no longer encounters extreme statements of the "tools makyth man" argument. Instead, most paleoanthropologists consider early tools to have been an important element in a cultural system. "In other words, it was not the tools themselves that were the key factors in successful evolutionary coping. Rather, the associated social, behavioral and cultural processes, directing such activities as tool-making, hunting and gathering, were basic." (Holloway 1981: 288) According to Tobias, other elements of this system linked to tool behavior included transport of meat, delayed consumption, sharing of food, and distribution of food to adult and juvenile members (Tobias 1983: 181-182), all of which had profound consequences for social organization and even intelligence. Some paleoanthropologists, especially archaeologists, who, as I mentioned earlier, tend toward materialist explanations, still view tools as the leading element of this system. "Humankind's ability to modify elements of its environment into a range of usable tools was undoubtedly one of the principal behavioral traits that contributed to the success of the genus *Homo* during the Pleistocene" (Toth & Schick 1986: 26). The emphasis is not so much on tools alone but on the immediate and evolutionary affects of tool behavior on other elements of behavior, especially social elements. Fifteen years ago this led Isaac (1987) to propose a model of early hominid behavior based on meat acquisition, home bases, and sharing, with sharing being the key element in hominization. This rosy interpretation has now been largely rejected (even by Isaac himself before his death) but the idea of a social reordering linked directly or indirecty to tool behavior is still the interpretive orientation of archaeologists of the early stone age.

I do not wish to imply that paleoanthropologists are ignorant of the concepts of modern ethology and evolution. These concepts, in fact, provide the primary framework for interpretations of early hominid evolution. By using concepts of foraging derived from modern ethology, paleoanthropologists are erecting a picture of the adaptive niche of our presumed ancestor. One of the issues of current debate, for example, concerns the foraging behavior of *Homo habilis*, a hominid that lived in East Africa between 2 million and 1.5 million years ago. We know what the local habitat was like, we have a fair knowledge of this hominid's anatomy, and we have stone tools and even some of the refuse produced by the tools, in this case fragmented bones of various species. Based on knowledge derived from the ethology of modern carnivores and scavengers and a knowledge of the local environment, paleoanthropologists have reconstructed a possible scavenging niche (e.g. Hill 1984; Blumenschine 1987). Based on taphonomic studies of bone attrition and damage caused by carnivores and also on knowledge of butchery techniques used by modern hunters and gatherers, archaeologists have been able to reconstruct some of the foraging episodes of these hominids (Bunn 1981; Potts

1988). The result is a picture that incorporates the concepts and processes emphasized by modern ethology.

However, this is a picture that draws relatively little from the primate literature, and almost nothing from the primate tool literature. Knowledge of primate foraging is occasionally introduced to modify the picture derived from carnivores and scavengers (e.g. Toth & Schick 1986) and knowledge of primate social organization is occasionally used to predict hominid social organization (e.g. Foley & Lee 1989), but the extensive literature on primate tool behavior is almost never used as a source for interpretations. Instead, archaeologists draw on ethnoarchaeological studies of modern Nunamiut Eskimo (Binford 1981) or the Hadza of Tanzania (O'Connell et al. 1988). Again, the reason for avoidance of the nonhuman tool literature appears to lie in the mismatch between the prevailing perspectives. When stone tools enter the picture, paleoanthropologists look to modern humans (and the superorganic perspective) for interpretive analogies.

The unfortunate result of this mismatch between nonhuman tool literature and the early hominid tool literature is that we actually understand very little about the transition from ape-like tool behavior to human-like tool behavior. Neither the natural historic perspective nor the superorganic perspective easily accommodates intermediate conditions. The natural historic approach so neutralizes the technology itself that it does not even allow a useful distinction between human and nonhuman tools. True, both human and nonhuman tools solve problems, but the difference between the two appears to encompass more than just the details of these problems. Exclusive focus on adaptive niche does not lead easily to characterizations of intermediate kinds of tool behavior. On the other hand, the superorganic approach so reifies technology that the idea of a nonhuman technology, even one intermediate between ape and human, simply has no place in the general conception. Because neither of the two approaches encompasses intermediate kinds of tool behavior we have no useful analytical concepts to apply to the archaeological record of early tools.

TOWARDS A REMEDY

I would be remiss if I were to give the impression that all discussions of tool behavior fall clearly into a "natural historic" camp or a "superorganic" camp. W.C. McGrew, for example, has consistently pursued the broader evolutionary implications of nonhuman primate tools. He has demonstrated, for example, that chimpanzee material culture varies geographically for what can only be termed cultural reasons.

> Within the set of flexible and resilient materials, why should chimpanzees at Mt. Assirik [Senegal] specialize in twigs and leaf-stalks while those at Gombe [Tanzania] prefer grass and bark? The choice seems arbitrary. We can find no functional reasons for these differences and are forced to conclude that they are based on cultural rather than environmental factors" (McGrew et al. 1979: 205-206).

Discussion of culture and choice is generally encountered only in the superorganic approach, yet McGrew uses it persuasively in an analysis of nonhuman tools, a domain generally excluded by the superorganic approach. However, he does not attach to his use of culture the contextual or progressive implications that are usually part of the superorganic approach; we see no discussion of technology reverberating through chimpanzee life. McGrew's analysis is not a "simple" natural historic account of adaptive niches - in fact, it focuses on characteristics that are not clearly adaptive - and has obvious implications for the evolutionary continuity of tool behavior (within the chimp-human clade, at least). However, despite being often cited, these implications have been almost universally ignored in the literature on early hominids and the idea of chimpanzee culture has been relegated to the realm of curious evolutionary convergence.

Parker & Gibson's (1979) paper on the evolution of language and intelligence has also had relatively little impact, despite extensive citation. At the core of their analysis is a persuasive case for continuity of one kind of tool behavior, something they term extractive foraging. They argue that early hominids continued a type of tool behavior practiced by modern chimpanzees (and presumably our common ancestor) - the use of tools to extract food encased in hard shells, embedded in nests or mounds, and so on. When hominids moved onto the savanna they came to rely more and more heavily on extractive foraging, eventually transferring the behavior to the extraction of marrow from bones of large mammals. This argument for evolutionary continuity is persuasive because it focuses on known behaviors. Unfortunately, this rather elegant point is obscured, ironically, by Parker & Gibson's' reliance on a rather extreme form of the superorganic perspective. From extractive foraging they derive selection for larger brains, sharing and its social correlates, and even language. In other words, they do see tool behavior reverberating throughout early hominid life, but not, apparently, chimpanzee life. By doing so they obfuscate the continuity they so carefully established.

These two examples point to a possible solution to the theoretical mismatch in studies of tool behavior. McGrew and, at least initially, Parker & Gibson, narrow their focus from tool behavior in general to a specific characteristic of tool behavior (cultural variation and extractive foraging). Part of the difficulty in comparing tool behavior (and establishing evolutionary continuity) may lie in the diversity of behaviors that come together in any episode of tool use - motor habit patterns, memory, availability of raw material, and social context, to name just four. One can hardly expect similarities in all of them, yet in comparative studies tool behavior is often treated as a unit. Chimpanzee tool behavior, generally considered, compared to human tool behavior, generally considered, will hardly inspire feelings of kinship. Yet if we break tool behavior into smaller components we can perhaps see some continuity. Such an approach would expand the natural historic perspective to include nonadaptive characteristics and dissect the superorganic perspective into analyzable units.

One can, for example, trace continuity in tool cognition. Primatologists have a fair idea of the cognitive abilities of modern apes (e.g. Premack 1976; Passingham 1982) and it is possible, using appropriate theories, to assess cognitive abilities from prehistoric tools (Wynn 1989). While various kinds of thinking come into play in tool behavior (including motor memory, sequence production, and several others), problem solving provides an especially informative example of evolutionary continuity. Modern apes use a style of problem solving that we can term trial and error. They have a goal in mind (puncturing a bee nest, for example), produce a tool and, if it fails, make successive modifications until it finally succeeds. The stone tools and remains from the earliest archaeological sites suggest a directly comparable kind of *ad hoc* technology. The hominids had a specific goal (breaking into long bone diaphyses, for example) and successively modified a stone tool until a usable edge was produced. In this narrow sense, at least, early hominid tool behavior was almost identical to that of chimpanzees. Indeed, when one examines many such specific components of tool behavior, including spatial concepts, carrying, and reuse of sites, all appear very ape-like, suggesting that the earliest stone tool technology was in fact more like ape tool behavior than it was like modern human technology (Wynn & McGrew 1990).

Using this more atomistic approach one can also examine the hypothesis that superorganic features are in fact the major characteristic distinguishing human from nonhuman technology. Modern human tools, for example, almost always carry social information about the user in addition to being mechanical devices (BMW's vs. Yugos). We know that modern apes use tools in agonistic encounters and for foraging, indicating that tools perform in the two separate behavioral realms. However, there is no evidence that the same tools perform in both realms *simultaneously*. When did tools first carry social as well as mechanical load? Nothing about the earliest stone tools suggests that they performed in any domain beyond foraging. But by 1.5 million years ago we have tools manufactured to an arbitrary standard; we find thousands of these bifaces in site after site. While we do not know why they produced this particular shape, the existence of a shared, arbitrary idea about appropriate shape suggests that the hominids considered factors other than just effective foraging when making tools. Once again, this is a very narrow component of overall tool behavior, but it is different from anything known for apes and suggests the first glimmerings of real superorganic characteristics.

Analyzing the evolutionary continuity and discontinuity of the many specific components of tool behavior will likely produce a complicated picture. We cannot expect that all aspects of tool behavior changed hand in hand. But such a mosaic of evolution is precisely what evolutionary theory predicts.

LITERATURE CITED

Beck, B.B. 1980. *Animal Tool Behavior.* New York: Garland Press.

Binford, L. 1981. *Bones: Ancient Men and Modern Myths.* New York: Academic Press.

Blumenschine, R. 1987. Characteristics of an early hominid scavenging niche. *Current Anthropology* 28, 383-408.

Boesch, C. 1978. Nouvelles observations sur les chimpanzes de la foret de Tai (Cote-D'Ivoire). *La Terre et la Vie* 32, 195-201.

Boesch, C. & Boesch, H. 1984. Mental map in wild chimpanzees: An analysis of hammer transports for nut cracking. *Primates* 25, 160-170.

Brewer, S. & W.C. McGrew. 1990. Chimpanzee use of a tool-set to get honey. *Folia Primatologica*.

Bunn, T.T. 1981. Archaeological evidence for meat-eating by Plio-Pleistocene hominids from Koobi Fora and Olduvai Gorge. *Nature* 291, 574-80.

Chavaillon, J. 1976. Evidence of the technical practices of early Pleistocene hominids. In *Earliest Man and Environments in the Lake Rudolf Basin*. (ed. by Y. Coppens, R. Howell, G. Isaac, and R. Leakey), pp. 565-573. Chicago, Illinois: University of Chicago Press.

Daugherty, J. & C. Keller. 1982. Taskonomy: a practical approach to knowledge structures. *American Ethnologist* 5, 763-774.

Ellul, J. 1963. Ideas of technology. In: *The Technological Order*. (ed. by Carl Stover). Detroit, Michigan: Wayne State University Press.

Foley, R. & P.C. Lee. 1989. Finite social space, evolutionary pathways, and reconstructing hominid behavior. *Science* 243, 901-906.

Gatewood, J. 1985. Actions speak louder than words. In: *Directions in Cognitive Anthropology*. (ed. by J.W.D. Daugherty), pp. 199-220. Urbana, Illinois: University of Illinois Press.

Gowlett, J. 1984. Mental abilities of early man: A look at some hard evidence. In: *Hominid Evolution and Community Ecology* (ed. by R. Foley), pp. 167-192. London: Academic Press.

_____. 1986. Culture and conceptualization: The Oldowan-Acheulean gradient. In: *Stone Age Prehistory*. (ed. by G. Bailey and P. Callow), pp. 243-260. Cambridge: Cambridge University Press.

Hamilton, W.J., Bushirk, R.E. & W. Bushirk. 1975. Defensive stoning by baboons. *Nature* 256, 488-489.

Harris, M. 1979. *Cultural Materialism*. New York: Random House.

Hill, A. 1984. Hyaenas and hominids: taphonomy and hypothesis testing. In: *Hominid Evolution and Community Ecology* (ed. by R. Foley), pp. 111-128. London: Academic Press.

Holloway, R. 1981. Cultural symbols and brain evolution: A synthesis. *Dialectical Anthropology* 5, 287-303.

Huffman, M. & D. Quiatt. 1986. Stone handling by Japanese macaques (*Macaca fuscata*): Implications for tool use of stone. *Primates* 27, 413-424.

Isaac, G.L. 1978. Food sharing and human evolution: Archaeological evidence from the Plio-Pleistocene of East Africa. *Journal of Anthropological Research* 34, 311-325.

Lane Fox/Pitt-Rivers. 1875. On the principles of classification adopted in the arrangement of his anthropological collection. *Journal of the Royal Anthropological Institute of Great Britain and Ireland* 4, 293-308.

Lemonnier, P. 1986. The study of material culture today: Toward an anthropology of technical systems. *Journal of Anthropological Archaeology* 5, 147-186.

McGrew, W.D., C. Tutin & P. Baldwin. 1979. Chimpanzees, tools, and termites: Cross-cultural comparisons of Senegal, Tanzania, and Rio Muni. *Man* 14, 185-214.

Menzel, E. 1978. Cognitive mapping in chimpanzees. In: *Cognitive Processes in Animal Behavior* (ed. by S. Hulse, H. Fowler, and W. Honig), pp. 375-422. New York: Wiley.

Mumford, L. 1961. History: Neglected clue to technological change. *Technology and Culture* 2, 230-236.

_____. 1966. *The Myth of the Machine: Technics and Human Development*. New York: Harcourt, Brace, Jovanovich.

O'Connell, J.,K. Hawkes & N. Jones. 1988. Hadza hunting, butchering, and bone transport and their archaeological implications. *Journal of Anthropological Research* 44, 113-161.

Oswalt, W. 1976. *An Anthropological Analysis of Food-Getting Technology*. New York: John Wiley and Sons.

Parker, S. & K. Gibson. 1979. A developmental model for the evolution of language and intelligence in early hominids. *The Behavioral and Brain Sciences* 2, 367-408.

Passingham, R. 1982. *The Human Primate*. San Francisco, California: W.H. Freeman.

Pfaffenberger, B. 1988. Fetished objects and humanized nature: towards an anthropology of technology. *Man* 23, 236-252.

Potts, R. 1988. *Early Hominid Activities at Olduvai*. New York: Aldine.

Premack, D. 1976. *Intelligence in Ape and Man*. Hillsdale, New Jersey: Lawrence Erlbaum Associates.

Somerville, J. 1967. *The Philosophy of Marxism: An Exposition*. New York: Random House.

Skolimowski, H. 1966. The structure of thinking in technology. *Technology and Culture* 7, 371-383.

Staudenmeier, J. 1985. *Technology's Storytellers: Reweaving the Human Fabric.* Cambridge, Massachusetts: MIT Press.

Strushsaker, T. & P. Hunkeler. 1971. Evidence of tool-using by chimpanzees in the Ivory Coast. *Folia Primatologica* 15, 212-219.

Tattersall, I., E. Delson & J. Van Couvering. 1988. *Encyclopedia of Human Evolution and Prehistory.* New York: Garland.

Thompson, M. 1977. *General Pitt-Rivers.* Bradford-on-Avon: Moonraker Press.

Tobias, P. 1983. Hominid evolution in Africa. *Canadian Journal of Anthropology* 3, 163-185.

Toth, N. & K. Schick. 1986. The first million years: The archaeology of protohuman culture. In: *Advances in Archaeological Method and Theory, Vol. 9.* (ed. M. Schiffer), pp. 1-96. New York: Academic Press.

Uehara, S. 1982. Seasonal changes in the techniques employed by wild chimpanzees in the Mahale mountains, Tanzania, to feed on termites (*Pseudocnathothermes spiniger*). *Folia Primatologica* 37, 44-76.

van Lawick-Goodall, J. 1970. Tool-using in primates and other vertebrates. In: *Advances in the Study of Behavior,* Volume 3 (ed. by D. Lehrman, R. Hinde, & E. Shaw), pp. 195-249. New York: Academic Press.

White, L. 1959. *The Evolution of Culture.* New York: McGraw-Hill.

Wynn, T. 1989. *The Evolution of Spatial Competence.* Urbana, Illinois: University of Illinois Press.

Wynn, T. & W.C. McGrew. 1990. An ape's view of the Oldowan. *Man.*

II
Method, Analysis, and Critical Experiment

Introduction

How do we come to know what we know? What is good evidence? How much of what we learn is constructed in the first place by our presuppositions or by the methods that we use to collect data? How much of what we learn is "really out there?" While the collection of behavioral data is often simple, its proper analysis and explanation can be very difficult. Early ethological studies were typically highly descriptive and nonquantitative, with little evidence of hypothesis testing. Perhaps this is what led de Solla Price (1960: 1830) to write that "biology is a system that proceeds from biochemistry to the associated subjects of neurophysiology and genetics. All else, as they used to say of the non-physical sciences, is stamp-collecting."

The chapters in this section go a long way towards making it clear that ethological research is not mere stamp-collecting. Some essays consider case studies, while others discuss the applications of various techniques to the study of animal behavior. They all begin with the foundational assumption, articulated by Charles Otis Whitman (1898/1985) and Oskar Heinroth (1910/1985), and later championed by Konrad Lorenz (1981), that behavior is something that an organism "has" as well as something that an organism "does:" Behavior is a measurable phenotype that evolves (Barlow 1977; Bekoff 1977; also see Gordon Burghardt & John Gittleman's chapter in this section) and is influenced by immediate environmental variables.

In their case study of the role of the Mauthner cell in avoidance and antipredatory behavior, particularly in zebra fish, Randolf DiDomenico & Robert Eaton show how rigorous study can inform arguments about the causation of this deceptively "simple" motor response. They use the term "neural positivism" to refer to "attempts to define neural functions by what fails to happen in the absence of neural structures, and by what occurs when neural structures are artificially stimulated." Based on their critical analysis of what is learned using the three main ways of studying relationships between the nervous system and behavior - *correlation* between the activity of a structure and observed output; *stimulation* and the production of a behavioral response; and *lesioning* or preventing the activity of a structure and observing whether a given response is eliminated - they reject the notion that a univocal (one function, one structure) causal paradigm can be applied to the nervous system. They show that even with respect to very "simple" behaviors in "simple" organisms, a given end is not always achieved by fixed means. This has important consequences for experimental design in neuroethology.

Jane Packard and her colleagues also consider methodological and analytical issues in their discussion of how artificial intelligence (AI) can be applied to the study of animal behavior. Although there is a great deal of popular and scientific interest in AI, there is little consensus about its utility in explaining mind and behavior (see for example Dreyfus 1979; Searle 1980, 1990; Minsky 1986; Harnad 1989; Penrose 1989; Churchland & Churchland 1990). Packard et al.'s essay is a good introduction for the uninitiated and a good review for the more sophisticated reader. Starting from the assumption that the four basic components of behavioral processes are perception, action, motivation, and learning, and that a detailed understanding of how stimuli and responses are paired "is central to all analyses of behavior," they argue that AI can be valuable in helping us to study and to understand complex behavioral patterns. They are optimistic that communication among scientists in different disciplines will improve with the application of AI to the analysis of behavior.

Comparative analyses of behavior are central to furthering our understanding of the evolution of behavioral phenotypes but, until recently, rigorous analyses using analytical techniques were restricted to comparative analyses of morphological phenotypes and often suffered from misapplications. Gordon Burghardt & John Gittleman provide a valuable discussion of the historical roots of the comparative study of behavior (see also Richard Burkhardt in section I of this volume) and show how different taxonomic methods stemming from phenetics and cladistics can be used in comparative ethological studies. Burghardt & Gittleman also discuss new methods for distinguishing and quantifying phylogenetic and adaptive effects and "outline how some new comparative methods might provide a rigorous means to incorporate the early Darwinians' dream of a comparative psychology of mental characteristics into modern evolutionary biology."

Donald Kroodsma emphasizes the fundamental role of experiment and the importance of experimental design in his essay on bird-song and the study of dialect variation. As he points out, "An experiment that is inadequately designed is severely limited in what it can tell us about the world." He convincingly argues that many studies of bird-song dialects have not succeeded in testing the hypotheses that they were intended to test. Rigorous concern with methodology may slow or reduce publication but, as Kroodsma rightly concludes, our goal should be to attain knowledge about animal behavior rather than to publish papers (see also Bekoff 1989).

Steven Lima's chapter on vigilance (antipredatory) behavior in birds is also skeptical about some commonly accepted results. Individual birds cannot scan for potential predators and feed at the same time. This has led to the "many eyes" hypothesis which holds that by living in a group an individual can benefit from others who scan while it feeds and feed while it scans. The area of vigilance also is one in which cognitive ethologists can find plenty of room for research. For example, does an individual monitor its behavior based on an awareness of what others have done, are doing, or are likely to do? A few data strongly suggest that it does. If this is the case, we can ask how this information processed.

While some will feel that Lima and Kroodsma overstate the case that perhaps very little is actually known about their areas of expertise, none could deny that the applications of rigorous thought that characterizes their two chapters will be valuable for their own areas of inquiry and also for the field of animal behavior in general. Proper experiments and more data are needed in these and other instances.

The final two chapters in this section also show how methodological and analytical procedures can influence our acquisition of knowledge. Walter Koenig & Ronald Mumme develop this thought in response to a challenge to functional explanations of helping behavior. This challenge suggests that helping "originated and is currently maintained nonadaptively as a result of its tight linkage with the clearly adaptive behaviors associated with normal parental care." On this view, helping is an unselected consequence of what is selected. Although the authors do not consider Sober's (1984) distinction between selection *for* (the causal concept) and selection *of* (the effects of a selection process), it is relevant here. The critic of functional explanations of helping behavior is suggesting that there is "selection of" this behavior but not "selection for" it. Koenig & Mumme reply that if four levels of analysis are distinguished (evolutionary origins, functional consequences, ontogenetic processes, and mechanisms [including cognitive and physiological processes]) there are ample data to support the notion that helping has current adaptive value. However many studies fail to control for confounding variables such as the "group-living effect" (perhaps helper effects are merely a consequence of living in groups) and Koenig & Mumme are aware of the fact that more rigorous analyses of data are needed as are more carefully stated hypotheses.

Gail Michener, in her analysis of the behavioral ecology and reproductive biology of (mainly Richardson's) ground squirrels, echos many of the concerns of the other chapters. She notes that, because researchers employ various methods of data collection and analysis, results are often difficult to compare. She shows how "date-based" analyses (which involves pooling data collected on the same date regardless of the reproductive status of the individuals comprising the sample) and "event-based" analyses (which involves pooling data from individuals of the same reproductive status without regard to the date on which the data were collected) produce different results. Thus, when comparing the results of different studies, one must examine in detail how the results were obtained.

The chapters in this section make clear that the study of behavior involves more than "stamp-collecting" (serious philatilists might say the same about "stamp-collecting"). If studies are really to give us information about animal behavior and if the results are to be used for meaningful comparative analyses, then we cannot forsake a detailed concern with method, analysis, and critical experiment in the interest of speedy publication.

LITERATURE CITED

Barlow, G.B. 1977. Modal action patterns. In: *How Animals Communicate* (ed. by T.A. Sebeok), pp. 94-125). Bloomington, Indiana: Indiana University Press.

Bekoff, M. 1977. Quantitative studies of three areas of classical ethology: Social dominance, behavioral taxonomy, and behavioral variability. In: *Quantitative Methods in the Study of Animal Behavior* (ed. by B.A. Hazlett), pp. 1-46. New York: Academic Press.

_____. 1989. Assessing publication impact. *BioScience* 39, 586.
Churchland, P.M. & Churchland, P.S. 1990. Could a machine think? *Scientific American* 262, 232-237.
Dreyfus, H.L. 1979. *What Computers Can't Do*, 2nd Edition. New York: Harper and Row.
Harnad, S. 1989. Minds, machines and Searle. *Journal of Experimental and Theoretical Artificial Intelligence* 1, 5-25.
Heinroth, O. 1910/1985. Contributions to the biology, especially the ethology and psychology of the Anatidae. In: *Foundations of Comparative Ethology* (ed. by G. Burghardt, transl. by D. Gove & C.J. Mellor), pp. 246-301. New York: Van Nostrand Reinhold Company Incorporated.
Lorenz, K.Z. 1981. *The Foundations of Ethology.* New York: Springer-Verlag.
Minsky, M. 1986. *The Society of Mind.* New York: Simon and Schuster.
Penrose, R. 1989. *The Emperor's New Mind: Concerning Computers, Minds, and the Laws of Physics.* Oxford: Oxford University Press.
Searle, J.R. 1980. Minds, brains and programs. *Behavioral Brain Sciences* 3, 417-424.
_____. 1990. Is the brain's mind a computer program? *Scientific American* 262, 26-31.
Sober, E. 1984. *The Nature of Selection: Evolutionary Theory in Philosophical Focus.* Cambridge, Massachusetts: MIT Press.
de Solla Price, D.J. 1960. Review of I. Asimov's *The Intelligent Man's Guide to Science. Science* 132, 1830-1831.
Whitman, C.O. 1898/1985. Animal behavior. In: *Foundations of Comparative Ethology* (ed. by G. Burghardt), pp. 141-194. New York: Van Nostrand Reinhold Company Incorporated.

6. Neural Positivism and the Localization of Function

Randolf DiDomenico and Robert C. Eaton

INTRODUCTION

One of the major goals of neurobiology is to discover the causal determinants of behavior. In neurobehavioral studies, the most widely accepted experimental paradigm for achieving this goal is generally referred to as the localization of function. This paradigm is broadly understood as the attempt to identify the specific role of a neural structure in producing a particular behavioral function.

As an empirical guide, the localization of function begins with an anatomically circumscribed neural structure which can range from an individual neuron to an entire region encompassing millions of neurons. Included in the paradigm are three basic methodological approaches. The first approach involves correlating the activity of a neural structure with some measure of an organism's physiological or behavioral performance. Second, the neural structure is artificially activated to determine if it produces the behavioral response in question. Third, the activity of the neural structure is prevented by lesion or other procedures to see if the behavior is eliminated. These procedures form the outline of a classical approach that will be called the localizationist program in this paper. This program has historically endured and gives us much of our contemporary understanding of neural functions.

There is ample evidence to demonstrate that the production of certain behavioral patterns are strongly associated with particular brain structures. As early as the nineteenth century, studies using cortical stimulation indicated that different parts of this structure were important for movement, vision and language (Brazier 1959). Although contemporary research continues to reinforce the idea that localized motor functions occur in the brain, many authors interpret these data cautiously (Luria 1966; Gazzaniga 1978; Kertesz 1983; Evarts et al.1984; Edelman

1987).

There are many reasons for caution when interpreting claims that behavioral functions originate from localized brain structures. First, the conceptualization of this paradigm depends on the definitions of ambiguous and sometimes arbitrary terms such as "function," "structure," "behavior," and "causality." (For a discussion of the definitions and implications of these terms see Eaton & DiDomenico [1985], Rosenberg [1985], and DiDomenico & Eaton [1988].) Second, since the methods typically include artificial stimulation and lesioning, they often yield results that are difficult to interpret because nervous systems show both resiliency and recovery in response to damage and have flexible or alternative strategies for performing essential functions (Evarts et al. 1984).

Despite these problems, there is a version of the localizationist program that claims that individual neurons or discrete groups of cells are depots for particular behavioral functions, suggesting a type of "one-structure, one-behavior" doctrine. The evidence of this is the use of one causal paradigm by many contemporary neurobehaviorists. Taking the results from the three localizationist methodologies (correlation, stimulation, and lesioning) they argue that particular structures or cells can be described as "both necessary and sufficient" for the behavioral phenomenon in question (Figure 1). In other words, "a neuron or a neurophysiological process is sufficient for a behavior or its modification if, when activated, it causes that behavior or its modification; a neuron or neurophysiological process is necessary for a behavior or its modification if, when eliminated, the behavior or its modification is eliminated, and so on" (Davis 1986: 288). This causal paradigm is both entrenched and unchallenged. Endorsements of it can be found throughout the recent literature. For example, in a lead article in *Science* on the neural basis of learning, R. F. Thompson (1986: 942) stated: "By 'essential memory trace circuit' I mean the neuronal circuitry from receptors to effectors that is necessary and sufficient for learning and memory in a given training paradigm." Or, to take another example from a highly cited target article on motor control in *The Behavioral and Brain Sciences*, "*responsibility for a given behavior should be attributed to a cell only if its activity is necessary and sufficient for the initiation of the behavior.*" (original authors' italics; Kupfermann & Weiss 1978: 7)

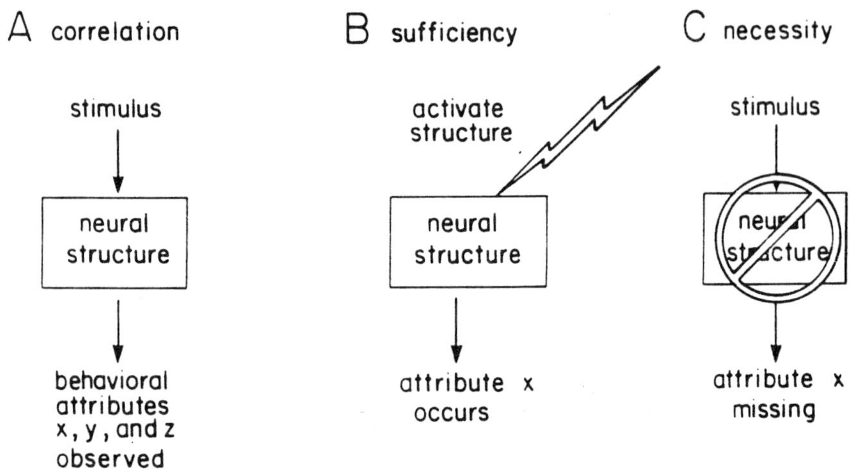

Figure 1. The three criteria for assigning functions to structures according to the strong localizationist view. They contend that tests B and C are equivalent to the causal criteria "sufficient and necessary". In this case, satisfaction of these tests leads the strong localizationist to conclude that the neural structure in question is the cause of behavioral attribute x. As explained further in the text, we refer to the advocates of this view as *neural positivists.*

According to Bunge (1979: 33) the clearest historical origins of this understanding of causation can be traced to Galileo who stated: "That and no other is to be called cause, at the presence of which the effect always follows, and at whose removal the effect always disappears." At first this definition seems satisfactory because it has a clear methodological meaning (stimulation and lesioning) and provides criteria for deciding whether a structure or neuron is the causal origin for a behavioral event. Although Galileo provided the "necessary and sufficient" causal definition, he was aware of the difference between ideal and experimental situations (Cohen 1985). However, contemporary advocates of this view of causation take it for granted that this approach is both a theoretical and empirical ideal that is unequivocal and universal (Eaton & DiDomenico 1985). In this paper we argue that the belief in an unequivocal causal paradigm has its origins in an important intellectual movement of the 1920s, known as logical positivism. This movement was developed by a group primarily comprised of prominent scientists, philosophers, and mathematicians, collectively referred to as the Vienna Circle, who advocated the formulation of rules having "the form of explicit definitions which provide necessary and sufficient observational conditions for the applicability of theoretical terms." (Suppe 1977: 18) This idealistic notion that logical laws of necessary connection "tower above the objects to which they apply" (Bunge 1979: 22) was eventually abandoned by its original creators and is no longer acceptable among knowledgeable scientists and philosophers (Suppe 1977; Rosenberg 1985; Churchland 1986). According to Rosenberg (1985), although its motives were laudable, the difficulties of logical positivism proved to be insurmountable.

Despite the inadequacies of logical positivism, the existence of strong localizationists demonstrate that a latent and naive form of it is very much alive in neurobehavioral science today. In this paper we introduce the terminology "neural positivism" to refer to strong localizationists who, in the influential spirit of logical positivism, advocate the "necessity and sufficiency" paradigm as shown in Figure 1. Neural positivism attempts to define neural functions by what fails to happen in the absence of neural structures, and by what occurs when neural structures are artificially stimulated. However, paradoxically the neural positivists have combined this faulty paradigm with a deterministic concept of causality that even the logical positivists considered untenable. We contend that neural positivist claims about structure and function are at best overstated and at worst wrong.

Very much like their predecessors in logical positivism, neural positivists resort to a radical operationalism to replace real theoretical constructs. Thus, to all positivists the meaning of a term is the method of its verification; then taking this to an extreme the neural positivist asserts that the operations to test "necessity and sufficiency" are "required" (Rock et al. 1981), and "exhaust" (Kupfermann & Weiss 1978) the meaning of the neural function in question. By borrowing the discredited ideas of the physicist Percy Bridgman (1927), the neural positivist reconstructs an operational empiricism in which neural concepts are synonymous with the corresponding set of operations used to reveal them. The operational approach failed in physics because it led to the denial of field theories of gravitation and electromagnetism because they are not directly observable (Bunge & Ardilla 1987). Moreover, operationalism resulted in anomalies in which theoretical terms such as *mass* have as many distinct definitions as experimental procedures used for determining them (Suppe 1977). Thus, there was on one hand an impoverishment of recognized theoretical constructs and on the other hand an unacceptable proliferation of concepts defining the same phenomenon.

In this paper we use both empirical and theoretical considerations to analyze neural positivism. We will show that neural positivism results in inconclusive and contradictory conclusions when applied to complex and probabilistic biological systems. Our argument is illustrated with an example from the neural network underlying control of the fish escape response. This behavior is the characteristic startle response seen when fish in an aquarium respond to a sudden tap on the glass. Under natural conditions, this response is used to avoid predators (Eaton & Bombardieri 1978; Webb 1986; Katzir & Intrator 1987). Experiments on this preparation include neurophysiological recordings from acute and freely-swimming animals and experiments involving various types of ablation and stimulation procedures. These studies introduce a comprehensive framework for discussing the general implications of neural positivism.

Our analysis will show that neural positivism is beguiling. It seems to apply a rigorous and logical decomposition of the neural mechanisms underlying behavioral acts. But, when experimentally tested, the causal criteria of the neural positivist result in inconsistent conclusions. This is because the extant neural circuitry is far more complex and adaptive than allowed for by background assumptions of the neural positivists. We conclude with a discussion in which we emphasize that the purpose of neurobehavioral experiments is to discover (and not to invent *a priori*) the causal neural mechanisms of behavior. The emancipation from preconceived causal paradigms means that diverse methodologies can provide insights into the causal explanations of behavior and allows the discovery of the natural adaptability of neural networks.

THE MAUTHNER CELL SYSTEM

The fish escape response is triggered by neurons of the reticulospinal formation. This system is phylogenetically old and its basic organization has been conserved over the course of vertebrate evolution (DiDomenico et al. 1988; Nissanov & Eaton 1989). A wide range of studies shows that this system is generally important in coordinating movements of the body (Lawrence & Kuypers 1968; Peterson 1984).

Our analysis concerns the function of an identified pair of bilateral reticulospinal neurons, the Mauthner cells (M-cells) of fish (Faber & Korn 1978; Eaton & Hackett 1984). M-cells are among the most well studied identifiable vertebrate neurons (Bullock 1978) and they share many similarities with mammalian reticulospinal neurons in general (Nissanov & Eaton 1989). M-cells are involved in the initiation of the C-start escape response shown in Figure 2A. This movement is produced by muscles of the trunk and also includes those muscles controlling movements of the eyes, jaw, operculum and fins. In the present study we focus on the trunk movements. These have two parts which result in a sudden and ballistic propulsion. During the initial preparatory phase of the C-start (stage 1) the animal bends its body in a C-like shape; during the propulsive phase (stage 2) the animal accelerates away from the stimulus. At issue is the role of the M-cell in the expression of this behavior.

A simplified representation of the system is shown in Figure 2B. The M-cell has a diverse array of sensory inputs, but of most importance to this discussion are the afferent fibers from the octavolateralis system. These afferents mediate perceptions of sound, water movement and body orientation. The axon of the M-cell synapses both on motoneurons and relay neurons that activate the complex and extensive muscular contractions associated with the escape movement (Hackett & Faber 1983). The extensive knowledge about this apparently simple system makes it well suited for an analysis of neural positivism.

The experimental question about the M-cell has centered on whether the firing of the M-cell is "sufficient" to trigger all the various muscle contractions associated with the C-start behavior and whether the cell's firing is "necessary" for those contractions to occur. In positivist terms, the issue was whether the M-cell could be shown to be "necessary and sufficient" or the "cause" of, the onset of the behavior (Kupfermann & Weiss 1978; 1986). We have previously referred to this translation of causal theory into an experimental paradigm as the Command Neuron Experiment (*CNE*; Eaton & DiDomenico 1985; DiDomenico & Eaton 1988). This issue developed from results on a number of systems, such as the crayfish escape network, where some experiments (e.g. Olsen & Krasne 1981) seemed to support the existence of such "command" neurons.

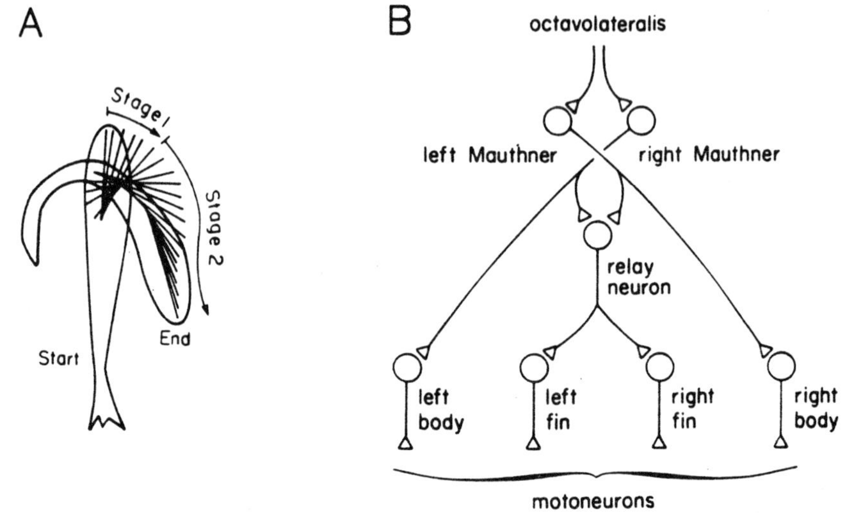

Figure 2. A. The C-start escape response of the goldfish. Shown are silhouettes of a fish in the beginning position (start) at the time of a stimulus on the left. This results in activation of the left M-cell and a 180-degree turn to the right in an interval of about 0.1 seconds (end). The midline segments are computed from high-speed imaging data obtained from a digital matrix camera. Analysis of the midlines is used to measure the parameters of the two major components (stages 1 and 2) of the movement (see text). B. A simplified diagram of the M-cell system. The M-cell excitatory sensory input is from the same side of the body whereas the M-cell axon contacts body motoneurons on the opposite side, thus resulting in contractions away from the side of the stimulus. Associated with the response are fin movements which are activated bilaterally by each M-cell (from Eaton & DiDomenico 1985).

M-Cell Activity Is Correlated with Escape

Because of the robust electrical properties of the M-cell it has been possible to implant an electrode in the brain and record the activity of one of these cells in the freely swimming animal. In these experiments, there is an action potential in one of the M-cells when the animal is presented with a threatening stimulus, such as an object dropped into the water (Eaton et al. 1981, 1982, 1988). The action potential is followed at a short and fixed interval by the onset of the C-start. When the stimulus is on the left, the left M-cell fires and the animal moves to the right. A stimulus on the right is followed by firing of the right M-cell and an escape to the left. Thus, because of its connections and its firing time, it was suggested that the M-cell might activate all of the motoneurons leading to the stage 1 contraction of the C-start.

M-cell "Necessity and Sufficiency" for Escape in Restrained Animals

The electrophysiological accessibility of the M-cell makes it seem technically possible to do experiments of the CNE type that directly test its "necessity and sufficiency" for the escape behavior. When sensory fibers to the M-cell are stimulated artificially by means of an electrical shock, the M-cell fires a single action potential which is followed within a few milliseconds by a contraction of the body musculature on the side opposite the activated cell. In these experiments on immobilized animals, the muscular contraction is monitored by recording an electrical event in the muscle. This electromyogram, or EMG, is an indication of the muscular activity associated with the escape behavior (Rock 1980). Such an experiment demonstrates that the firing of the M-cell has a powerful and direct influence on contractions of the body musculature.

Because of the large size of the M-cell it is possible to use a glass micropipette electrode to penetrate the M-cell membrane for artificial activation. Since M-cell firing results in a body movement, this experiment must be done on a restrained, anesthetized animal. With this method the experimenter can selectively turn on the M-cell and its post-synaptic followers without turning on any additional pathways involved in movement production. When the M-cell is stimulated in this way, the large EMG, mentioned above, is elicited and the animal moves its tail to one side (Rock 1980; Rock et al. 1981; Hackett & Faber 1983). From such studies it was concluded that the firing of the M-cell is a sufficient condition for the onset of the escape behavior (Rock et al. 1981; Hackett & Greenfield 1986).

Is the M-cell required for the EMG seen in these experiments? This can be readily tested by inactivating the cell during the presentation of the electrical shock to its sensory input fibers. The inactivation was also done with a micropipette electrode which, in this case, was used to pass a small hyperpolarizing current to increase the cell's threshold to the electrical stimulus (Hackett & Faber 1983; Hackett & Greenfield 1986; Rock et al. 1981). When this experiment was performed, the EMG, formerly elicited by the sensory stimulus, was prevented.

From such CNE experiments the positivist conclusion was that the firing of the M-cell is necessary and sufficient for the first stage of the escape response (Rock et al. 1981). According to one popular conclusion based on these views: "To wiggle the tail of a worm may take a network of thousands of neurons, but to flick the tail of a fish, only one, the Mauthner cell" (Edelman 1987: 23).

Is the M-Cell Indispensable for Escape Behavior?

If the positivist conclusion is correct, it should be consistent when tested in a variety of ways. For instance, removal of the M-cell should prevent the escape behavior when the animal is given a threatening environmental stimulus. However, the evidence does not support this prediction.

Lesions of the M-cells do not disrupt motor performance of C-starts in freely-swimming animals. Individual M-cells were lesioned by passing a small current from a metal microelectrode inserted into the M-cell (DiDomenico et al. 1988). This coagulates the neuron and causes a small lesion of about 0.1 mm in diameter. Following destruction of one (or both) M-cells, we analyzed C-starts by comparing the responses to those seen in the same animals before the destruction of their M-cells. To limit the possibility of overlooking subtle deficits in motor performance following the lesion, we used a high-speed and sensitive analysis method involving a computerized matrix camera that operates at 500 frames per second (Eaton et al. 1988). This instrument system allows quantitative analysis of multiple trials using four biomechanical (performance) measures.

As shown in Figure 3, fish give C-starts that are very similar whether or not the M-cells are intact. The upper pair of midlines (1 and 2) in Figure 3 show control responses before experimental treatment: the lower set (3 through 8) are responses after both M-cells were removed. Responses in the absence of the M-cell are referred to as non-M responses. No statistically significant differences were found between M-responses and non-M responses when we analyzed the movement according to 1) stage 1 angle, 2) stage 2 angle, 3) angle of the center of mass (CM) relative to the starting position and 4) distance covered. The large number of trials, the excellent temporal resolution, and the continuous ranges of the response parameters, make it very unlikely that we missed important subcomponents that clearly distinguish the M-responses from the non-M movements. This result does not imply that differences do not exist, only that the M-responses and non-M are so similar that their differences cannot be easily detected even with very sophisticated methods.

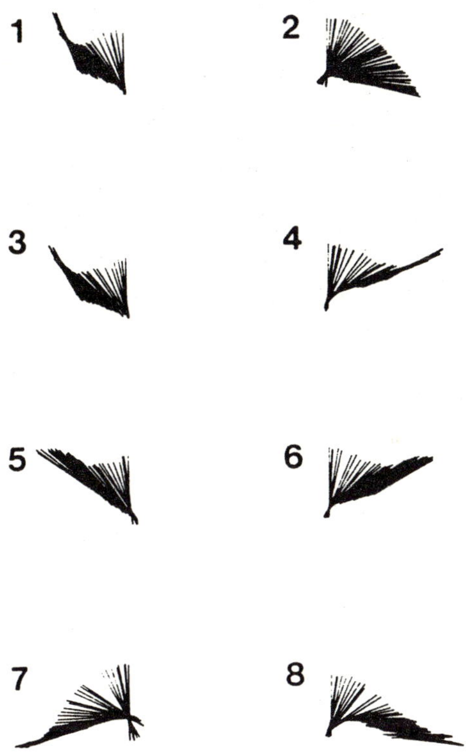

Figure 3. Comparison of the midlines of two control responses (1 & 2) in an animal in which the M-cells were intact and six non-M responses (3 through 8) in the same animal after both M-cells were lesioned. Both sets of responses are within the range of normal varlability and are statistically indistinguishable from each other when evaluating motor performance according to all measures (from DiDomenico et al. 1988).

M-cell presence appears to result in responses with a slightly faster reaction time to the stimulus. This 0.004 second difference between the average of the M-responses and the non-M was statistically significant (DiDomenico et al. 1988). Despite this difference, 77% of the non-M and M-responses were in the same range of latencies (0.012 - 0.022 seconds). Within this range one could not distinguish non-M from M-responses. Elsewhere we have discussed the behavioral implications of this difference in latency (DiDomenico et al. 1988).

Functional Substitution of Mauthner by Non-Mauthner Circuits

A neural positivist who knew only about the CNE tests on restrained animals would reach a conclusion opposite that of a positivist who knew only about the lesion test on free-swimming animals. In the CNE experiment the M-cell seems to be necessary and sufficient for escape and in the other it is dispensable. Even for behavior as seemingly simple as the escape response of fish, neural positivists run into trouble.

Part of the origin of the contradiction is that the various experiments used different methodological approaches in their tests of M-cell function. Thus, although the terms *necessity* and *sufficiency* appear incontrovertible, different methodological approaches result in different conclusions about the underlying neural network.

To illustrate this point, consider the following thought experiment based on the model network in Figure 4A. In visualizing how the model works, the starting condition is that the M-cell is the only neuron that triggers escape to the sensory stimulus. In the terms of neural positivism, the firing of the M-cell is the neural cause of the onset of the behavior. The positivist makes two claims for the adequacy of tests for "necessity and sufficiency." First, conclusions based on experimental tests for "necessity and sufficiency" can accurately account for the causal chain under normal conditions (if the conclusions cannot inform us about the normal condition, then there is little point to the experiment). Second, the positivist claims that the conclusions from "necessity and sufficiency" are unequivocal, even though the complete set of system interactions is unknown (if the experimenter had complete knowledge in advance, again there would be no need to perform the tests).

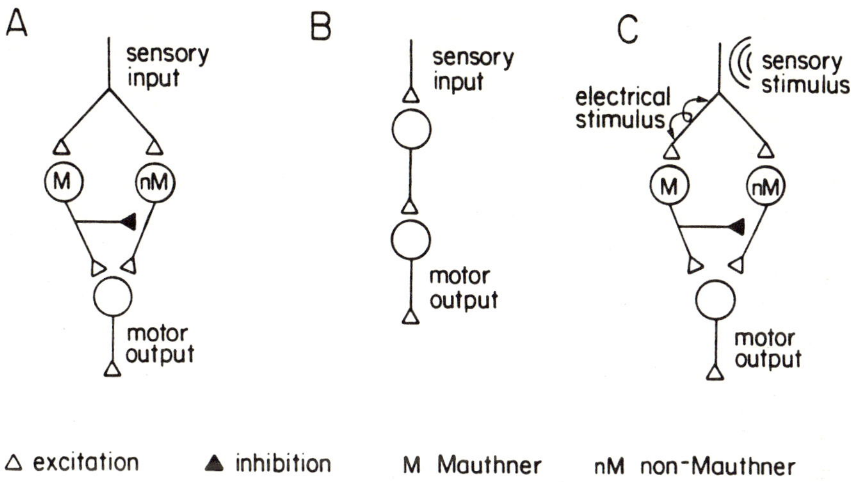

Figure. 4. A. A model network with non-linear response properties in which the non-M circuit can substitute for the M-cell. Networks with such properties result in false-negative conclusions when the neural positivist attempts to assign function to structure (see text for details). B. A linear model network in which the tests for "necessity and sufficiency" provide consistent and correct evaluations. Here triggering of the intermediate neuron results in the motor response and blocking this neuron prevents it. However, the conclusion is the trivial one that the intermediate neuron is necessary and sufficient, or the cause of, its own output. C. A model showing how different stimulus tests for "sufficiency" can activate the network in different ways and thereby result in different conclusions about network connections and characteristics. In one case, a naturalistic sensory stimulus was used and activated both the M-cell and the non-M branches of the circuit. In the other case, an artificial electrical stimulus activated sensory fibers in a way that only triggered the M-cell and not the non-M.

However, imagine in our model that there is one other cell (a non-M) whose presence is unknown to the experimenter. Both the M-cell and non-M cell drive similar or identical motoneurons and produce the same movement. The M-cell has an inhibitory effect on the non-M cell and prevents it from firing whenever the M-cell is activated. Under normal circumstances, because of its faster response time, the M-cell is activated first by the sensory stimulus, it turns off the non-M cell, and it activates the behavior. Because of the noisiness of the motor connections, the difference in reaction time between the M-cell and non-M cell is not evident.

Now, consider the conclusions reached when we test for M-cell function using the logic of "necessity and sufficiency." According to the model, whenever we present a sensory stimulus the M-cell fires and we observe the behavior. Thus, the firing of the M-cell is correlated with the presence of the behavior. Whenever we artificially activate the M-cell by itself we observe the behavior. Thus, the M-cell appears to be sufficient for the initiation of the movement. However, if we block the M-cell artificially, and present a sensory stimulus the behavior continues to be expressed. This is because the sensory stimulus activates the non-M branch of the circuit and there is no inhibition from the M-cell. Thus, the neural positivist, not knowing about the non-M, wrongly concludes that M-cell firing is sufficient but not necessary for the onset of the behavior.

The neural positivist assigns responsibility for the triggering of the behavior only if the cell can be shown to be both necessary and sufficient for the behavior to occur. Therefore, according to this logic, the M-cell is not the cause of the onset of the behavior under normal conditions.

The essence of the problem is that the positivist conclusion violates the starting assumption that the M-cell in the model is the only neuron that triggers the behavior. The reason for the contradiction is that the experiment itself alters the condition of the system in two ways, one known to the experimenter and the other unknown. Blocking the M-cell prevents not only its excitatory effect on the motoneurons but also its inhibitory effect on the non-M. Thus, unknown to the positivist the system has been altered by the experiment in such a way that its internal conditions are no longer the same. Their assumption that only one isolated neuron, the M-cell, is all that is in question ignores other cells or circuits that may affect the behavior. Most investigators recognize that such ancillary effects are the consequence of any experimental manipulation, but the neural positivist is blinded by adherence to the operational criteria of "necessity and sufficiency." This case is an example of a positivist reaching a false negative conclusion about the functional significance of a neural structure in a simple network. Elsewhere we have discussed other three-neuron networks with interactions that lead the positivist to false positive conclusions (Eaton & DiDomenico

1985). (There is however one case, shown in Figure 4B, where the positivist approach is appropriate.)

The above thought experiment suggests that the discrepancy between the two types of experiments on the M-cell system might be due to extant mechanisms that are intrinsic to the reticulospinal system and that allow it to continue to function despite disruption. The system may be organized in such a way that individual cells do not have indispensable functional roles. If this is the case, then the question of indispensability is irrelevant. Rather, by virtue of the nature of the connections, such as the lateral inhibition mechanism of our model, there are *computational* means (interconnectivity of the system) by which the correct output of the system is automatically produced despite the native unreliability of its individual components, and whether or not the M-cell has been lesioned (Nissanov & Eaton 1989).

If a computational mechanism is the explanation for the appearance of the behavior after the lesion, the neural positivist approach becomes irrelevant because claims of dispensability and indispensability, the cornerstones of their mechanistic explanations, are of no importance in this context. The positivist might escape this dilemma by claiming that the M-cell is indispensable, but after the lesion the animal learns a new way to perform the behavior or other cells grow new connections and acquire the role of the lesioned cell. Such mechanisms have been previously described in many other systems (e.g. Finger 1978). Is the compensation due to a temporal recovery mechanism that takes place *because* of the lesion? Can this account for the presence of the non-M responses and thus allow the neural positivist to escape the dilemma?

Non-Mauthner Responses are Mediated by Extant Circuits

Available evidence shows that learning or re-growth of new connections cannot account for the non-M responses. Instead, they are probably mediated by extant circuits. This conclusion comes from the fact that non-M responses require neither recovery time for neuronal growth, nor practice for learning; they can be elicited on the very first trial after the lesion (Eaton et al. 1982). This is as soon as the animal can be tested (within two hours). Further, there is no change in threshold of the non-M responses following the lesion even after a period of several months (DiDomenico et al. 1988). *De novo* compensatory mechanisms should be accompanied by changes in threshold after the lesion. Thus, the non-M responses are due neither to regrowth of neuronal connections nor the learning of a new way to perform the behavior. It appears that existing circuitry is capable of mediating the behavior in the absence of the cell thought to be necessary and sufficient for its onset.

We have evidence that non-M responses can be triggered in normal animals by other cells that are activated by a variety of types of threatening stimuli (Eaton et al. 1984). In the zebrafish, vibrational stimulation of the head usually elicits a M-initiated escape response whereas non-M responses can be readily obtained in response to stimulation of the tail. In this respect, the M-cell and non-M circuits are functionally analogous to the systems of parallel fibers used for escape in various invertebrates (Eaton 1984). In intact animals, it appears that the M-cells and non-M are triggered by different and overlapping sensory stimuli. In this sense, the non-M circuits are not functional copies of the M-cell circuit. They function in different behavioral contexts but substitute for the M-cell when it is absent.

Because non-M circuits co-exist with the M-cell, it should be possible to identify them anatomically. Several M-cell analogues have been identified by Kimmel and colleagues (Kimmel et al. 1982; Metcalfe et al. 1986) and more recently ourselves (Lee et al. 1988; Lee & Eaton 1989). These neurons reside in the reticulospinal formation near the M-cell and have similar dendritic and axonal projections and thus provide a correlate for the non-M trigger cells in intact animals.

The explanation we have provided based on the model in Figure 4A also requires the presence of some inhibitory effect between the M-cell and non-M circuit. Such an inhibitory effect has been long known to exist between the two M-cells (Furukawa & Furshpan 1963) and our lesion experiments provide evidence for it between the M-cell and non-M circuits as well (DiDomenico et al. 1988).

Why is it that non-M neurons did not activate the EMG when the M-cell was hyperpolarized in the CNE experiments (Hackett & Faber 1983; Hackett & Greenfield 1986; Rock et al. 1981)? This probably results from the differences in the method used to activate the escape network (Figure 4C). Major inputs to the M-cell are from sensory afferents supplied by various branches of the octavolateralis nerves. Only one of these nerves, the posterior branch, was electrically stimulated to show the "necessity" of M-cell firing. In contrast, the escape eliciting stimulus for free-swimming fish emulated many aspects of a predatory attack. This stimulus consisted of dropping a ball into the aquarium or suddenly displacing the water around the fish (Eaton et al. 1982; DiDomenico et al. 1988). These stimuli turn on a complex combination of octavolateralis afferents and other sensory fibers.

In the acute experiment, activation of escape was restricted to a particular set of afferents; in the chronic experiment, activation was multimodal. Furthermore, the electrical stimulus instantaneously turns on not only excitatory afferents but inhibitory afferents and efferents as well (Eaton & DiDomenico 1985). Thus, the electrical stimulus excites the system in a very different way from the environmental stimulus. On the one hand, the M-cell seemed necessary for activating the EMG because shocking only one branch of the octavolateralis nerve fails to activate the non-M circuit. On the other hand, the non-M cells are activated by environmental stimuli when the M-cell is absent.

CONCLUSIONS

Neural positivism is based upon the assumption that in the nervous system a given end is achieved from fixed means. This view ignores the real issue of neural function and instead pursues the cosmetic satisfaction of artificial logical criteria. In the guise of "necessity and sufficiency", neural positivism requires procedures that interfere with the operation of the system to draw inferences about the undisturbed system. This approach disregards the fact that neural systems have a variety of homeostatic processes allowing them to continue to operate despite disruption or intervention.

Few would doubt that tests to artificially block and emulate neural functions are highly successful methods for gaining insights into the neural basis of behavior. However, although such tests seem to be appealing, they cannot provide infallible answers in a biological context. "What is often a methodologically sound hypothesis may, if exaggerated, lead to ridiculous extremes." (Bunge 1979: 133)

Even when properly implemented, the criteria of "necessity and sufficiency" can only tell us about certain linear processes (such as in Figure 4B). In addition, the experimental conditions may be so specific that we gain little information, or even incorrect information, about how neural networks operate when exposed to the spectrum of variables present in the environment. Worse still, the statement that a neuron is "necessary and sufficient" for a behavioral function can lead to the false impression that this conclusion is incontrovertible.

Neural positivism elevates claims of "necessity and sufficiency" to the status of empirical truth when in fact such claims are subjective, and often unreliable, interpretations of experimental interventions (DiDomenico & Eaton 1988). We have shown that conclusions based on "necessity and sufficiency" change with the experimental approach and therefore cannot result in either true or false interpretations that are "unequivocal" (Kupfermann & Weiss 1978). Whether intended or not, the belief that unequivocal empirical tests can be performed and verified is a metaphysical position. Thus, "instead of eradicating metaphysics from the empirical sciences, positivism leads to the invasion of metaphysics into the scientific realm" (Popper 1959: 37).

Biological systems are probabilistic, whereas claims of "necessity and sufficiency" are determinative and ignore contingencies important in the analysis of system operation. This does not mean that tests to emulate or ablate functional capacity must be abandoned. But, it does mean that neural theories of behavior cannot be based upon explicit and preconceived causal criteria. Neural functions must be defined independently of any such causal paradigms.

Moreover, there is no scientific reason to tie neural function to preconceived causal schemes. Because causation is an inference, no one causal explanation has precedence in all experimental contexts. Rather the mechanistic explanations are the causal explanations (Eaton & DiDomenico 1985). Likewise, the purpose of neurobehavioral experimentation is to discover the causal neural mechanisms of behavior. This principle enables the discovery of new causal explanations which will become the basis of neurobehavioral theories (DiDomenico & Eaton 1988).

The liberation of neural function from preconceived causation opens neurobehavioral concepts to enrichment by diverse methodological approaches. What is required is a multidisciplinary approach that directs a variety of questions at the same subject, thereby verifying that the emerging principles are consistent with different experimental paradigms and techniques. This approach acknowledges the organizational principles of contemporary biological science.

ACKNOWLEDGEMENTS

We thank Marc Bekoff, Carol Cleland, Dale Jamieson, Phyllis O'Connell, Jim G. Canfield, Mark B. Foreman, and Roy E. Ritzmann for their comments on an earlier version of this manuscript. However, they are not to be held responsible for our conclusions. Our research described in this paper was supported by grants to R.C.E. from the National Institutes of Health (NS22621) and the National Science Foundation (BNS-12423) and N.I.H. grants BRSG SO7RR07013-23 (Division of Research Resources) to the University of Colorado.

LITERATURE CITED

Brazier, M.A.B. 1959. The historical development of neurophysiology. In: *Handbook of Physiology, Section 1: Neurophysiology,* Volume I (ed. by J. Field, H.W. Magoun and V.E. Hall), pp. 1-58. Washington, D.C.: American Physiological Society.

Bridgman, P.W. 1927. *The Logic of Modern Physics.* New York: Macmillan.

Bullock, T.H. 1978. Identifiable and addressed neurons in the vertebrates. In: *Neurobiology of the Mauthner cell* (ed. by D.S. Faber and H. Korn), pp. 1-12. New York: Raven Press.

Bunge, M. 1979. *Causality and Modern Science.* New York: Dover.

Bunge, M. & Ardilla R. 1987. *Philosophy of Psychology.* New York: Springer-Verlag.

Churchland, P.S. 1986. *Neurophilosophy.* Cambridge, Massachusetts: MIT Press.

Cohen, I.B. 1985. *Revolutions in Science.* Cambridge, Massachusetts: Harvard University Press.

Davis, W.J. 1986. Memory: Invertebrate model systems. In: *Learning and Memory: A Biological View* (ed. by J.L. Martinez and R.P. Kesner), pp. 267-297. Orlando, Florida: Academic Press.

DiDomenico, R. & Eaton, R.C. 1988. Seven principles for command and the neural causation of behavior. *Brain, Behavior and Evolution* 31, 125-140.

DiDomenico, R., Nissanov J. & Eaton, R.C. 1988. Lateralization and adaptation of a continuously variable behavior following lesions of a reticulospinal command neuron. *Brain Research* 473, 15-28

Eaton, R C. (ed.) 1984. *Neural Mechanisms of Startle Behavior.* New York: Plenum Press.

Eaton, R.C. & Bombardieri, R.A. 1978. The behavioral functions of the Mauthner neuron. In: *Neurobiology of the Mauthner Cell* (ed. by D.S. Faber and H. Korn), pp. 221-244. New York: Raven.

Eaton, R.C. & DiDomenico, R. 1985. Command and the neural causation of behavior: A theoretical analysis of the necessity and sufficiency paradigm. *Brain, Behavior and Evolution* 27, 132-164.

Eaton, R.C. & Hackett, J.T. 1984. The role of the Mauthner cell in fast-starts involving escape in teleost fishes. In: *Neural Mechanisms of Startle Behavior* (ed. by R. C. Eaton), pp. 213-266. New York: Plenum Press.

Eaton R.C., DiDomenico, R. & Nissanov, J. 1988. Flexible body dynamics of the goldfish C-start: Implications for reticulospinal command mechanisms. *Journal of Neuroscience* 8, 2758-2768.

Eaton, R.C., Lavender, W.A. & Wieland, C.M. 1981. Identification of Mauthner-initiated response patterns in goldfish: evidence from simultaneous cinematography and electrophysiology. *Journal of Comparative Physiology* 144, 521-531.

_____. 1982. Alternative neural pathways initiate fast-start responses following lesions of the Mauthner neuron in goldfish. *Journal of Comparative Physiology A* 145, 485-496.

Eaton, R.C., Nissanov, J. & Wieland, C.M. 1984. Differential activation of Mauthner and non-Mauthner startle circuits in the zebrafish: Implications for functional substitution. *Journal of Comparative Physiology A* 155, 813-820.

Edelman, G.M. 1987. *Neural Darwinism.* New York: Basic Books.

Evarts, E.V., Shinoda Y. & Wise, E.P. 1984. *Neurophysiological Approaches to Higher Brain Functions.* New York: Wiley.

Faber, D.S. & Korn, H. 1978. Electrophysiology of the Mauthner cell: basic properties, synaptic mechanisms and associated networks. In: *Neurobiology of the Mauthner Cell* (ed. by D.S. Faber and H. Korn), pp. 47-131. New York: Raven.

Finger, S. 1978. *Recovery from Brain Damage.* New York: Plenum Press.

Furukawa, T. & Furshpan, E.J. 1963. Two inhibitory mechanisms in the Mauthner neurons of goldfish. *Journal of Neurophysiology* 26, 140-176.

Gazzaniga, M.S. & LeDoux, J.E. 1978. *The Integrated Mind.* New York: Plenum Press.

Hackett, J.T. & Faber, D.S. 1983. Mauthner axon networks mediating supraspinal components of the startle response. *Neuroscience* 8, 317-331.

Hackett, J.T. & Greenfield, J.L. 1986. The behavioral role of the Mauthner neuron impulse. *Behavioral and Brain Sciences* 9, 725- 764.

Katzir, G. & Intrator, N. 1987. The striking of underwater prey by a reef heron *Egretta gularis schistacea*. *Journal of Comparative Physiology* 160, 517-523.

Kertesz, A. 1983. Issues in localization. In: *Localization in Neuropsychology* (ed. by A. Kertesz), pp. 1-20. New York: Academic Press.

Kimmel, C.B. & Powell, S.L. & Metcalfe, W.K. 1982. Brain neurons which project to the spinal cord in young larvae of the zebrafish. *Journal of Comparative Neurology* 205, 112-127.

Kupfermann, I. & Weiss, K.R. 1978. The command neuron concept. *Behavioral and Brain Sciences* 1, 3-39.

_____. 1986. Command performance. *Behavioral and Brain Sciences* 9, 736-737.

Lawrence, D.G. & Kuypers, H.G.J.M. 1968. The functional organization of the motor system in the monkey. II. The effects of lesions of the descending brainstem pathways. *Brain* 91, 15-36.

Lee, R.K.K. & Eaton, R.C. 1989. Segmental template for reticulospinal organization. *Society for Neuroscience Abstracts* 15, (in press).

Lee, R.K.K., Zottoli, S.J. & Eaton, R.C. 1988. Mauthner-like reticulospinal neurons in the adult goldfish. *Society for Neuroscience Abstracts* 14, 54.

Luria, A.R. 1966. *Higher Cortical Functions in Man.* New York: Basic Books.

Metcalfe, W.K., Mendelson, B. & Kimmel, C.B. 1986. Segmental homologies among reticulospinal neurons in the hindbrain of the zebrafish larva. *Journal of Comparative Neurology* 251, 147-159.

Nissanov, J. & Eaton, R.C. 1989. Reticulospinal control of rapid escape turning maneuvers in fishes. *American Zoologist* 29, 103-122.

Olsen, G.C. & Krasne, F.B. 1981. The crayfish lateral giants as command neurons for escape behavior. *Brain Research* 214, 89- 100.

Peterson, B.W. 1984. The reticulospinal system and its role in the control of movement. In: *Brain Stem Control of Spinal Cord Function* (ed. by C.D. Barnes), pp. 27-86. New York: Academic Press.

Popper, K. 1959. *The Logic of Scientific Discovery.* New York: Harper and Row.

Rock, M.K. 1980. Functional properties of Mauthner cell in the tadpole *Rana catesbeiana. Journal of Neurophysiology* 44, 135-150.

Rock, M.K., Hackett, J.T. & Brown, D.L. 1981. Does the Mauthner cell conform to the criteria of the command neuron concept? *Brain Research* 204, 21-27.

Rosenberg, A. 1985. *The Structure of Biological Science.* New York: Cambridge University Press.

Suppe, F. 1977. The search for philosophic understanding of scientific theories. In: *The Structure of Scientific Theories* (ed. by F. Suppe), pp. 3-232. Chicago, Illinois: University of Illinois Press.

Thompson, R.F. 1986. The neurobiology of learning and memory. *Science* 233, 941-947.

Webb, P.W. 1986. Effect of body form and response threshold on the vulnerability of four species of teleost prey attacked by largemouth bass (*Micropterus salmoides*). *Canadian Journal of Fisheries and Aquatic Science* 43, 763-771.

7. Applications of Artificial Intelligence to Animal Behavior

Jane M. Packard, L. Joseph Folse, Nicholas D. Stone, Merry E. Makela, and Robert N. Coulson

INTRODUCTION

The conceptual tools used to investigate animal behavior not only limit our abilities to explain reality, they also reflect the ways in which human thought processes are structured by biological organs and cultural legacy. Two styles of thinking, "linear" and "synthetic" (Table 1) represent diverse approaches to the study of animal behavior. Conceptual tools developed in the field of artificial intelligence (AI) provide ways of integrating these two styles. AI tools match the way biologists think about behavioral processes. In essence, AI programming is a reflective "backside of the mirror", shared in common by scientists from different disciplines.

Miscommunication regarding interpretation and explanation of animal behavior has occurred when researchers differed in their definition of and style of addressing a problem (e.g. Beach 1950; Lehrman 1953; Lorenz 1965). In other realms of human interaction, two individuals functioning under different styles of thinking are likely to interpret a phenomenon differently and disagree over an explanation of the phenomenon, even though each may be accurate from their own limited perception of the world (Harrison & Bramson 1982; Tanenbaum 1989).

The integration of linear and synthetic styles of thinking in AI programming resulted from a stimulating and lengthy debate over how (and whether) computers can be programmed to mimic the human mind (Dreyfus 1979; Waltz 1982; Searle 1984; Denning 1986). Our purpose in this chapter is to explain that even though AI may not have yet achieved the goal of making computers act as if they think, AI does provide powerful tools for integrating linear and synthetic approaches to interpreting, explaining and predicting animal behavior.

Table 1. Characteristics of two styles of thinking, which often result in miscommunication regarding interpretation and explanation of animal behavior.

Source	Linear style	Synthetic style
Lorenz1977	"objective"	"subjective"
	experimental approach	comparative approach
	test hypotheses	develop hypotheses
	analysis	understanding
	deduction; logic	induction; intuition
Harrison & Bramson	"analyst"	"synthesist"
1982; pp 98-99	general rules	concepts; opposing viewpoints
	describes things systematically	speculative arguments
	offers substantiating data	ignores data
	structured, rational examination	qualifying phrases
	reason, logic	humor, sardonic
Searle 1984;	"scientific explanation"	"common sense explanation"
pp 71-75	physics and chemistry	natural science
	prediction	explanation
	deduce what will happen	deduce what has happened
	regularity within a population	individual variation
	straight forward laws of	complex interrelated networks
	statistical mechanics	different paths lead to same
	same result from same path	result
Tanenbaum 1989	"male"	"female"
pp 58, 85, 95, 109	sequential ideas	random information and ideas
158-159	slow process	rapid insight
	focused on one idea and problem	several simultaneous problems
	impose limitations, follow rules	ignore preset rules and
	to maintain order	regulations to address problem
	start with component parts	start with overview
	interruptions resisted	interruptions not a problem
	continuity important	change/ improvement welcome
	define concepts	explore and play with ideas
Application to	linear programming	parallel process programming
models of animal	stepwise flowchart	modular structure
behavior in present	create one model	create alternative models
paper	cause and effect	cyclical
	population-level rules	individual level rules
	prespecified parameters	dynamic linkages
	rules of behavior do not change	changing rules of behavior
	confounding variables controlled	variables not controlled
	spatially homogeneous,	suitable for heterogeneous,
	constant environment	fluctuating environment
	large populations	small populations
	forward chaining	backward chaining

As long as explanation is an acceptable goal for the study of animal behavior, no synthesis is needed. However, if scientists seek to apply knowledge of animal behavior to managing populations affected by human activities, the integration of linear and synthetic approaches is imperative. This integrative task is a challenge, because when we talk about a synthetic structure of behavior, we tend to lose those readers who operate from primarily a linear style. Likewise, synthetic thinkers have difficulty following a logical development of AI applications to modeling behavior. However, despite evidence that differences in the predominant style of thinking may be as deeply rooted as biological differences between individuals (Tanenbaum 1989), we retain faith that each reader can comprehend both styles when the difference is made explicit (Harrison & Bramson 1982).

Thus, this chapter is intended as an overview to provide entry to the animal behavior literature for those who are well versed in AI and an introduction to "soft" AI programming tools for those who study animal behavior. Since the jargon in both fields has become dense, references to semi-popular as well as technical works are included. Wherever possible, we refer readers elsewhere for definition of concepts. Our purpose is to give an overview, in typical synthetic style, exploring relations among ideas by discussing concrete examples.

In this chapter, we seek (a) to define a basic integrative structure used by ethologists in analyzing behavior, (b) to identify some of the limitations of linear approaches to modeling behavior in heterogeneous environments, and (c) to illustrate some ways that AI concepts can be used to overcome such limitations. This is quite a different approach from the usual discussion of how artificial intelligence relates to the cognitive capacities demonstrated by human and nonhuman animals. Thus, we first provide some background so readers have a better perspective on the experiences that have shaped our view of the question "how have conceptual tools influenced the interpretation and explanation of behavior."

BACKGROUND

When Konrad Lorenz sat next to the moor pond and lectured his student assistants on the basic concepts of ethology, he used common mechanical analogies that even the untutored mind could understand, such as the description of motivation subsequently criticized as the "flush-toilet model" (Nisbett 1976; Lorenz

1981; Toates 1986). Working on his philosophical treatise titled *Behind the Mirror* at the time (Lorenz 1977), Lorenz was intensely interested in how our own perceptual capabilities shape our views of reality and the way we approach science. He was interested in explanation, not prediction. He argued that the linear and synthetic approaches were both necessary for good science, citing the productive collaboration when Niko Tinbergen was able to test the hypotheses that arose from Lorenz's intuitive knowledge base.

Our research group has been stimulated to readdress this issue from quite a different perspective as we tried to communicate across disciplines in which people formulate the basic question of "knowledge representation" in quite different ways (Table 2). For the modern ethologist, computer models have become an extension of perceptual capabilities, generating predictions against which reality can be tested. However, it is difficult to communicate with nonspecialists about the structure and content of such models. Mechanical analogies such as flush toilets and thermostats were common knowledge of an earlier era, unlike the sophisticated gaming, optimization, homeostatic, time-sharing, and contingency models of today.

Our research group has been interested in modeling population dynamics and animal movements in patchy environments that are not stable over time (Coulson et al. 1987; Makela et al. 1988; Saarenmaa et al. 1988; Folse et al. 1989; Roese 1989). This set of problems required a synthetic as well as a linear perspective, because it violated the assumption of a homogeneous, constant, environment containing animals that behave like gas particles or unintelligent robots. We needed to test alternative models to be able to deduce rules of behavior based on how the animal rather than the investigator perceives the environment.

Using a synthetic style, we sought to develop and share a basic "shell program", or common structure, containing modules that could be expanded to model the content of both invertebrate and vertebrate behavior. We encountered the practical problem that the computer programmers on our team had little experience with animal behavior or behavioral concepts. Their first inclination was to generate their own structures for behavioral processes.

For effective interdisciplinary communication, we needed to address the differences in knowledge base as well as different styles of thinking. These two components of "knowledge base" and

Table 2. Questions raised by separate disciplines involved with creating models of knowledge representation.

Discipline	Question	References
Computer engineer	How can machines be designed to operate beyond human capabilities?	Albus 1981; Raibert & Sutherland 1983; Poggio 1984 Denning 1985
Knowledge engineer Van Horn	How can computers be programmed to solve similar to human experts?	Kurzweil 1985; Bobrow & Stefik problems 1986; Davis1986; 1986
Philosopher	How does human knowledge match the real world?	Dennett 1984; Tennant 1984; Denning 1986
	Can computers think?	Searle 1984
Experimental psychologist	What is the most parsimonious way of representing human interactions with the real world?	Rolls & Rolls 1982; Staddon 1983; Toates 1986
Comparative psychologist	To what extent do other species solve problems the way humans do?	Dewsbury 1978; Pearce 1988; Colgan 1989
Ethologist	How does an animal's knowledge match the real world?	Tinbergen 1951; Lorenz 1981; Boden 1984; Alcock 1989
Behavioral ecologist	How are animals designed to solve ecological problems?	Krebs & McCleery 1984; Stephens & Krebs 1986; Alcock 1989
Simulation modeler	How can computers be programmed to behave like animals, populations and ecosystems?	Hassell & May 1984; Starfield & Bleloch 1986; Reiter 1986; Saarenmaa et al. 1988; Folse et al. 1989
Resource manager	How can a computer help people solve problems involving management of biological resources?	Rykiel et al. 1984; Marcot 1984; Starfield & Bleloch1986; Coulson et al. 1987

"procedures" are basic to the manner in which AI programming mimics human thought processes involved in knowledge representation (Davis 1986).

As a result of a linear style of analysis, the knowledge base in behavioral studies has been separated into components of motivation, learning, perception and coordination of action patterns. A synthetic overview of these components is provided in the following section. The examples used below to illustrate the concepts represent a common knowledge base for anyone who has had a basic course in animal behavior. They were chosen so readers unfamiliar with the behavioral literature could understand the knowledge base from which the overview was derived.

THE BASIC STRUCTURE OF BEHAVIORAL PROCESSES

Dennett (1984) warned computer programmers against rediscovering "cognitive wheels" already designed in other disciplines. The cognitive structure used in the study of animal behavior is outlined in Figure 1. The input and output of the system is defined as stimulus and response messages, respectively. The four components of this basic structure include: perception, action, motivation, and learning systems.

This basic structure of behavioral processes has been similar throughout the works of the classical ethologists to the present day. Admittedly, the ways in which this structure is described have varied among disciplines over the years. We choose to describe this structure using terms common to AI, to better suggest the appropriate application of techniques described in the last section of this chapter.

We think of this structure as a basic "shell" for knowledge, which could be made operational in a work environment such as that provided by an artificial intelligence style program shell (e.g. Glymour et al. 1987; Richmond et al. 1987). These four components of the basic shell are described in more detail in the following subsections.

Stimulus and response messages

The basic structure of behavior typically is represented as a series of stimuli and associated responses, as if there are sets of external objects and internal objects, which "pass" messages to one another (Figure 1). In this subsection, we discuss the

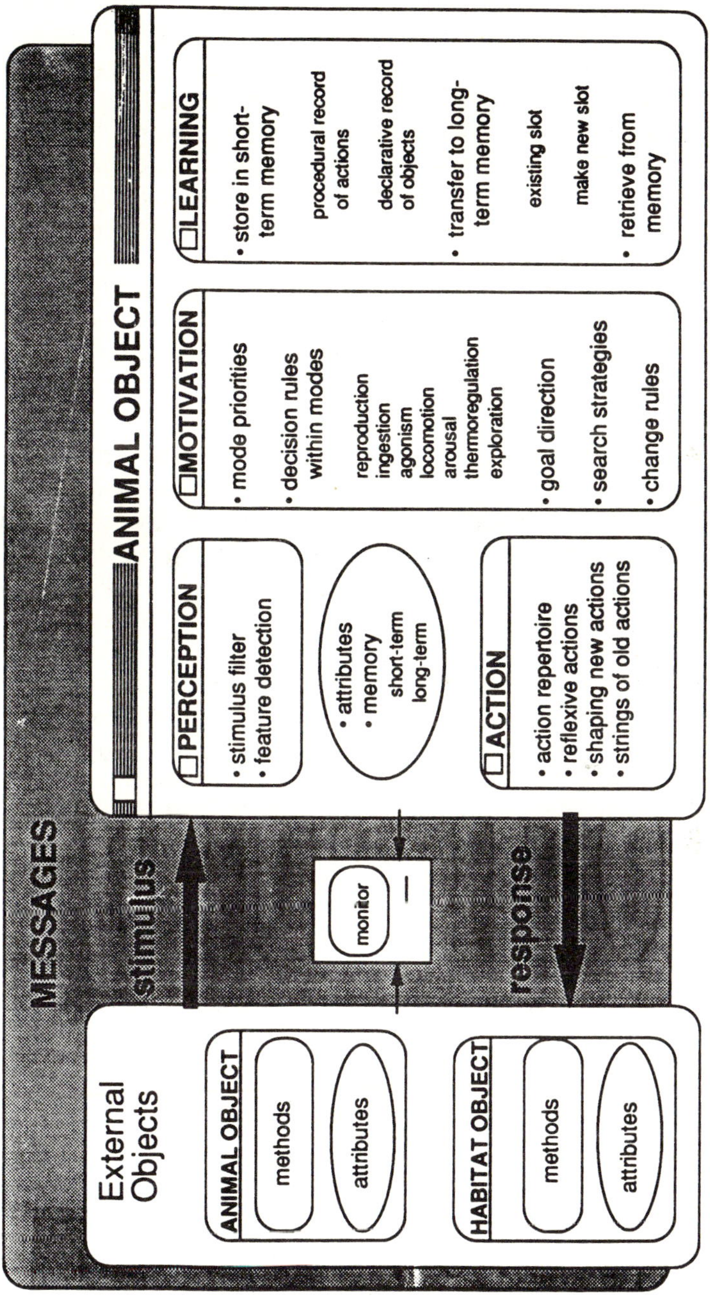

Figure 1. Basic shell structure of animal behavior.

concepts of a dichotomy between internal/external information and the message versus the meaning of a stimulus.

The pairing of stimulus and response is central to all analyses of behavior and must be clarified prior to looking at the separate components. When the stimulus message always elicits the same response message, the process is considered to be relatively hard wired, or innately determined by developmental processes resistant to environmental perturbance (Heuer & Sanders 1987). When the correspondence between stimulus and response is not one to one, intervening variables influencing the meaning of the message are important. The intervening variables (processes) are the components we identified above as motivational systems and learning (Staddon 1983; Toates 1986).

The internal/external dichotomy. Our human perceptual apparatus makes an artificial distinction between processes within the body, relatively hidden from view, and visible processes outside the body. An obvious analogy is the input and output of a computer program, the contents of which remain a "black box" for the outside observer who can only monitor what goes in and what comes out.

This dichotomy sometimes is confusing to the introspective synthesist who accepts mental states as part of reality (see Dupré Volume I). Are you not aware of the internal messages sent by a gnawing, empty stomach? The major issue is that an external observer is unaware of your internal messages, unless that observer has some way of monitoring what happens inside. A linear style of thinking requires special tools to collect data on what goes on inside other organisms. Thus, linear thinkers reserve the concept of "stimulus" to refer to information originating outside the organism and reject the notion of mental states. To the synthetic thinker, the information exchanged between two objects is a "message", whether it is from outside or inside the organism. To avoid confusion in the general structure of animal behavior, we reserve the term "stimulus message" to refer to information of external origin.

In the modeling of internal systems, a synthesist would treat a system as a series of components, or interactive objects. For example, the internal processes of the stomach appear relatively hidden from the conscious processes of the brain until a certain threshold is reached. A threshold can be modeled as a decision point at which a message is passed between internal objects. This is the approach used by object-oriented programming as we will describe later.

From a synthetic approach, the basic structure of behavior is a way of representing knowledge and does not imply a direct correspondence to physiological systems. In contrast, computer representations of information processing in neuronal networks have been developed (Poggio 1984; Ullman 1986; Tank & Hopfield 1987) and criticized as inadequate to ever represent the functions of a mind (Searle 1984). Whether computer models of neuronal networks represent a mind, or predict behavior, may be irrelevant to the neurophysiologist seeking better ways to represent interactions among neurons that are known to be branched rather than linear processes. Our brains are inadequate to understand much more than the input and output of such complex neuronal network computer models.

Message vs. meaning. How simple the study of animal behavior would be if the same stimulus message always resulted in the same response message. However, the meaning of a male songbird's song differs for another territorial male, a nonterritorial floater male, or a nesting female. In response to the same song, the neighboring territorial male approaches and displays, the floater male leaves quietly, and the nesting female remains nearby the singing male. A linear approach to determining general rules of behavior for animals is quite different than that required for billiard balls because the internal state of the animal changes.

The difference between the message and the meaning lies in the way the stimulus information is processed (see Smith 1977 and in Volume I). Not only is the information processed differently by intervening factors described as motivational systems, the roles of learning and developmental history of individuals are also important. For this reason, motivational systems and learning processes are two of the components containing rules within the structure of the basic animal behavior shell (Figure 1). Although individuals may initially share a common set of information-processing rules characteristic of the species, those rules change with age (e.g. young male songbirds do not sing) and with experience (e.g. floaters start singing when they acquire a territory). The specific information stored in memory also influences how an individual processes information from a stimulus (e.g. how vigorously it was chased).

If our basic shell structure of behavior is to accommodate diverse species, we need to be able to represent sensory capabilities and response repertoires (Figure 1), which differ characteristically among species. Developmental processes and learning influence the variation in response to stimuli expressed among individuals of the same species. However, with regard to differences between species, variations in perceptual systems and activation systems also need to be considered (e.g. the messages received by a wasp are different than those received by a songbird).

Perceptual systems

Two major concepts in the study of perceptual systems suggest it is appropriate to model incoming information as discrete messages. First, the sensory system of a given species is tuned into certain channels of information better than other channels (stimulus filters). Second, specific features of incoming information are often recognized in the central nervous system, which detects information that has been biologically important in the evolutionary history of the species (feature detection). These two concepts are illustrated below by the classic example of the Griffin and Roeder study of bats and moths (Roeder 1965; Alcock 1989).

Stimulus filters. The hearing system of noctuid moths is tuned with maximum sensitivity to the frequency range of their predators, bats. Unimportant information is thus filtered out by the sensory system. Moths receive messages within a narrow acoustic channel as if tuned into the emergency channel of a CB radio. In contrast, larvae of the moth are tuned to the frequency range of the sounds made by their main predators, wasps. The hearing of bats is maximally sensitive to ultra-high frequencies corresponding to their sonar system, but also includes a broad range of pitches as low as squeaks audible to the human ear.

Feature detection. Up to a certain threshold of sound intensity, moths turn and fly away from a source of sound that mimics sonar pulses from a distant bat. In response to very loud sonar pulses, moths fall in a fluttering dive that confuses a bat closing in for a kill. This feature detection is encoded in two sensory cells within the hearing system of the moth. The A1 cell fires in response to low intensity sound. The A2 cell fires in response to high intensity sound, sending a neural message that temporarily blocks coordination of the wing movements and makes the moth fall.

Thus, it is very appropriate to model incoming stimuli as discrete messages that are mapped to characteristic (and usually adaptive) responses.

<u>Action systems</u>

The responses of animals are coordinated in certain recognizable action patterns that reoccur (Lorenz 1958; Tinbergen 1960). Such action patterns can be modeled as discrete response messages. The set of response messages typical of a species is called its "action repertoire". Those actions that are paired reliably and rapidly with specific stimulus messages are called "reflexive". New response messages may be acquired by "reshaping" old actions, or combining "sequences" of old actions. The concepts of action repertoire, reflexive action, shaping actions and action sequences are illustrated below with examples that are common knowledge of practicing ethologists (Ridley 1987; Alcock 1989).

<u>Action repertoire</u>. A species' repertoire of action patterns is like a list of preprogrammed messages ready to be sent. For example, all herring gulls perform certain recognizable action patterns such as the "long call", "choking", or "forward display" (Tinbergen 1960). The action repertoires of duck species differ quite a bit from those of gulls, and comparisons among duck species provide insight as to their evolutionary history (Lorenz 1958; Ridley 1986; see also Burghardt & Gittleman this volume). The number of actions in a species' repertoire is, for practical purposes, a finite set (Fagen 1978).

Certain subsets of action patterns are likely to be highly correlated in time (Bekoff 1977; De Ghett 1978; Packard & Ribic 1982), giving rise to the notion that a motivational system (e.g. hunger, reproduction, migration) controls their expression. However, a particular action pattern may occur in the context of several motivational systems (Lorenz 1981: 198; Colgan 1989: 41); for example, a digger wasp may sting an enemy (defensive mode) or a caterpillar (reproductive mode).

<u>Reflexive actions</u>. Reflexive actions solve predictable problems that have been encountered by many individuals in the past history of a species. For example, herring gull chicks run to cover and crouch to the ground in response to the raucous warning call of adults. When a covey of quail is flushed, each individual scatters

in an unpredictable direction. Those individuals that did not possess such hard wiring must have been eaten in previous generations of the species. Reflexive actions occur when split-second timing of an appropriate response(s) has been adaptive.

Although reflexive actions are typical of simple organisms, they also occur in complex vertebrates responding to reliable cues in the environment. For example, many single-cell organisms, insects and fish follow simple rules of orienting toward a light, thermal or chemical gradient. Kittens orient toward warmth and a familiar odor. Dogs salivate in response to the taste of food. Thus, reflexes are analogous to rules that map stimulus messages directly to response messages without intervening variables.

Shaping new actions. For those preoccupied with an anthropocentric view of behavior, the list of potential actions appears to be a *tabula rasa* (clean tablet ready to be written upon). However, even in humans, new actions arise as a modification of actions existing previously in the repertoire. The process of acquiring a new action is called "shaping"; for example, pigeons can be taught to turn in circles by delivering a reward each time the pigeon makes a move that closely resembles turning (Gardner & Gardner 1988). After an individual has learned a new action, it is as if a new message has been added to the action repertoire.

Sequences of old actions. Sequences of old actions also can become new messages. Anyone who has learned to play golf remembers a time when each separate component of the golf swing required conscious effort. However, after the sequence of actions was repeated many times, it became an "unconscious" discrete motion, interrupted only if the player thought too much about it. Even the complex show routines of marine mammals are built on previously existing sets of behaviors.

Motivational systems

Ethologists think of the rules governing behavior as occurring in clusters with a hierarchical organization (Dawkins 1976; De Ghett 1978). The components of this model include rules for implementing (1) priority of modes, (2) decision rules within modes, (3) goal direction, and (4) search strategies (Figure 1). To the computer programmer or neurophysiologist, this may appear as an unnecessary or even incorrect structure, probably because it reflects the organization of human minds (analogous to software) more than the body (analogous to hardware). Nevertheless, the components have provided a useful approach for analysis of the dynamics of behavior within a species and for comparisons within a motivational system across species (Colgan 1989).

Motivational systems are like modules of a computer program, which contain "action rules" and "stop-action rules" relative to some "goal" state (Dawkins 1976). Examples of motivational systems (Figure 1) include: reproduction (social interaction, mate choice and parental care), ingestion (hunger, thirst and foraging), agonism (fight and flight), locomotion (movement and migration), arousal (activity and rest), thermoregulation, and exploration (search and play). Although behaviorists may quibble over a basic set of motivational systems common to all organisms (Toates 1986), this list has worked fairly well for us. Some of these modules may be unimportant for certain taxonomic groups or research questions. However, this basic structure is consistent with most models of animal behavior.

Mode priorities. The relations among motivational systems may be complex, involving inhibition, facilitation, neutrality or time sharing (McCleery 1983). From an analytical perspective, animals "decide" which motivational system has highest priority at a specific moment. From a synthetic perspective, each system decides its own priority and the highest priority system takes control of the animal.

Each motivational system is not totally independent from other systems. For example, a deer that finishes ruminating at daybreak may arise and forage. However, as air temperature rises, it may seek a cooler location to feed. If a coyote approaches, it may flee; or the deer may chase the coyote if its fawn is hidden nearby.

Interactions between motivational systems have been quite elegantly determined for the simple nervous system of the sea slug (*Pleurobranchia*). Escape takes precedence over all other systems, egg laying inhibits feeding and supersedes mating and reorientation, but feeding is more likely than mating, reorientation and withdrawal from touch (Kovac & Davis 1980; Alcock 1989; Colgan 1989).

Decision rules within modes. Within a particular mode (motivational module), an animal acts as if it chooses from a set of action patterns in the repertoire (messages). The probability of one action occurring given that another has occurred previously has been analyzed for diverse invertebrates and vertebrates (Dawkins 1976; Sustare 1978).

This probabilistic approach to understanding internal behavioral processes has been extended to include response contingencies relative to external messages. For example, the probability that a digger wasp digs a new burrow or enters an existing burrow has been analyzed relative to the action probability of others in the population (Dawkins 1982: 118-132).

Goal direction. Acting as independent modules, some motivational systems appear to monitor the state of the individual relative to a set point (template or tolerance range) and assume high priority when the difference (between the actual state and the set point) passes a threshold (Staddon 1983; Toates 1986). Internal processes (e.g. changes in hormones, gut fill, or blood glucose) may change the actual state relative to the goal state, as well as external information (e.g. photoperiod, temperature, or ingestion of food). Although goal directed behavior may appear to the synthesist to imply intentionality (Dennett 1988), the linear analyst would argue there is no need to construct intervening variables that cannot be measured (Toates 1986).

In the simplest model of goal directed action (adapted from Staddon 1983), four functions are involved (Figure 2). The "set point function" evaluates the goal state relative to the actual state and sets the probability of action. The "action transfer function" decides on the appropriate action to reduce the discrepancy. The "incentive function" may override (inhibit or facilitate) the action probability, depending on input from memory and other modes. The resulting response message is the input to an external "contingency function". This function of external objects produces the stimulus message that is the input to internal objects at the "set point function". Thus, contingency functions are part of the external objects that send messages to organisms in our basic shell structure of behavior (Figure 1).

Search strategies. The complexity of search patterns modeled for animals ranges from random movement to insight (Schöne 1984; Pearce 1988). We use the term "insight" to refer to apparent internal manipulation of information yielding an appropriate response not previously executed. Between these extremes are

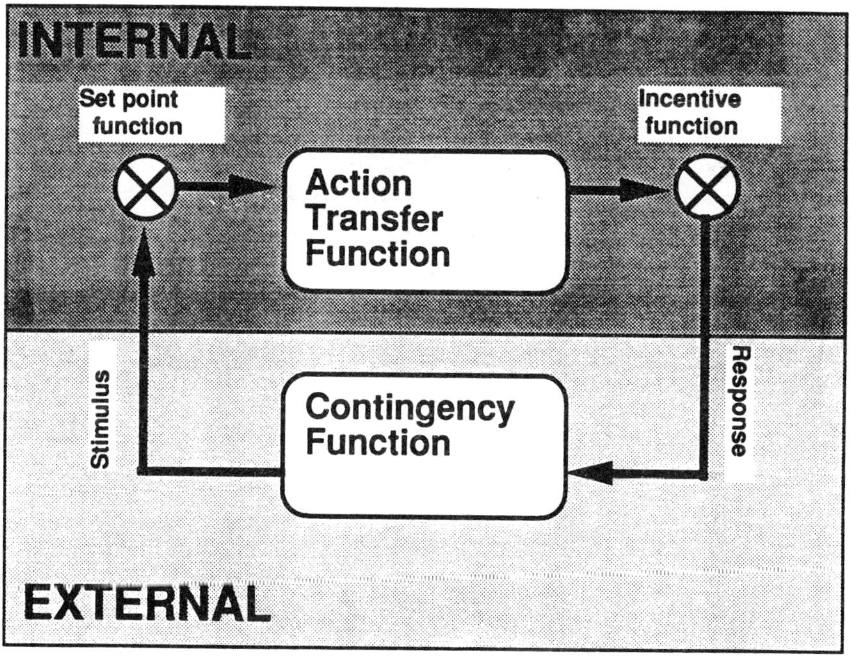

Figure 2. Basic homeostatic representation of a motivational system.

intermediate types of search strategies, including: specific rules that increase probability of encounter (Shettleworth 1983; Schöne 1984); repeating previous sequences that were successful (Gardner & Gardner 1988); and evaluating changes in the rate of return within patches (DeVries et al. 1989). Although we do not yet know of examples of invertebrates showing insightful behavior, the other types of search strategies occur in both vertebrates and invertebrates.

These strategies may be categorized as sets of discrete decision rules (algorithms), continuous functions, backward chaining based on a memory of previous solutions to a problem, and forward chaining based on projection of probable causal relationships. Each of these strategies has a direct counterpart in AI programming (Waltz 1982; Charniak & McDermott 1985), as described in the last section of this chapter. We do not imply that the behavioral strategies were independently derived from the AI strategies, rather both reflect processes of the human mind.

Learning systems

The debate over learning theory (Fox 1983) illustrates the difference between a linear style of analyzing brain structures and the synthetic style of studying cognitive capacities. The components of learning processes (Figure 1) include storage and retrieval of information from the knowledge-base called memory (Kendrick et al. 1985; Squire 1986; Thompson 1986). Conscious and unconscious decisions influence whether information enters short-term memory and then whether (and in what form) it is stored in long-term memory (Atkinson & Shiffrin 1971). Retrieval is a separate process.

Memory. Short- and long-term memory are involved in building an internal representation of external processes (Narayanan 1986; Squire 1986). The capacity to store information from messages received and delivered would be quickly overwhelmed if there was not some way of erasing irrelevant information.

Memory appears to involve explicit declarative as distinct from implicit procedural knowledge (Alper 1986; Squire 1986). Procedural knowledge is acquired during processes involving motor responses (e.g. rehearsal of skills, priming, classical conditioning) and is thought to occur in a primitive portion of the brain, the cerebellum (Alper 1986). Storage of procedural knowledge need not be conscious and is basic even to invertebrates that do not have the specific brain structure (hippocampus) responsible for declarative knowledge (Squire 1986; Mishkin & Appenzeller 1987).

Declarative knowledge involves facts, episodes, lists and routes of everyday life, including events specific to a time and place (episodes) and reference to associated facts from the context of an experience (semantics) (Squire 1986). Declarative knowledge allows an animal to record and access the particular encounters that lead to behavioral change. It stores the attributes of objects and relationships among objects and is only found in organisms (e.g. mammals) with a well developed hippocampus, which supports conscious decisions in humans.

As an initial model, we suggest procedural memory involves storing information about response messages and internal messages regarding the state of the organism (attributes), whereas declarative memory involves storing information about patterns of stimulus messages, memory networks, and attributes of external objects. The former appears adequate for backward chaining based on previous experience; however, the latter appears necessary for planning or projecting future results of alternative actions.

Storage and retrieval of information. Learning can be viewed as the process by which an organism or machine decides which information to store as an internal representation of the the external world (Forsyth 1986). Learning theorists have found it useful to distinguish between nonassociative (habituation, sensitization) and associative learning (classical conditioning) categories (Dewsbury 1978; Thompson 1986).

The "quick" learning that occurs in biologically adaptive circumstances appears distinct from the "slow" learning that occurs in the novel laboratory environment (Gardner & Gardner 1988). For example, Shettleworth (1983) observed that a marsh tit cocks its head after hoarding a seed. Later, the bird uses visual information to return to holes where it hid seeds. The bird behaves as if there is an "open slot" for information associated with each seed. In contrast, repeated pairing of stimulus and reward is required for a rat to learn to press a bar.

For synthetic thinkers, the process of storing information appears to involve pattern matching to a preexisting "slot" (Tennant 1984; Gardner & Gardner 1988). The concept of cognitive structures explains that if the incoming information can be mapped into an existing slot, it will be stored and retrieved rapidly. If the incoming information does not match existing slots, the process of constructing a new slot takes time or repetition. Organisms differ in the ability to "make new slots" as a result of their evolutionary histories. However, from the perspective of a linear style, it is very difficult to collect data on the intangible intervening variable of a memory slot or template (Johnston 1988) even though the concept is used in AI models of memory (Narayanan 1986).

Certain sensory channels involve more rapid learning than others (Alcock 1989: 45). For example, a rat will not learn to avoid the taste of water associated with a shock to the tail, even though it learns after one experience to avoid the taste of water it drank prior to vomiting. Such species-specific constraints have defied the linear approach that seeks to derive general laws of learning applicable to all organisms.

Evolutionary biologists have recognized a continuum ranging from "restricted developmental programs" that are closed to irrelevant information, to "open developmental programs" that store information about external processes (Mayr 1974; Alcock 1989). The concept of an open versus restricted developmental program is better suited to a linear style of analysis than the previous distinction of genetic versus learned control of behavior (e.g. Johnston 1988).

Invertebrates tend to fit the restricted model and vertebrates to fit the open model; however, the variation is actually related to life history traits (Horn 1978). Restricted developmental programs are fast and reliable in solving predictable problems when many expendable copies of the organism are produced. Although an "expensive" open program that picks up specific information about a problem is more flexible and durable in dealing with unpredictable problems, it suffers the limitations of slower operation and potential disruption due to an inappropriate developmental history. An optimal degree of flexibility may involve sensitive periods in which an "innate school marm" (Lorenz 1965) tells the system what kind of information to store from incoming messages (Bateson 1979).

LIMITATIONS OF PREVIOUSLY APPLIED TOOLS

The synthesis of components of behavior into a common structure probably has not been explicitly outlined previously due to the prevalence of the linear style of thought that focuses on universal laws rather than providing an overview of how components interrelate. Historically, mechanical models for describing the complexity of motivational processes were criticized when systems analysis, optimization models and game theory approaches were borrowed from disciplines other than ethology (Krebs & McCleery 1984; Maynard Smith 1984; Toates 1986; Mangel & Clark 1988; Colgan 1989). However well-matched to the questions they address, such analytical tools are based on the assumption of a stable, homogeneous environment (or at least one in which stochastic fluctuations are bounded and represented by mean values; Mangel & Clark 1988).

Tools based on a linear style of thinking have limited our ability to represent the following: (1) an environment that is spatially heterogeneous; (2) changing contingency rules due to environmental processes; (3) changing response rules due to internal processes; (4) incomplete rather than global knowledge; (5) goal-directed search patterns; (6) interaction between motivational systems; (7) individual differences resulting from historical differences during development; and (8) "transparency" of the content of computer programs. Each of these limitations are briefly outlined below and illustrated by criticisms of optimal foraging behavior as reviewed by Pyke (1984), Stephens & Krebs (1986) and Pierce & Ollason (1987). Applications of object-oriented programming as a synthetic approach to overcoming such limitations are discussed afterwards.

Spatial heterogeneity

The legacy of MacArthur & Pianka (1966) led many modelers to represent resources as homogeneously distributed "prey" or disjunct "patches" within which resources were homogeneously distributed (reviewed by Schoener 1971). An elegant application of the marginal value theorem (Charnov 1976) allowed modelers to use optimization algorithms to predict behavior (Pyke 1984; Stephens & Krebs 1986). Optimization is a linear style because it involves a search for general laws, a deduction of what will happen in the future, and a model to be tested analytically.

Although homogeneous or clumped distributions of resources may be an appropriate representation for nectar feeding insects and birds (Pyke 1984), resource distributions for most other species are more complex (Lessels & Stephens 1983). To simplify such complexity, the probability of encountering prey has been modeled as if a forager is stationary and the habitat "flows" past its senses in one dimension (Owen-Smith & Novellie 1982; Arditi & Dacorogna 1988). However, the practical difficulties of translating from the real environment to the model environment are complex and some have argued that patchy distributions can only be determined *a posteriori* in terms of the forager's behavior (Hassell & May 1984).

For many larger vertebrates, researchers collect information about the location of individuals in relation to vegetation patches. However, we still understand very little about how animals perceive discontinuities in their environments compared to patches perceived by researchers.

Providing more of a synthetic perspective, Senft et al. (1987) have stressed the importance of a hierarchical approach to understanding ungulate foraging patterns. They found it inappropriate to model the landscape as discrete patches. For example, elk feeding in mountain meadows encounter a series of decisions. For ungulates, hierarchical decisions may involve seasonal movements among pastures, then decisions made within pastures (e.g. the number of bites per step and the choice of plant parts at a feeding station). At decision nodes higher in the hierarchy (earlier in the sequence), input from other motivational systems may be more important than at nodes lower in the hierarchy.

Changing contingency rules

Optimization and game theory approaches assume that the environment is stable over time. If this assumption is violated, the usual approach is to take an average of the expected fluctuations or to limit tests of the model to a time frame that can be considered invariant. To accommodate such limitations, "dynamic optimization" procedures have been developed, involving a sequence of decisions (Mangel & Clark 1988).

Few models have allocated to patches the "rules" for changing over time, although seasonal changes in vegetation and prey distribution are well documented. Animals that deal with seasonal variation may not be optimally designed for all seasons. One approach has been to argue that animals merely "satisfy" their resource requirements, performing at what appear to be less than optimal levels during resource-poor periods (Bekoff et al. 1989), but nevertheless minimizing the risk of dying (Krebs & McCleery 1984).

Changing response rules

A major difficulty in testing optimality models has been in detecting when a behavior that appears not to be optimal is in the process of change, tracking changes in the environment (Pierce & Ollason 1988). Game theory (Maynard Smith 1984) provides a stimulating intellectual approach when the optimal behavior for one individual depends on the actions of others in the environment. One solution has been to model conditional strategies as sets of tactics (subroutines) that are activated by specified conditions in the environment. For example, the optimal time for a territorial bird to switch to flocking in winter has been analyzed by comparing the cost/benefit ratio of the three strategies: territoriality, flocking, and switching (Davies & Houston 1984).

In contrast to models based on discrete strategies, traits subject to natural selection are likely to result in incremental responses to some environmental cue. A modeling approach that provides for variation in response rules observed in a real population is needed. Furthermore, rules of response change during development (Chalmers 1987). If we are truly to examine parent-offspring conflict, such rules for changing rules need to be included in models.

Incomplete knowledge

Many optimality models of foraging (and other behaviors) assume that the organism has perfect knowledge of the environment and can calculate when the rate of return within a patch falls below alternative patches (DeVries et al. 1989). However, in the real world, individuals only have access to (not perfect knowledge of) information available within their habitual range of movement. The information that is actually used may depend not only on their perceptual capabilities, but also on their ability to store information in memory (reviewed by Roese 1989).

Goal directed search

Most models of optimal foraging assume patterns of movements with encounter rates that are random or drawn from an *a priori* probability distribution even though many foragers search systematically (Pyke 1984). Simulations of movement patterns based on a random or *a posteriori* analysis of turn angles and distances traveled have been used in analysis of radio-tracking data (Siniff & Jensen 1969; Bekoff & Mech 1984; Brody 1988). However, such approaches ignore what is known about motivational systems and goal-directed search.

Interaction between motivational systems

One of the assumptions of an optimality approach to modeling foraging behavior is that natural selection influencing evolution of foraging traits has been independent of selective factors influencing other behavioral traits. Likewise, systems approaches to modeling motivational systems typically account for so many details within one system that interactions with other motivational systems have been excluded (Rolls & Rolls 1982: LeMagnen 1986; Toates 1986). Successful analysis of the interactions among systems (McCleery 1983) has been limited to the laboratory environment.

During field work, researchers typically follow individuals for long periods during which the animals switch among several modes of behavior. For such data, transitions between activities organized by different motivational systems can be difficult to analyze.

Individual history

The search for general principles is easily obscured by the history of living organisms and their ecosystems (Hull 1984; May & Seger 1986). Field studies of long-lived animals in a fluctuating environment have demonstrated that the behavior and reproductive success of an individual may be dependent on demographic and environmental characteristics at the time of its birth (Clutton-Brock 1988). May & Seger (1986) called for modeling populations as aggregates of individuals. Łomnicki (1988) reviewed some analytical approaches to representing variation in a population due to categories of individuals.

Few models have incorporated constraints on an individual's behavior, which arise because it happened to be born into a particular territory that differed from other territories in the population, or to a particular mother whose behavior influenced the manner in which it learned to forage. However, such complexities of the real world influence the variance in success of individuals and the complications of testing the validity of models (e.g. Chepko-Sade & Halpin 1987; Clutton-Brock 1988).

Transparent computer programs

At the end of his review of motivational systems, Toates (1986) rather plaintively commented that soon scientists would need to exchange programs rather than reprints. If the complexity of biological models reaches such a level, the contents of programs need to be accessible to people other than those who built them. Unfortunately, the structure and content of programs often are so complex that only the behavior is reported. The result has been a proliferation of models that are untested, providing the function of illustrating a conceptual framework but not fitting into the scientific method of testing alternative hypotheses and rejecting those that yield invalid predictions (Thompson 1981).

ALTERNATIVE AI APPROACHES APPLYING OBJECT-ORIENTED PROGRAMMING

The limitations described above are common frustrations expressed by scientists who take a synthetic rather than reductionist approach and who seek to apply conceptual models to real world problems. Linear thinkers who accept a universal model "as reality" are likely to perceive the problems we identified above, not as limitations, but rather as the assumptions that give elegance and clarity to a model. In contrast, resource managers tend to take a very different view of the world, perceiving reality to be the "problem-in-the-field" and modeling to be one step in finding solutions to complex problems.

Resource management questions led us to take another look at the tools available in AI programming (Coulson et al. 1987). For example, these questions spanned micro- to macro- geographical, temporal, and biological scales, such as the following. (1) How can outbreaks of pine bark beetles be predicted and controlled (Rykiel et al. 1984)? (2) How does the size and mosaic of cotton fields influence biological control of pests by a parasitoid wasp (Makela et al. 1988)? (3) How can damage by moose foraging on pine seedlings be reduced (Saarenmaa et al. 1988) and habitat quality be assessed (Roese 1989)? (4) How can brush treatments in cattle pastures be designed to minimize impact on deer habitat (Folse et al. 1989)? (5) How do adaptations to a fluctuating environment influence individual reproductive success (Coulson et al. 1987) and population dynamics (Wilber 1987) in an aseasonal ungulate? (6) How can large herbivore populations be managed (Starfield & Bleloch 1985)? (7) How does social structure influence population dynamics (Graham 1986)? (8) How can wildlife habitat be classified (Marcot 1984)? (9) How can interaction between large carnivores and visitors be reduced in parks (Ruth & Packard unpublished data)?

In the following subsections, we provide examples of how AI programming concepts have been used to address the limitations identified in the previous section. The particular style of programming is described as "object-oriented" (Bobrow & Stefik 1986; Stefik & Bobrow 1986) in contrast to the procedural style typical of other approaches. The application of object-oriented programming to animal behavior is still in the initial exploratory stage; however, the following examples provide insights for future elaboration.

Spatial heterogeneity

Habitat patches (spatial heterogeneity) can be represented as objects in the environment, each object having a set of descriptive attributes and arranged within a hierarchical class (Saarenmaa et al. 1987). For example, Makela et al. (1988) represented a wasp world as one type (class) of habitat object (cotton field) with three member units (field 1, field 2, field 3). Each field contained a different set of interconnected parcels. The attributes of a cotton field were defined once and all units of the class inherited those attributes. Likewise, the attributes of parcels were defined once and each instance inherited the "information slots" for attributes. Examples of attributes of parcels included: list of neighbors, number of egg-laying sites, number of immature females, number of mature females.

Several features of this object-oriented structure are intuitively appealing to biologists. First, a continuous environment is represented by units that correspond to our conceptual framework of foraging relative to patches. This appeals to a basic "Gestalt" recognition of discontinuities in the environment (Lorenz 1981: 44; Allman 1986). Second, we tend to lump certain units as more similar than others, organizing information in hierarchical classes (Dawkins 1976). Third, we attach similar attributes to all members of a class (e.g. all vertebrates have forearms), while recognizing that each member may have a slightly different form of a particular attribute (e.g. the forearm of a bat differs from an elephant; Lorenz 1981: 94).

Different types of habitat patches can be represented via object-oriented programming. For example, Saarenmaa et. al. (1988) mentioned three classes: forest compartment, farmland, and water. Folse et al. (1989) created one class of habitat object that included three types of vegetation patches in a cattle pasture (e.g. untreated-shrub, root-plowed-shrub, sprayed-shrub) defined by the values of attributes. The vegetation attributes were determined by field data from a sample in each shrub type. The model environment was complex, representing 265 instances of vegetation patches in the pasture.

The linkages between habitat objects can be represented in a dynamic manner (Folse et al. 1989). Travel routes for deer moving between patches were represented as a network; however, not every patch was linked to all its neighbors. For example, deer do not readily move across root-plowed shrub patches, so habitat patches of this vegetation type were not in the network and functionally blocked movement between neighboring patches. The linkages among habitat patches were changed between subsequent runs of the model to examine the effect of vegetation pattern on simulated deer movements under three conditions: no shrub treatment, many small, or several large root-plowed patches of equivalent total area.

The spatial representation of the habitat objects can be stored in a series of map layers in a computer database commonly referred to as a Geographic Information System (GIS) (Coulson et al. 1987). This approach was used to represent the habitat of cougars in Big Bend National Park (Ruth & Packard, unpublished data.). Habitat patches were digitized using a GIS; based on characteristics of slope, aspect and soil type. The resulting list of patches and corresponding attributes were used as input to the object-oriented model and served as a network for simulated movement of the cougars (Folse et al. in press). To validate the model, simulated movements can then be compared with data acquired via radio-telemetry.

This direct interface between landscape-based spatial data and object-oriented models may not seem terribly exciting to the analytical modeler. However, the breakthrough will be clear to anyone who has struggled with analyzing radio-telemetry data on a grid-point basis, which is divorced from the biological and topographical realities of the landscape. Previous tools for analyzing home range were based on assumptions that animal use of space fits a distribution similar to an ellipse (Ford & Krumme

1979). Such assumptions are difficult to accept when a cougar's home range includes a massive cliff, lake, or campground.

Furthermore, the tools provided by GIS and object-oriented modelling provide the flexibility to switch between grid-based analysis at the micro-level, as is suitable for studying foraging paths (Roese 1989) and the coarse-grained analyses over larger patches, as is suitable for studying seasonal changes in animal distribution (Folse et al. 1989). Within the GIS data base, the environment can be represented as both points and patches. This flexibility provides options for redefining patches as meaningful to the animal in contrast to the researcher. For example, a researcher may perceive the difference between untreated-shrub and root-plowed shrub patches to be important, whereas the deer may respond in terms of edges and interiors.

Changing response rules

The animal object that interacts with its environment can receive stimulus messages and send response messages to external objects (Saarenmaa et al. 1988). External objects may be habitat objects or they may be other animal objects (Coulson et al. 1987; Folse et al. 1989). The exciting aspect of this structure is that it provides for dynamic modification of the program rules and memory content during simulations.

Use of object-oriented programming to simulate changes in response rules due to nutritional state were illustrated by Saarenmaa et al. (1988). In this prototype model, the moose-object had a specified daily intake requirement, monitored by a state variable "nutritional-balance". It met this goal state by searching for a patch containing young trees, entering the patch and consuming pine (protein and energy) and hardwoods (minerals). When the nutritional balance was positive, the moose did not consume vegetation. An internal function simulated metabolism of food for each day, reducing the nutritional balance as a function of time.

In contrast, external events in the moose model (Saarenmaa et al. 1988) drove the motivational system corresponding to "activity". A clock sent a message when it was daylight, resetting the "shelter balance" of the moose-object to a threshold level where the moose-object sought the nearest mature forest stand for resting.

One value of the object-oriented programming environment is the ease in which rules for changing responses can be attached to objects. Each object has a set of attributes and a set of procedures (Bobrow & Stefik 1986; Saarenmaa et al. 1988). The attributes and rules of one object are relatively hidden from another object. The language in which the program is written is at a higher level, such that it "automatically" takes care of bookkeeping details (e.g. how objects exchange messages). The user is thus freed to focus more on the content of the model. Furthermore, the model can be "event-driven" rather than "time-driven", more closely approximating the real world (Folse et al. 1989).

An object-oriented program can exchange messages with other chunks of programs (Denning 1985), making it possible to link complex models of motivational systems to the object-oriented model. For example, Folse (unpublished data) has expanded the deer model to include a complex chunk of programming simulating rumination processes.

In an AI program (LISP language), the types of dynamic modifications are limited only by the programmer's imagination, so flexibility need not constrain a modeler in making the best formal representation suitable for a certain question. The program can modify motivational rules dependent on age, setting bounded rules for changing rules. For example, the rules controlling movements of a cougar kitten will differ in certain motivational systems from those of its mother (Figure 3). The transition from one developmental stage to another may be modelled by transferring the attributes of an individual (an "instance" in AI jargon; Saarenmaa et al. 1988) of an immature class to a new instance of a mature class of object and erasing the immature instance in species with discrete life stages (Makela et al. 1988). For slowly maturing vertebrates whose behavior changes as a result of individual history of development, rules for changing rules could be triggered by certain threshold values, such as weight.

Changing contingency rules

The values of attributes in each instance of habitat-objects can be updated periodically, unknown to the animal-objects. Thus, contingency rules may change due to processes within the habitat-object or by an external model of ecological processes. Changes induced by procedures within a habitat-object were illustrated by Saarenmaa et al. (1988). Feeding by a moose object was simulated by a message sent from the moose-object to

SETS OF METHODS	OBJECT CLASSES	
	Adult female	**Juvenile male**
MODE PRIORITIES	1. agonism 2. ingestion 3. thermoregulation 4. arousal 5. reproduction/social	1. agonism 2. ingestion 3. thermoregulation 4. arousal 5. reproduction/social
MOTIVATION Agonism	• hide from intruders • flee if there is an escape • attack if cornered	• hide from intruders • flee if there is an escape • freeze if cornered
Ingestion	• hunger rises to an asymptode at five days as an inverse function of time since last meal • when hungry search for prey • assess vulnerability when prey is encountered • attack vulnerable prey • eat from kill	• hunger rises to an asymptode at 24 hours as an inverse function of time since last meal • when hungry search for mother • suckle from mother when she responds by nursing • follow mother when no nursing • feed from mother's kill
Thermoregulation	• monitor body temperature • move to shade when hot • move to breeze if no shade • move to sun when cold • move to shelter from wind when cold in sun	• monitor body temperature • move to crevice when cold • approach warm body when cold • move to shade when hot
Arousal	• movement potential declines to nil in 12 hr after last sleep • duration of sleeping bout is direct function of duration of previous activity bout	• movement potential declines to nil in 6 hr after last sleep • duration of sleeping bout is a direct function of duration of previous activity bout
Reproduction/ Social	• approach offspring at least once a day • nurse offspring if it requests • leave offspring when it finishes nursing	• approach mother when near • approach siblings when mother is not near
Exploration	• visit nearby patches that have not been encountered recently	• approach new objects within 5 m of den • practice action repertoire with sibs
LEARNING	• store new patches in memory • update attributes of memory-net • actions leading to prey encounter • actions leading to social encounter	• store new patches in memory • update attributes of memory-net • actions leading to prey encounter • actions leading to social encounter

Figure 3. Preliminary set of rules for an adult and juvenile cougar, illustrating how several motivational systems may be represented in a rule of thumb model.

the patch-object (patch-objects were one class of habitat-objects). The patch-object contained procedures to reduce its attribute values for pine and hardwood by an amount corresponding to what the moose-object ate.

When more than one animal-object is in the population, the attributes of a patch-object may thus change from the time an individual first visits a patch to the time it returns. Such options were useful for modeling movements of nectar feeding birds (Triono et al. unpublished data), based on observations that one bird may drain a flower of nectar, reducing the value of a patch for individuals that follow (Pyke 1984). Internal rules may also change the attribute values of habitat objects, e.g. by simulating nectar production within a flower. Folse (unpublished data) has expanded the deer model to monitor the effects of several deer feeding in habitat patches.

In concept, seasonal changes in attributes of patches could be simulated by linking with an external model. For example, expert systems have been developed to choose an appropriate model of bark-beetle population dynamics, transfer input from the user to the model and transfer the model's output to the user (Rykiel et al. 1984). This approach was developed at a time when several complex models of bark beetle and forest dynamics had been developed, each with slightly different yet complementary characteristics. The concept was a pioneering step in integrating programming techniques developed in the field of artificial intelligence (e.g. expert systems) with traditional systems modeling techniques. There is no reason why the concept cannot be extended to include geographical data bases and input from external models that change the attributes of patches within the geographical data base (Coulson et al. 1987).

Incomplete knowledge

The discrepancy between what an animal "knows" about its environment and the actual environment may be modeled by an "internal representation" of the environment. Folse et al. (1989) illustrated how a deer-object can be programmed to build a memory network of the habitat objects that it passes through during an initial phase when it explores its model environment. The model environment can be represented as a network of habitat patches. When the deer-object enters a patch, it searches its memory and adds a new memory-patch if the habitat-patch is not in memory. Thus, the linkages among memory-patches are dynamic and represent a subset of habitat-patches known to an individual deer. Another individual might have a slightly different representation of the environment in its memory-net.

Incomplete knowledge of the environment does influence behavior of a model that simulates movement among habitat patches. Folse et al. (1989) modeled a simple goal directed search, representing a deer looking for water. They found that the distance traveled by a deer-object with a memory net was always shorter (by a factor of 100) than a deer-object with no memory (which moved randomly among patches). Furthermore, when the attribute values of the patches were changed, the path to the goal was shortened (by a factor of 3) after the deer-object "learned" of the modifications and updated its memory-net.

Within a patch, limited information is available to a browsing ungulate as it moves from shrub to shrub. Using a "rule-of-thumb" modeling approach, Roese (1989) illustrated how the distance of detection can influence the path of a moose-object moving during a foraging bout. As a moose moves from one feeding station to another, only a subset of plants are detected (perceptive field), of which a smaller subset are within reach (consumptive field). Roese (1989) found that a larger perceptive field significantly reduced foraging efficiency, as the moose-object spent more time deciding what to eat rather than eating.

Goal directed search

Search processes are well-defined techniques in AI programming (Charniak & McDermott 1985). For example, a robotic arm given the task of stacking blocks may use a "backward-chaining" or "forward-chaining" strategy (Waltz 1982). In the backward-chaining strategy, a solution to the problem is found by following the decision path of previous solutions that were successful. In the forward-chaining strategy, the solution is found by an internal representation of subsequent moves, discarding projected decision paths that were not successful in the internal analysis. There are many options for searches within these two basic approaches. For example, in a "depth-first" search strategy, one path is explored at a time, whereas in a "breadth-first" search strategy, several paths are explored in parallel (Saarenmaa et al. 1988). Certain search strategies (e.g. A* used by Folse et al. 1989) utilize heuristic (i.e. rule-based) information to improve search efficiency. All of these methods are well-tested modules that can be inserted into models developed by different laboratories.

Saarenmaa et al. (1988) compared the behavior of a moose-object using three different search strategies. They found that a random search (the strategy most often used in foraging models) resulted in a greater nutritional deficit and more evenly distributed frequency of visiting patches, compared to a local search or global search strategy. The local search strategy differed from the global search in that the moose object decided to which adjacent patch it would move rather than evaluating all patches within the model environment. Both the local and global search resulted in frequent visitation to "good" patches and infrequent visitation to "poor" patches and similar performance in terms of mean nutritional deficit. However, the local search strategy resulted in a greater variance in the index of nutritional deficit. Where risk-sensitive foraging is important, variance in intake may be a critical factor (Stephens & Krebs 1986).

Therefore, the type of search strategy can influence the behavior of an animal-object. To devise more predictive models, it will be necessary to identify ways of testing which search patterns most closely represent the actual behavior of animals in the system to be modeled. AI tools provide specific definitions of alternative search strategies, potentially taking the mystery out of vague notions of animal intelligence (Pearce 1988), which have defied operational definitions in the past. For example, a researcher might test actual behavior against the outcome predicted by alternative models based on random search, forward chaining, or backward chaining.

From our synthetic perspective, it does not matter whether animals actually use the search processes in the model, only whether one of several alternatives better matches the behavior of animals observed in the field. Of course from the perspective of a linear style, the task of developing an algorithm that provides a one-to-one match with biological mechanisms and accurately predicts individual behavior appears insurmountable. At issue is the recognition that any model, no matter how complete, will not be able to predict where you will eat dinner tonight. However, based on statistical data, a model may be able to identify which options are more likely than others.

Interaction between motivational systems

The object-oriented programming approach encourages modeling at a level where interactions between motivational systems can be considered. By using a modular style, it is easier to modify the "rules-of-thumb" for one motivational system without disturbing the rest of the program. Thus, sensitivity testing is made easier in exploring unusual behavior of the model when rules from different systems interact in unanticipated ways. We have set up the structure for simulating interactions among motivational systems and have a research tool to begin exploring the pattern of emergent properties resulting from interactions of behavioral decision rules internally consistent within each system.

The degree of elaboration within a motivational system and in interactions between systems depends, in part, on the scale of the research question. For example, in modeling foraging paths of moose, Roese (1989) considered only one motivational system, essentially considering a moose to be an "eating machine". Empirical data on foraging paths of moose can be obtained by following individual animals or trails in the snow. However, information on animal movements obtained from radio-telemetry accumulates over a much longer time span, broader geographic scale and less frequent sampling schedule. The questions of moose movements addressed by Saarenmaa et al. (1988) were at this broader level, involving movements between forest stands. Thus, it was important to consider whether moose could be discouraged from foraging in a patch of optimal nutritional value if no daytime shelter was nearby.

In their prototype model, Saarenmaa et al. (1988) took the parsimonious approach of assuming a moose to be an "eating/resting machine". The moose-object chose between two basic behavioral modes, foraging and resting. The nutritional balance directed the moose to seek a patch suitable for feeding. Daylight raised the priority of the resting mode, directing the moose-object to seek shelter. When the nutritional deficit was extremely severe, the moose-object fed during the day in the most sheltered patch available.

The deer-object modelled by Folse et al. (1989) was an "exploring/drinking machine". Although it only contained rules for the motivational systems of exploring and finding water, the program was expanded subsequently to include three additional motivational systems (feeding, thermoregulation, resting) with little effort due to the modular nature of the program (Folse unpublished data).

In developing the most parsimonious set of rules they thought were needed to explain cougar movements, researchers included four motivational states (Figure 3). However, simulated movements based on these rules did not approach a realistic pattern of movement based on telemetry data (H. Mueller personal communication). The exercise forced the field researchers to reexamine their database and assumptions.

Individual history

Object-oriented simulation procedures can accommodate variation among individuals due to learning, developmental history, geographic variation or fluctuations in environmental factors. The memory of hardware used in AI modelling is large enough to store at least several thousand instances of objects, representing individuals in a population (Folse, unpublished data). This magnitude is adequate for complete representation of all known individuals in many existing populations of large vertebrates. For example, the cougar population of Big Bend National Park is estimated at less than 100 individuals and the threatened southern sea otter population consists of less than 2,000 individuals (Brody 1988).

Conceptually, it should be possible to allow properties of populations to emerge from the combined decisions of individuals within the population (Reiter 1986; Huston et al. 1988). Often it is easier to test individual decision rules than population processes, particularly with long-lived vertebrates.

A major programming break-through was needed to solve the problem of synchronizing individual instances of objects with processes that are event-driven rather than time-driven. In the model of Makela et al. (1988), the decisions of each individual were made sequentially at each time step. In contrast, the event-driven model of Folse et al. (1989) has been expanded to include ten deer with memory-maps based on their individual experience during exploratory trips. As long as the deer-objects do not send messages to each other, their decisions are made in parallel and it does not matter if one individual finds water after 25 decision events and another finds it after 100 events. However, to model the interaction between a cougar and her cub, the parallel processes of decisions will have to be synchronized periodically. AI solutions to such problems are currently being explored for managing computer communication networks (Denning 1985) and complex systems (Gelernter 1989). One solution has been developed involving sliding windows of time in which individuals are resynchronized (Folse & Schnase unpublished data).

Transparent computer programs

One of the basic principles in AI programming is to make the content of programs "transparent" to the user (Bobrow & Stefik 1986). Transparency refers to the use of natural language interfaces or subroutines that allow the user to examine the rule-base contained in the program. For example, an expert system for diagnosis of an illness can be programmed such that the user can ask why the system made a certain diagnosis. In traditional programming approaches, this principle is implemented via program documentation. Like program documentation, transparency is often overlooked even in AI programming.

The basic method involved in AI programming is one that increases the communication between the scientist who knows a lot about the real world and the computer programmer who knows how to implement the model representation of that world. To construct an expert system, people in three roles are involved: the expert, the computer programmer and the knowledge engineer (Van Horn 1986). The role of the knowledge engineer is to package the knowledge of the expert in a form compatible with the information that can be put into the computer by the programmer.

In modeling animal behavior, the modeler assumes the role of knowledge engineer. We have found that the basic model of behavior illustrated in Figure 1 aided immensely in defining the role of the knowledge engineer in three steps required in developing an object-oriented model of behavior (Figure 4).

In the first step, the modeler interacts with the biological expert to arrive at a clear definition of the biological problem. This involves listing the relevant types of animals and habitat characteristics, the state variables important to monitor, and behavioral rules. Each box in Figure 1 and linkage between boxes is fully examined at this stage. Decisions are made regarding which boxes are to be left empty. The biological expert specifies the user input and the output of the model.

In the second step, the modeler restructures the information in an AI compatible form. This involves defining animal- and habitat-objects in the model world. The number of instances (individuals) in each object class is specified, as are the

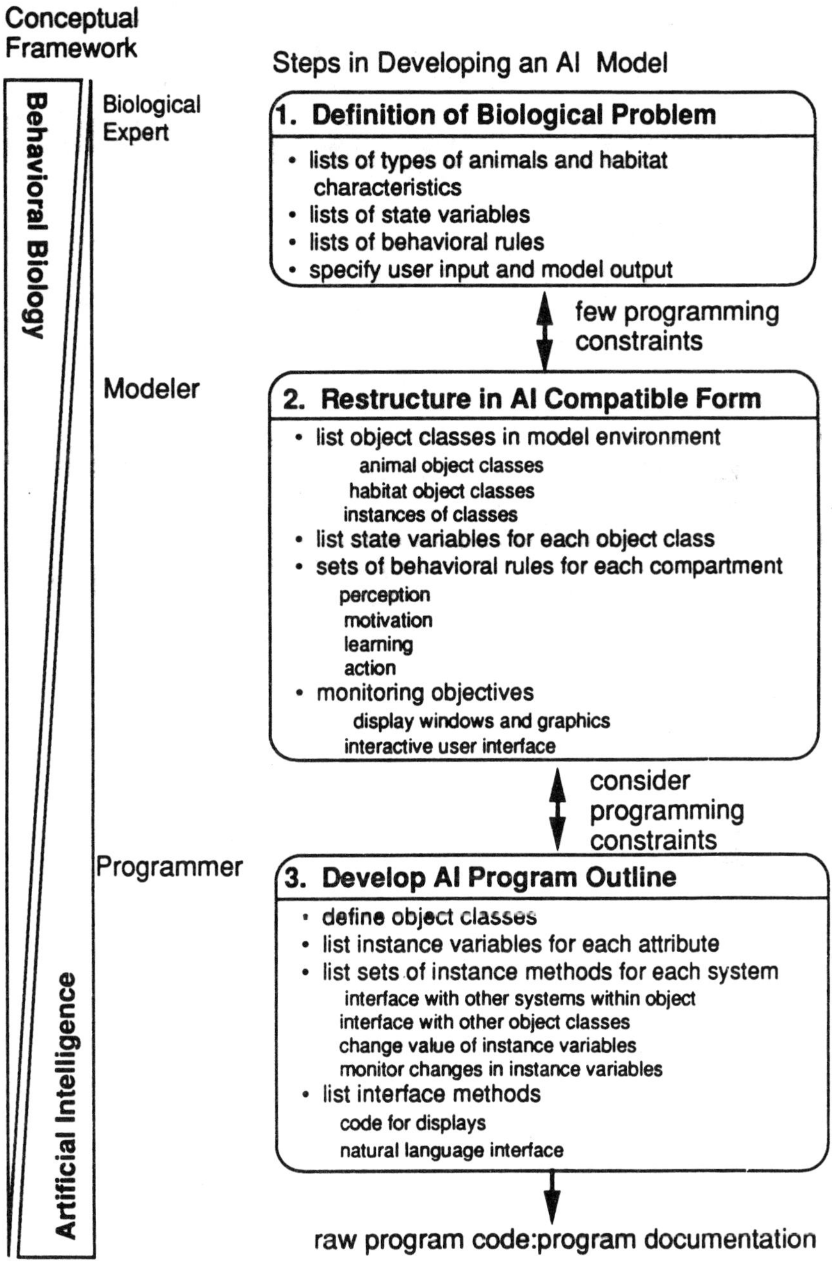

Figure 4. Steps involved in developing an object oriented model using the basic shell structure of behavior.

attributes (state variables) of each object class. For each object class, sets of behavioral rules are developed within the following subsystems: message sending (input/output corresponding to perception and activation), motivation, and learning. Finally, objectives for monitoring the behavior of the model are implemented by defining the options for windows, guages, dynamic graphic displays and user interfaces.

In the third step, the modeler interacts with the programmer in developing an AI program outline that documents the program in modules corresponding to the way the information was organized in steps 1 and 2. The programmer's job is to define object classes (including attribute variables and their methods), set up main simulation drivers, and create the user interfaces. For each object class, the programmer creates code for the methods specifying interfaces with other internal classes and external objects, the changes in attribute values due to internal processes, and the displays that monitor changes in attribute values and make the program transparent to the biological expert.

Ideally, an untutored user should be able to interact with an AI program to learn of its structure and content. The program should be modularized such that packages of information can be substituted with little disturbance to operations in the rest of the program (Denning 1985). Only when technology reaches such achievable functions will behavioral biologists from different laboratories truly be able to place confidence in the internal validity of models and proceed with the empirical testing required to reject alternative models.

CONCLUSIONS

The tools we use to communicate about complex behavioral processes influence the way we think about, interpret and explain those processes. Early ethologists used analogies based on mechanical, hierarchical or feedback mechanisms to describe behavior. Such linear processes have been adequate for analysis of linear causality resulting in a reductionist approach to science. However, the synthetic approach required in resource management is based on what has previously been criticized as subjective, intuitive thought, in which the emergent properties of the whole are more than the sum of the parts.

The programming tools developed in the field of artificial intelligence allow users to capture knowledge in a form that represents synthetic thought. For example, animals and patches in the environment can be represented as objects. Each object has slots for information about state variables (attributes) and behavioral rules (methods). Since the bookkeeping details of structure and communication between objects are handled by the higher level programming language, the user is freed to elaborate more regarding the content of the knowledge base and rules. We illustrated some ways in which this approach has been used to address questions involving spatial and temporal heterogeneity, interacting motivational systems, individual history, incomplete knowledge and goal-directed search.

In the past decade, analytical approaches to modeling behavior involved optimality theory and game theory. Empirical approaches to simulating behavioral processes involved elaborate homeostatic, contingency or compartment models. Although researchers who built their own computer programs were satisfied with the tools available to them, such models are opaque to researchers in other laboratories who seek to duplicate results.

We envision that the next decade will benefit from a proliferation of transparent programs based on artificial intelligence programming techniques, in which the structure and content of behavioral models will be accessible to even untutored users. Such accessibility will depend on appropriate structure as well as technology developed in expert systems to allow users to interact using command language phrases, menus, or function keys. Communication among laboratories will be greatly facilitated if the structure of such programs is similar, providing for modules to be easily replaced to simulate the results of alternative variants of a behavioral trait.

We presented a basic structure representing four components of behavior: perceptual systems, action systems, motivational systems, and learning systems, and discussed the major conceptual framework of each component. This basic framework not only aids in intellectual communication across disciplines, it also could provide the basis for a program shell that would provide a work environment with tools to integrate knowledge regarding individual decision rules, population dynamics and population genetics.

ACKNOWLEDGEMENTS

We are very grateful to colleagues who have engaged us in stimulating and thought-provoking discussion of the ideas presented in this paper. Valuable reviews were provided by M.

Bekoff. M. Childress, W. E. Grant, D. Jamieson, H. Mueller, and E. Rykiel. Members of our working group, W. E. Grant, J. Roese, and J. Schnase challenged us to evaluate whether artificial intelligence really provides a fundamentally new approach to modeling. The senior author is grateful to N. Thompson for clarification of some philosophical issues. We appreciate the contagious enthusiasm of students who developed prototype programs, including J. Brown, G. Triono, H. Mueller and J. Shoshank. The cooperation of L. A. Graham, E. Dougherty, A. Bunting, R. Flamm, and C. Lovelady in the Knowledge Engineering Laboratory greatly aided T. Ruth in exploring ways to integrate GIS information into movement models. We are grateful for support from the Entomology Department, the Department of Wildlife and Fisheries Sciences and the Texas Agricultural Experiment Station. This is a publication of the Texas Agricultural Experiment Station.

LITERATURE CITED

Albus, J. 1981. *Brains, Behavior and Robotics.* New York: McGraw Hill.

Alper, J. 1986. Our dual memory. *Science* 7, 44-49.

Alcock, J. 1989. *Animal Behavior: An Evolutionary Approach.* 4th Edition. Sunderland, Massachusetts: Sinauer Associates.

Allman, W.F. 1986. Mindworks. *Science* 7, 22-31.

Arditi, R. & Dacorogna, B. 1988. Optimal foraging on arbitrary food distributions and the definition of habitat patches. *American Naturalist* 131, 837-846.

Atkinson, R.C. & Shiffrin, R.M. 1971. The control of short-term memory. *Scientific American* 225, 82-90.

Bateson, P.P.G. 1979. How do sensitive periods arise and what are they for? *Animal Behaviour* 27, 470-486.

Beach, F.A. 1950. The snark was a boojum. *American Psychologist* 5, 115-124.

Bekoff, M. 1977. Quantitative studies of three areas of classical ethology: Social dominance, behavioral taxonomy, and behavioral variability. In: *Quantitative Methods in the Study of Behavior* (ed. by B. A. Hazlett), pp 1-46. New York: Academic Press.

Bekoff, M. & Mech, L.D. 1984. Simulation analyses of space use: Home range estimates, variability and sample size. *Behavior Research Methods Instrumentation and Computers* 16, 32-37.

Bekoff, M., Scott, A.C., & Conner, D.A. 1989. Ecological analyses of nesting success in evening grosbeaks. *Oecologia* 81, 67-74.

Bobrow, D. & Stefik, M.J. 1986. Perspectives on artificial intelligence programming. *Science* 231, 951-957.

Brody, A. 1988. A simulation model for assessing the risks of oil spills to the California sea otter population and an analysis of the historical growth of the population. In: *Population Status of California Sea Otters* (ed. by D. B. Siniff & K. Ralls), pp. 191-274. Washington, D.C.: Minerals Management Service Report No. 88-0021.

Chalmers, N.R. 1987. Developmental pathways in behaviour. *Animal Behaviour* 35, 659-674.

Charniak, E. & McDermott, D. 1985. *An Introduction to Artificial Intelligence.* Reading, Pennsylvania: Addison Wesley.

Charnov, E.L. 1976. Optimal foraging, the marginal value theorem. *Theoretical Population Biology* 9, 129-136.

Chepko Sade, B.D. & Halpin, Z.T. (eds.). 1987. *Mammalian Dispersal Patterns: The Effects of Social Structure on Population Genetics.* Chicago, Illinois: University of Chicago Press.

Clutton-Brock, T.H. (ed.) 1988. *Reproductive Success: Studies of Individual Variation in Contrasting Breeding Systems.* Chicago, Illinois: University of Chicago Press.

Colgan, P. 1989. *Animal Motivation.* New York: Chapman & Hall.

Coulson, R.N., Folse, L.J. & Loh, D.K. 1987. Artificial intelligence and natural resource management. *Science* 237, 262-267.

Davies, N.B. & Houston, A.I. 1984. Territory economics. In: *Behavioral Ecology: An Evolutionary Approach* 2nd Edition. (ed. by J. R. Krebs & N. B. Davies), pp. 148-169. Sunderland, Massachusetts: Sinauer Associates.

Davis, R. 1986. Knowledge-based systems. *Science* 231, 957-963.

Dawkins, R. 1976. Hierarchical organisation: A candidate principle for ethology. In: *Growing Points in Ethology.* (ed. by P. P. G. Bateson & R. A. Hinde), pp. 7-54. Cambridge: Cambridge University Press.

_____. 1982. *The Extended Phenotype: The Gene as the Unit of Selection.* Oxford: Oxford University Press.

De Ghett, V.J. 1978. Hierarchical cluster analysis. In: *Quantitative Ethology* (ed. by P. W. Colgan), pp. 115-144. New York: John Wiley & Sons.

DeVries, D.R., Stein, R.A., & Chesson, P.L. 1989. Sunfish foraging among patches: The patch departure decision. *Animal Behaviour* 37, 455-464.

Dennett, D. 1984. Cognitive wheels: The frame problem of AI. In: *Minds, Machines and Evolution* (ed. by C. Hookway), pp. 129-151. Cambridge: Cambridge University Press.

_____. 1988. Précis of *The Intentional Stance. Behavioural and Brain Sciences* 11, 495-546.

Denning, P. J. 1985. The evolution of parallel processing. *American Scientist* 73, 414-416.

_____. 1986. Will machines ever think? *American Scientist* 74, 344-346.

Dewsbury, D. A. 1978. *Comparative Animal Behavior.* New York: McGraw Hill.

Dreyfus, H. 1979. *What Computers Can't Do.* New York: Harper & Row.

Fagen, R.M. 1978. Repertoire analysis. In: *Quantitative Ethology* (ed. by P. W. Colgan), pp. 25-42. New York: John Wiley & Sons.

Folse, L.J., Packard, J.M., & Grant, W.E. 1989. AI modelling of animal movements in a heterogeneous habitat. *Ecological Modelling, Special Issue, Artificial Intelligence and Expert Systems in Ecology and Natural Resource Management* 46, 57-72.

Folse, L.J., Mueller, H.E., & Whittaker, A.D. In press. Object-oriented simulation and geographic information systems. *AI Applications in Natural Resource Management 4.*

Ford, R.G. & Krumme, D.W. 1979. The analysis of space use patterns. *Journal of Theoretical Biology* 76, 125-155.

Forsyth, R. 1986. Machine learning. In: *Artificial Intelligence: Principles and Applications* (ed. by M. Yazdani), pp. 205-225. London: Chapman and Hall.

Fox, J.L. 1983. Debate on learning theory is shifting. *Science* 222, 1219-1222.

Gardner, R.A., & Gardner, B.T. 1988. Feedforward versus feedbackward: An ethological alternative to the law of effect. *Behavioral and Brain Sciences* 11, 429-493.

Gelernter, D. 1989. The metamorphosis of information management. *Scientific American* 261, 66-73.

Glymour, C, Scheines, R., Spirtes, P., & Kelly, K. 1987. *Discovering Causal Structure: Artificial Intelligence, Philosophy of Science, and Statistical Modeling.* New York: Academic Press.

Graham, L. A. 1986. *HAREMS: A Generalized Data Base Manager and Simulator for Barrier Island Feral Horse Populations.* Atlanta, Georgia: Cooperative Park Studies Unit Technical Report 32.

Harrison, A.F. & Bramson, R.M. 1982. *Styles of Thinking: Strategies for Asking Questions, Making Decisions, and Solving Problems.* Garden City, New York: Anchor Press/Doubleday.

Hassell, M.P. & May, R.M. 1984. From individual behaviour to population dynamics. In: *Behavioural Ecology* (ed. by R. M. Sibly & R. H. Smith), pp. 3-32. Oxford: Blackwell Scientific Publications.

Heuer, H. & Sanders, A.F. (eds.) 1987. *Perspectives on Perception and Action.* Hillsdale, New Jersey: Lawrence Erlbaum Associates.

Horn, H.S. 1978. Optimal tactics of reproduction and life history. In: *Behavioral Ecology: An Evolutionary Approach* 1st Edition. (ed. by J.R. Krebs & N.B. Davies), pp. 411-429. Sunderland, Massachusetts: Sinauer Associates.

Hull, D. 1984. Historical entities and historical narratives. In: *Minds, Machines and Evolution* (ed. by C. Hookway), pp. 17-41. Cambridge: Cambridge University Press.

Huston, M., De Angelis, D., & Post, W. 1988. New computer models unify ecological theory. *BioScience* 38, 682-691.

Johnston, T.D. 1988. Developmental explanation and the ontogeny of birdsong: Nature/nurture redux. *Behavioral and Brain Sciences* 11, 617-663.

Kendrick, D.F., Rilling, M.E., & Denning, M.F. (eds.) 1985. *Theories of Animal Memory.* Hillsdale, New Jersey: Lawrence Erlbaum Associates.

Kovac, M.P. & Davis, W.J. 1980. Neural mechanisms underlying behavioral choice in *Pleurobranchia*. *Journal of Neurophysiology* 43, 469-487.

Krebs, J.R. & McCleery, R.H. 1984. Optimization in behavioral ecology. In: *Behavioral Ecology: An evolutionary approach* 2nd Edition. (ed. by J.R. Krebs & N.B. Davies), pp 91-121. Sunderland, Massachusetts: Sinauer Associates.

Lehrman, D.S. 1953. A critique of Konrad Lorenz's theory of instinctive behavior, *The Quarterly Review of Biology* 28, 337-363.

Lomnicki, A. 1988. *Population Ecology of Individuals.* Princeton, New Jersey: Princeton University Press.

Lorenz, K.Z. 1958. The evolution of behavior. *Scientific American* 199, 67-78.

_____. 1965. *Evolution and Modification of Behavior.* Chicago, Illinois: University of Chicago Press.

_____. 1977. *Behind the Mirror: A Search for a Natural History of Human Knowledge.* New York: Harcourt, Brace & Jovanovitch.

_____. 1981. *The Foundations of Ethology.* New York: Springer Verlag.

MacArthur, R.H. & Pianka, E.R. 1966. On optimal use of a patchy environment, *American Naturalist* 100, 603-609.

Makela, M.E., Stone, N.D., & Vinson, B. 1988. Host parasitoid population dynamics in a heterogeneous environment. In: *Artificial Intelligence and Simulation: The Diversity of Applications* (ed. by T. Henson), pp. 228-223. San Diego, California: Simulation Council Incorporated.

Mangel, M. & Clark, C.W. 1988. *Dynamic Modeling in Behavioral Ecology.* Princeton, New Jersey: Princeton University Press.

Marcot, B.G. 1984. Use of expert systems in wildlife habitat modeling. In: *Wildlife 2000: Modeling Habitat Relationships of Terrestrial Vertebrates* (ed. by J. Verner, M.L. Morrison & C. J. Ralph), pp. 145-150. Madison, Wisconsin: University of Wisconsin Press.

May, R.M. & Seger, J. 1986. Ideas in ecology, *American Scientist* 74, 256-267.

Maynard Smith, J. 1984. The evolution of animal intelligence. In: *Minds, Machines and Evolution* (ed. by C. Hookway), pp. 63-71. Cambridge: Cambridge University Press.

Mayr, E. 1974. Behavior programs and evolutionary strategies, *American Scientist* 62, 650-659.

McCleery, R.H. 1983. Interactions between activities. In: *Animal Behaviour 1. Causes and Effects* (ed. by T. R. Halliday & P. J. B. Slater), pp. 134-167. Oxford: Blackwell Scientific.

Mishkin, M. & Appenzeller, T. 1987. The anatomy of memory. *Scientific America* 256, 80-89.

Narayanan, A. 1986. Memory models of man and machine. In: *Artificial Intelligence: Principles and Applications.* (Ed. by M. Yazdani) pp. 226-259, New York: Chapman and Hall.

Nisbett, A. 1976. *Konrad Lorenz.* London: J.M. Dent & Sons, Ltd.

Owen-Smith, N. & Novellie, P. 1982. What should a clever ungulate eat? *American Naturalist* 119, 151-178.

Packard, J.M. & Ribic, C.A. 1982. Classification of behavior of sea otters (*Enhydra lutris*). *Canadian Journal of Zoology* 60 1362-1373.

Pearce, J.A. 1988. *An Introduction to Animal Cognition.* Hillsdale, New Jersey: Lawrence Erlbaum Associates.

Pierce, G.J. & Ollason, J.G. 1987. Eight reasons why optimal foraging theory is a waste of time. *Oikos* 49, 111-118.

Poggio, T. 1984. Vision by man and machine. *Scientific American* 250, 106-116.

Pyke, G.H. 1984. Optimal foraging theory: A critical review. *Annual Review of Ecology and Systematics* 15, 523-575.

Reiter, C. 1986. Toy universes. *Science* 7, 54-59.

Richmond, B., Peterson, S., & Vescuso, P. 1987. An Academic Users Guide to Stella. Lyme, New .Hampshire: High Performance Systems

Ridley, M. 1986. *Animal Behaviour: A Concise Introduction.* Oxford: Blackwell Scientific Publications.

Roeder, K.D. 1965. Moths and ultrasound. *Scientific American* 212, 94-102.

Roese, J.R. 1989. *A Simulation Model of Ruminant Foraging Strategies.* Ph.D. Dissertation, Texas A & M University.

Rolls, B.J. & Rolls, E.T. 1982. *Thirst.* Cambridge: Cambridge University Press.

Rykiel, E.F. Saunders, M.C., Wagner, T.L., Loh, D.K., Turnbow, R.H., Hu, L.C., Pulley, P.E., & Coulson, R.N. 1984. Computer-aided decision making and information accessing in pest management systems with emphasis on the southern pine beetle, *Dendroctonus frontalis*, (Coleoptera: Scolytidae). *Journal of Economic Entomology* 77, 1073-1082.

Saarenmaa, H., Stone, N.D., Folse, L.J., Packard, J.M., Grant, W.E., Makela, M.E., & Coulson, R.N. 1988. An artificial intelligence modelling approach to simulating animal/habitat interactions, *Ecological Modelling* 44, 125-141.

Schöne, H. 1984. *Spatial Orientation and the Control of Behavior in Animals and Man.* Princeton, New Jersey: Princeton University Press.

Schoener, T. 1971. Theory of feeding strategies. *Annual Review of Ecology and Systematics* 2, 369-404.

Searle, J. 1984. *Minds, Brains and Science.* Cambridge, Massachusetts: Harvard University Press.

Senft, R.L. Coughenour, M.B., Bailey, D.W., Rittenhouse, L.R., Sola, O.E. & Swift, D.M. 1987. Large herbivore foraging and ecological hierarchies. *BioScience* 37, 789-799.

Shettleworth, S.J. 1983. Memory in food-hoarding birds. *Scientific American* 248, 102-110.

Siniff, D. & Jessen, C. 1969. A simulation model of animal movement patterns. In: *Advances in Ecological Research* (ed. by J.B. Cragg), pp. 185-219. New York: Academic Press.

Smith, W.J. 1977. *The Behavior of Communicating.* Cambridge, Massachusetts: Harvard University Press.

Squire, L.R. 1986. Mechanisms of memory. *Science* 232, 1612-1619.

Staddon, J.E.R. 1983. *Adaptive Behavior and Learning.* Cambridge: Cambridge University Press.

Starfield, A.M., & Bleloch, A.L. 1986. *Building Models for Conservation and Wildlife Management.* New York: Macmillan.

Stefik, M. & Bobrow, D.G. 1986. Object-oriented programming: themes and variations, *AI Magazine* 4, 40-62.

Stephens, D.W. & Krebs, J.R. 1986. *Foraging Theory.* Princeton, New Jersey: Princeton University Press.

Sustare, B.D. 1978. Systems diagrams. In: *Quantitative Ethology* (ed. by P. W. Colgan), pp. 275-311. New York: J. Wiley & Sons.

Tanenbaum, J. 1989. *Male & Female Realities: Understanding the Opposite Sex.* Sugar Land, Texas: Candle Publishing Company.

Tank, D.W. & Hopfield, J.J. 1987. Collective computation in neuronlike circuits. *Scientific American* 257, 104-114.

Tennant, N. 1984. Intentionality, syntactic structure and the evolution of language. In: *Minds, Machines and Evolution* (ed. by C. Hookway), pp. 73-103. Cambridge: Cambridge University Press.

Thompson, N.S. 1981. Towards a falsifiable theory of evolution. In: *Perspectives in Ethology,* Volume 4 *Advantages of Diversity.* (ed. by P.P.G. Bateson, & P.H. Klopfer), pp. 51-74. New York: Plenum Press.

Thompson, R.F. 1986. The neurobiology of learning and memory. *Science* 233, 941-947.

Tinbergen, N. 1951. *The Study of Instinct.* Oxford: Oxford University Press.

_____. 1960. The evolution of behavior in gulls. *Scientific American* 203, 118-130.

Toates, F.M. 1986. *Motivational Behavior.* Cambridge: Cambridge University Press.

Ullman, S. 1986. Artificial intelligence and the brain: computational studies of the visual system. *Annual Review of Neuroscience* 9, 1-26.

Van Horn, M. 1986. *Understanding Expert Systems.* New York: Bantam Books

Waltz, D.L. 1982. Artificial intelligence. *Scientific American* 247, 118-133.

Wilber, J.P. 1987. *Effects of Seasonally Varying Dietary Crude Protein Levels on Collared Peccary Population Dynamics - A Simulation Study.* M. Science Thesis. Texas A&M University.

8. Comparative Behavior and Phylogenetic Analyses: New Wine, Old Bottles

Gordon M. Burghardt and John L. Gittleman

INTRODUCTION

Comparative issues deriving from an evolutionary framework have been an explicit concern in psychology and ethology since the onset of Darwinian biology in the late nineteenth century (Bornstein 1980). The fate of this "comparative imperative" has had a distinctly checkered history. Comparative psychology and classical ethology have been repeatedly dissected and compared (e.g. Burghardt 1973; Beer 1980; Dewsbury 1989), and comparative methodologies outlined (e.g. Gittleman 1989a; S. Mitchell this volume; Thornhill 1984 this volume;).

Our aim in this chapter is to: (1) trace the historical development of comparative and phylogenetic concerns in psychology and ethology; (2) apply to behavior recently developed taxonomic methods for classification and evolutionary reconstruction by reanalyzing a classic study (Lorenz 1941/1970); (3) present a new methodology for distinguishing and quantifying phylogenetic and adaptive effects; and (4) outline how some new comparative methods might provide a rigorous means to incorporate the early Darwinians' dream of a comparative psychology of mental characteristics into modern evolutionary biology. We hope to stimulate critical interest, using new tools, in problems prematurely considered archaic, dead, empirically too tedious, or in principle impossible.

A BRIEF HISTORY OF COMPARATIVE STUDIES: WHO, WHAT, AND HOW

Comparative ethology can be characterized as the use of behavior patterns in the analysis of behavioral evolution emphasizing the processes underlying the mechanisms and evolution of behavior, adaptive function, and phylogenetic reconstruction. *Comparative* psychology was originally the parallel study of the cognitive, motivational, and affective processes underlying behavioral performance and their phylogenetic parameters and reconstruction (Crawford 1989). Issues of physiology, ontogeny, and experience have played important roles in both fields and there always has been some overlap.

Most comparisons are performed at the species level. This comes straight from the comparative anatomy and taxonomic approaches of pre-Darwinian biologists. In his chapter on instinct in *The Origin of Species* (Darwin 1859/1967) and in his posthumous chapter on instinct (Darwin 1883), Darwin compared closely related species of animals, such as bees and ants, to obtain insight into the possible ways in which a given remarkable ability, such as comb construction or slave-making could have evolved out of less dramatic precursors by the natural selection of minor variants.

Darwin also compared across much wider taxonomic groups such as orders, classes, and phyla. Possible variation within them was set aside to draw attention to evolution above the species level (here termed macroevolution). These trends were often viewed as progressive (anagenesis - Gottlieb 1984). This led to attempts (e.g. Romanes 1883) to organize a hierarchical, progressive scheme of psychological, rather than strictly behavioral, evolution based on large taxonomic groupings. Specific "higher" mammals were elevated, with our own species sole occupant of the top.

Comparative Psychology

The interest of the early comparative psychologists in making comparisons across major taxonomic groups dictated that the specifics of behavior were primarily treated as reflections (ambassadors) of underlying abilities constrained by the organization of the mind or nervous system. Certainly differences among species within higher taxa would occur, but they were superficial adaptations to peculiar or specific conditions, the "animals' ecology" in more modern terms. Thus the comparative psychologists tried to compare animals and humans on a number of cognitive and emotional traits such as reason, communication of ideas, morality, sympathy, affection, rage, and remorse. Unfortunately the database often consisted of anthropomorphic anecdotes collected by amateurs or professionals unfamiliar with a species' natural history and without the benefit of modern methods of recording and analyzing behavior. The most influential example was Romanes' (1882) classic volume on animal intelligence. His subsequent analytical volume on mental evolution (Romanes 1883) attempted to order this material in an evolutionary scheme. Both his taxonomic and mental hierarchies are actually quite modern in spite of improved methods and quantum leaps in amount of data (Burghardt 1985a).

Efforts such as Romanes were criticized both by those sympathetic with his aims (e.g. Morgan 1894) as well as by the laboratory experimental schools represented by figures such as Thorndike, Pavlov, Loeb, and Jennings. The culmination was Watsonian behaviorism that ruled out both the study of mental traits and, because of an antipathy to instinct and genetics, any role for evolutionary comparison. Thus motivation was just a physiological process demoted to a variable in conditioning experiments. Emotion was ignored or considered of no particular comparative importance. Cognition or intelligence was the only behavioral trait worth measuring and eventually this was demoted to the study of learning mechanisms in a few species.

Some psychologists remained interested in species diversity in behavior (Dewsbury 1984, 1985; Demarest 1987). Species differences were emphasized in studies of sensory and brain function, but comparative analysis was typically limited to gross trends within vertebrates such as the perfection of hearing as we ascend the "scale" (see Jerison 1973 for a recent statement). Brain studies focused on structure and behavior (largely learning, sensory and motor control, motivation, and emotion). Textbooks variously compared fish, frog, turtle, alligator, pigeon, rat, cat, dog, monkey, and human to trace evolutionary trends in sensory, neural, and hormonal control, and in the importance of experience and learning. Such comparisons of "typical" animals reflect pre-evolutionary typological, idealistic, even Platonic views of not only species (Mayr 1982) but even orders and classes. These are still common in the neuroanatomy section of psychology and biology textbooks and permeate to varying extent all current texts in comparative anatomy, comparative physiology, physiological psychology, and comparative endocrinology. Historically this may reflect the marginal role played by morphology in the evolutionary synthesis (Coleman 1980; Ghiselin 1980; Hull 1988).

Cross sectional slices may be the most heuristically practical way to convey large amounts of comparative data. But such stereotypical presentations, especially when arranged to look progressive, are potentially misleading, not only to students but, by the blinders they encourage, to working scientists themselves. By being part of an explicitly anthropocentric field, early comparative psychology was "preadapted" to move away from the study of naturalistic behavior patterns, to eliminate species level differences, and to focus on underlying principles and progressive general trends in learning, perception, brain organization, and other mechanisms.

Romanes' scale of mental evolution and its successors in physiological psychology and animal learning were sincere attempts to map the broad brush of animal phylogeny onto mental processes. But they were also meant to convince skeptics that human beings were products of evolution and that an essential continuity existed. When this goal was largely accomplished, attempts to test the mental phylogeny and work out the details lost appeal and added to the relative decline of comparative psychology. But in spite of the behavioristic *Zeitgeist*, such essentially intuitive evolutionary comparisons remained in animal behavior as in Maier & Schneirla (1935) and down to popular texts used in the 1960s (e.g. Dethier & Stellar 1964). But such broad taxonomic sweeps were not based on a sophisticated methodology of comparison and were vulnerable to the well-known critiques by Hodos & Campbell (1969), Lockard (1971) and other psychologists who, having digested some of the main tenets of comparative ethology, pointed out the narrow database, few and ill-chosen species, and unnatural behaviors usually measured.

Comparative Ethology

In contrast to the trend in comparative psychology, the early ethologists stayed with naturalistic behavior patterns and gloried in species differences (Burghardt 1985b; Barlow 1989). They emphasized stereotyped instinctive movements and their stimulus control, especially of consummatory acts and displays (fixed or modal action patterns, Barlow 1977; Chiszar 1986), which they considered on a par with morphological characters (Bekoff 1977). Thus Heinroth used wing position in swans to show the taxonomic value of a simple overt behavior. Many species can be distinguished only, or most clearly, through behavior. For example, the gray treefrogs of North America are composed of two identical appearing species (*Hyla chrysoscelis* and *H. versicolor*) that can be distinguished only by their mating calls and preferences and then only if ambient temperature is known because trill rate of the males' calls are differentially temperature dependent.

Konrad Lorenz certainly had this comparative taxonomic approach in mind when he anointed C. O. Whitman and Oskar Heinroth as the post-Darwinian founders of ethology (Lorenz 1950) and chose comparative anatomy as the most appropriate model for ethology. Thus, for the early ethologists, "comparative" was explicitly tied to an evolutionary framework based on nineteenth century morphology. But ethology only took from morphology the need for detailed description and looked to systematics to see how morphological characters were used in phylogenetic reconstruction. Not only should a variety of species, especially nondomesticated ones, be studied in lab and field, but comparative studies should involve similar biologically relevant behavior patterns across species and higher taxa. *In other words, to be comparative is to adopt a taxonomically oriented evolutionary framework based on characters with common origins (homologies).*

The ethological use of discrete stereotyped action patterns led to an emphasis on identifying and tracing homologies in the evolution of behavior (Wickler 1961, 1973; McKinney 1975; Bekoff 1977; Lauder 1986), using behavior both to tap on to existing phylogenies and to supplement and sometimes force their revision. The high point of this effort was Lorenz's (1941/1970) comparative study of ducks. The focus on topographically discrete behaviors made some behaviorists happy since motivational and mental factors were bypassed. This use of behavior as an objective "thing" to be compared was, however, opposed by students of the Lehrman-Schneirla school (e.g. Atz 1970; see the largely sympathetic review by Beer 1980) who variously argued that behavior was an evanescent process, or too variable and experientially influenced to allow its equation with "real" things such as bones. Furthermore, it was argued that behavioral homologies were especially unlikely above the genus level since most were really superficial similarities based on divergent underlying mechanisms.

While the critics did point out some conceptual and methodological pitfalls, behavioral homologies have been shown in virtually every group of multicellular animals. The point is that *some* behavioral patterns, like some morphological characteristics, are indeed "good" characters and others are too variable or nondiscriminating. As we shall show, modern comparative methods use empirical data to evaluate the usefulness of any and all characters at different taxonomic levels.

Although ethologists have concentrated on behavioral comparisons among closely related species, species and genera are relatively large units and evolutionary change is generally recognized as being derived from variation among individuals within species. Darwin pointed out such differences and compared domestic with nondomestic species and breed differences within domesticated species such as dogs and pigeons. Early ethologists such as Whitman and Heinroth also pursued intraspecific variation (microevolution). So too did the early comparative psychologists, such as Yerkes (1907) in his study of the dancing mouse and, more recently, Boice (1973) in his comparison of wild and domestic rodents. Both Heinroth and Lorenz compared hybrids with parental stock.

However, most of the early ethological comparative work focused on uncovering taxonomic relationships or confirming with behavioral evidence phylogenies based on more traditional characters. The *process* of behavioral evolution was later looked at in terms of ritualization and the evolution of displays (Tinbergen, 1952).

Nonetheless, it is remarkable how little focus there was on *individual variation* in the early studies of animal behavior by both psychologists and ethologists. There were several reasons for this. First was the fear of confounding motivational, sensory, hormonal, and other factors in animals compared under less than fully controlled conditions. Second was the view, especially among behavioristic psychologists, that as most behavior was experientially based to begin with, individual differences were created solely by prior learning events or ontogenetic history and the issue was trivial in any evolutionary sense. Third was the impact of Mendelian genetics that emphasized large single gene effects and which made small subtle differences less easy to relate to different genetic backgrounds. Quantitative and population genetics were little known prior to the 1930s and after to those outside animal breeding.

This picture has radically changed in recent years, although convincing data have been around for some time. For example, the greater stereotypy of displays within than between individuals has been shown by Dane & van der Kloot (1964) in ducks, by Jenssen

(1971), Ferguson (1971), and Stamps & Barlow (1973) in lizards, and in other species (Bekoff 1977). Variation in response to biologically meaningful sensory cues occurs in response to chemical cues from prey in naive garter snakes (*Thamnophis sirtalis*) from different populations (Burghardt 1970) and even within litters (Burghardt 1975). These findings were followed by the quantitative genetic studies of Arnold (1981) on population differences and the inheritance of prey recognition in *Thamnophis elegans*. Developmental approaches to individual differences have also become more sophisticated (e.g. Bekoff 1988).

The Decline of Phylogenetic Comparisons in Ethology

An obvious conclusion is that if intraspecific variation is so great in behavior, then macroevolutionary comparisons must be suspect. Although we feel such comparisons are not only feasible but essential, it does means that variation within taxa must be evaluated and not ignored. Species and population level adaptations must be distinguished from phylogenetic phenomena that constrain the solutions that natural selection can reach at the species, population, or individual level. This insight has not always been recognized by theorists. Thus, by the 1960s, many commentators considered attempts to compare behavior across both narrow and wide taxonomic chasms (such as phyla and classes) impractical or impossible.

The ethologists also wanted to go beyond narrow comparative phylogenies of closely related species. Initially they identified as homologous only topographically similar movements across vertebrate classes such as head scratching, yawning, and other basic generally maintenance type behaviors (Heinroth 1930/1985). But to show the power of a comparative method with profound implications for human beings, Lorenz (1966), Morris (1967), and others wrote provocative popular works that often relied on motivational constructs such as sex, territoriality, dominance, and aggression as much or more than on detailed topographical analysis of motor patterns. To some extent these works mimicked the earlier psychologists who moved from specific behavior to instincts, drives, and propensities such as James (1890), McDougall (1908), and Freud (1915). Critics focused on the problematic similarities in ethological writings, and ignored innovations such as the analytic use of the distinction between appetitive behavior and consummatory acts in sequential analysis.

Thus even as such writings created a popular audience for ethology, many ethologists in America and especially in England became critical of the comparative phylogenetic approach. This wasn't based on direct attacks so much as an unhappiness with classical ethological notions such as "innateness," a concern shared by many psychologists. Genetic specification of aspects of behavior were viewed as essential to the existence of behavioral homologies. And if one were skeptical about innate behavior it follows that one would not be accepting of studies on the phylogeny of behavior patterns. Lorenz was forced to refine and defend his position (Lorenz 1965). He did so by conflating the approach and nationality of his critics in an unnecessarily provocative and stereotyping manner. Regardless, Hinde's (1966) well-known text limited the presentation of phylogeny material to a short chapter. The implicit message was that tracing phylogenies was a tired, unimaginative business. Studies of behavioral evolution should focus on general processes (eerily reminiscent of what earlier happened in the study of learning in comparative psychology). Thus, during the 1960s the comparative perspectives of both comparative psychology and comparative ethology were being conceptually undermined and empirically ignored.

The formal challenge came in John Hurrel Crook's social ethology program unveiled in a plenary address at the 1969 International Ethological Conference in France (Crook 1970). It was a response to the growing trend in primate and bird studies to classify and hierarchically arrange different kinds of social organization. Crook felt that to approach this endeavor with the classical comparative methods based on individual action patterns and releasers was an error. He felt the study of social behavior should look to links with population biology and ecology, on the one hand, and, on the other, to sociology, social psychology, and social systems research. His basic assumption was that intraspecific variability in social and ecological responses was much greater than that of individual behavior and morphology, and that different levels of complexity were involved.

Crook's model paradigm was his own work with both weaver birds and primates. The former showed that different species of weaver birds living in forest or savanna habitat had a different suite of adaptations involving food type, foraging mode, group size, and antipredator strategies that seemed much more related to the current ecology of the species than to any phylogenetic heritage. One of his explicit goals was "to bypass the never ending sterility and unreality of the nature-nurture controversy when applied to social life" (Crook 1970: 198). The replacement was a phylogeny - ecology dichotomy in which current ecological adaptations were viewed as either ontogenetically shaped or facultatively induced responses to specific ecological settings, which in turn shaped social organization. This view was buttressed by the finding that social organization in given primate species was highly variable across different ecological conditions, presumably due to intraspecific plasticity. Thus was paved the way for viewing phylogenetic or ancestral characteristics as weak constraints as compared to the plasticity of populations when faced with given ecological realities. And it was these malleable details that were of most interest and importance, not universal patterns of response.

The impact of this research program resulted in even further diminution of the perceived value of a comparative phylogeny of behavior in the manner of Whitman, Heinroth, Lorenz, even Darwin (McLennan et al. 1988) or the anagenesis of comparative psychology. Nonetheless, social ethology (also termed social ecology and socioecology) had difficulty answering such questions as why all species living in a given habitat did not converge on the same solutions. But Crook's social ecology approach was soon overshadowed and displaced by the growth of sociobiology (Wilson 1975) which utilized the modern population biology and ecology that Crook emphasized, but ignored the emergent levels of social complexity derived from sociology that Crook also considered critical. Other aspects of social ethology influenced the growth of behavioral ecology, however.

Sociobiology, behavioral ecology, and new methods of comparative biology have now led to a systematic focus on whether or not evolved features of animals are due to adaptations to current or recent ecological circumstances or whether they are based on phylogenetic or historical constraints from ancient ancestral stocks. Comparative analyses in biology allow evaluation of hypotheses across taxa, either with respect to evolutionary function (selective value) or phylogeny (the evolutionary history of a lineage).

Evolutionary biologists are sometimes so taken with the power of optimization and design arguments for adaptations that they seem to view any population as putty that could be moved around in an adaptive space. Gould & Lewontin (1979) and others challenged this view with arguments on constraints of adaptation.

Regardless of natural selection, crabs are not going to evolve wings or bats feathers. The evolutionary moment for such innovations is long past and, in any event, are based on what happened to long extinct animals that were neither crabs, birds, or bats.

But how do we determine whether given genetic adaptations are due to current utility (adaptation) or phylogenetic heritage? In both cases we may be dealing with genetically based characters. Ironically, before such questions can be dealt with the taxonomic relationships among species, genera, and families must be known with some degree of certainty. Thus classification schemes assume great importance, and behavior, as the early ethologists held, may provide crucial information. Clearly, detailed comparative ethograms with a common language and approach (Drummond 1983; Schleidt et al. 1984) are needed.

SYSTEMATIC METHODS AND BEHAVIOR

Unfortunately, since the beginning of biology the establishment of consistent systematics has been a hotbed of debate (see Hull 1988). Recently, with the advent of new biometrical techniques, questions regarding the functional significance of specific characters for classification (e.g. morphology versus molecular), and philosophical critiques of the relationship between systematics and evolutionary theory, the debate has rekindled over which methodologies should be used for constructing taxonomies (Ridley 1986; Hull 1988). For the most part, this discussion has left out behavioral contexts. But classification, *per se*, is a necessary preliminary to all scientific and scholarly pursuits regardless of how the field deals with relationships or natural origins, be it linguistics, chemistry, art history, or cultural anthropology.

There are three modern schools of classification: evolutionary systematics, phenetic (or numerical) taxonomy, and cladism (Hull 1988). In brief, *evolutionary classification* posits that similarities and differences among taxa are distributed according to their pattern of evolutionary change (Simpson 1961; Mayr 1969). That is, evolution produces natural groups of taxa because all evolutionary change occurs in the same phylogenetic tree: All character change takes place similarly within the same pattern of lineages. The primary difficulty in evolutionary classification is to sort out convergent or *analogous* characters (wings in bats and bees) from nonconvergent or *homologous* characters (canine teeth in bears and wolves) derived from common ancestors. Because evolutionary classification includes multiple and diverse characters, the method is always subject to change and improvement as new data accumulate, including new species and "key" characters. Critics charged that subjective impressions guided the application and weighting of specific characters.

Two schools of taxonomy, aided by the computer, arose as responses to the subjective nature of the evolutionary school. *Phenetics*, or *numerical taxonomy* (Sokal & Sneath 1963), attempted to eliminate decisions about character weighting. Essentially, all taxonomic characters are equally measured, weighted and aggregated in a quantitative measure of similarity across taxa. The pheneticist is only striving to estimate overall similarity, irrespective of phylogeny, in an objective and repeatable manner.

Cladistics aims to rule out the infinity of classifications produced by numerical taxonomists and to highlight classifications that best represent the hierarchy of phylogenetic branching. The cladistic technique hinges on the ability to distinguish ancestral from derived characters (Hennig 1966), with the "outgroup comparison", being particularly relied on in comparison to other methods (Felsenstein 1983; Sober 1988), although ontogeny and especially heterochrony (differential rates of character development) also provide important data (Kluge & Strauss 1985); controversy over methods and philosophy are rampant within cladistics (e.g. Hull 1988). Outgroup comparison uses the logic of parsimony. This states that a minimum of evolutionary events have taken place between character states and that shared characters across taxa are more likely to be due to common ancestry than convergence (Maddison et al. 1984; see Gittleman 1981 for a behavioral example).

Cladism, in its initial form, appears to use the strengths of various methods while avoiding many of their problems (particularly the selection and weighting of characters). Nonetheless, comparative studies of behavior must eventually point out the effects of the classification used and, better yet, model varying conclusions based solely on changes in classification.

As an example we will reconsider Lorenz's (1941/1970) treatment of the Anatidae. His seminal figure has often been

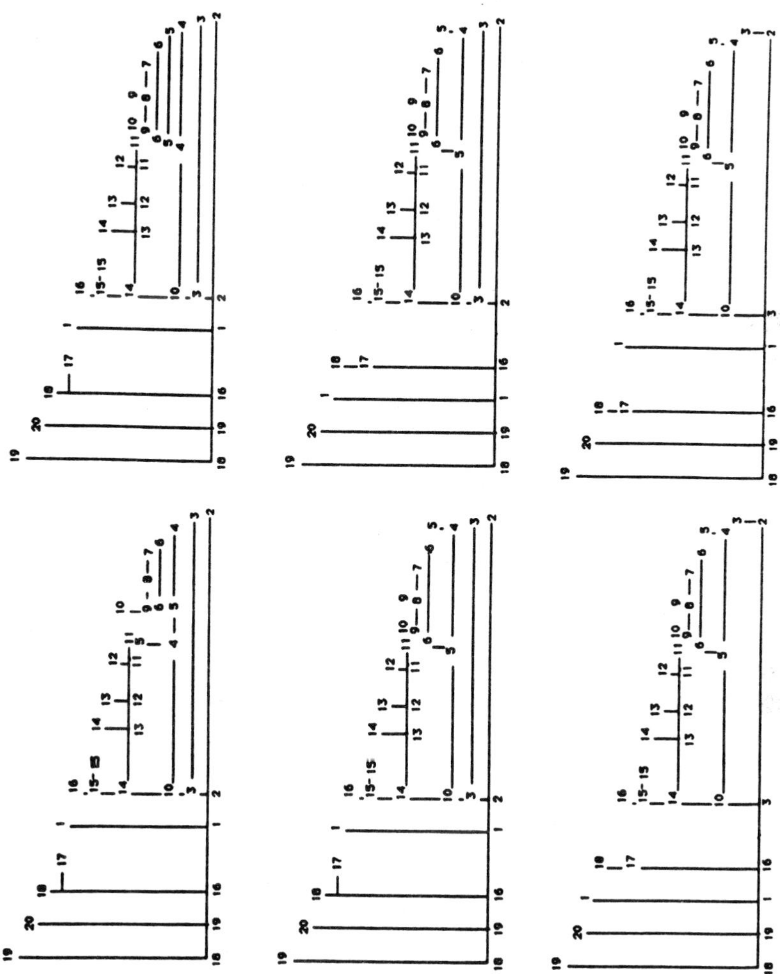

Figure 1

Figure 1. Six equally parsimonious phylogenies based on a statistical analysis from PHYLIP (see Felsenstein 1985a). Numbers represent comparative data on waterfowl displays originally presented in Lorenz 1941. Details concerning assumptions and methodology of the analyses are in the text. Lorenz's species code is as follows: (1) *Cairina moschata*, Muscovy duck; (2) *Lampronessa sponsa*, Carolina wood-duck; (3) *Aix galericulata*, mandarin duck; (4) *Mareca sibilatrix*, Chiloe wigeon; (5) *Mareca penelope*, wigeon; (6) *Chaulelasmus streptera*, gadwall; (7) *Nettion crecca*, teal; (8) *Nettion flavirostre*, South American teal; (9) *Virago castanea*, chestnut-breasted teal; (10) *Anas Genus* including mallard, spot-billed duck, Meller's duck; (11) *Dafila spinicauda*, South American pintail; (12) *Dafila acuta*, pintail; (13) *Poecilonetta bahamensis*, Bahama duck; (14) *Poecilonetta* (?) *erythrorhyncha*, red-billed duck; (15) *Querquedula querquedula, garganey*; (16) *Spatula clypeata*, shoveller; (17) *Tadorna tadorna*, shelduck; (18) *Casarca ferruginea*, Ruddy shelduck; (19) *Anser* genus; (20) *Branta* genus; (? = Lorenz's uncertainty).

reprinted (e.g. Burghardt 1973; Bekoff 1977; Grier 1984). His comparative study of motor patterns across the Anatidae is exemplary in showing the technique, strengths and weaknesses of evolutionary classification. He used 18 species among 14 genera, with two outgroups (see legend to Figure 1). Across these taxa Lorenz indicated the distribution of 48 characters. He concluded that some characters (e.g. monosyllabic 'lost-piping'; display drinking) were good primitive characters and that others (polysyllabic gosling social contact call of Anserinae) were not. More importantly, not only were these characters distinguishing features, but they could be assessed in similar fashion with morphological traits (e.g. bill type; pelage coloration).

How valid is Lorenz's arrangement? Obviously after 50 years we know more about these species and missing data points have been filled in and others revised (Johnsgard 1965; McKinney 1975, 1978). Furthermore, much more is known about the general ecology and field behavior of many of the species, which were exotic and known to Heinroth and Lorenz only from captivity. What we have done is to reanalyze Lorenz's data, using a phylogenetic program (part of PHYLIP, see Felsenstein 1985a), to illustrate the many permutations of phylogenies. Essentially the trees we generated are derived from adding species to the topology of a tree, rearranging species at about $2n^2$ different topologies, that will at least guarantee that no nearby topology is better. It is essential to realize, however, that this tree-searching process is contingent upon the order of inputting species -the ordering of species directly follows Lorenz, moving from left to right in his seminal Figure. Clearly, with 20 species involving 48 characters the total number of possible tree topologies is enormous. To erect parsimonious trees we establish the following assumptions that will permit a set of working phylogenies (from Wagner parsimony method): (1) ancestral states are unknown; (2) different characters and lineages evolve independently; (3) changes in characters are equally probable; (4) rates of evolution in different lineages are sufficiently low that two changes in a long segment of the tree are far less probable than one change in a short segment (validity of these assumptions may be found in Felsenstein 1978, 1979).

Given these assumptions, six equally parsimonious trees are found for Lorenz's data (Figure 1). The branch lengths do not represent evolutionary time or the amount of evolution. However, the primary rearrangements through the six trees occur in species 1, 3, 4, 5, and 16. Some of these taxa were discriminating for Lorenz and the trees are overall quite similar to the arrangement Lorenz constructed out of paper and wire. The purpose of this exercise, though, is to establish that phylogenetic construction *per se* will shape our taxonomic comparisons in behavior. If we were to carry through with this reanalysis of Lorenz' data it would be necessary to alter some of the assumptions above and the input of species ordering to complete all possible trees. Only through this process could we be certain of the validity of our comparative test.

Another issue is to compare Lorenz's conclusions with the current status of waterfowl systematics and phylogeny. This is difficult because many of the species used by Lorenz have been lumped into *Anas* and the relationships among them are in a state of flux (F. McKinney personal communication). However, at the generic and tribe levels a recent cladistic analysis (see below) of standard morphological characters (Livezey 1986) and one of single-copy DNA (Madsen et al. 1988) gave roughly similar results.

The validity of comparative studies on phylogenetic and adaptive sources of variation is contingent upon the taxon level chosen as well as on using an accurate and systematic classification. But note that a null hypothesis in a comparative study must initially extirpate the natural structure of species (taxonomic) differences presented solely by the classification (Felsenstein 1985). In the following, we present an example of the kind of taxonomic problem that faces many comparative analyses of behavior. Following a brief introduction to various ways of addressing this problem, we then present a new methodology which is a "null model".

A Comparative Problem

Consider a simple comparative test of adaptive explanations for the evolution of social behavior (imagine animals ranging from extreme solitary living, rarely interacting except during a brief mating season, to large social groups, constantly interacting for everyday needs). Following some descriptive statistic showing the distribution of social behavior across a given taxonomic group, we might search for various ecological associations and find that solitary species are strictly carnivorous and social species are mainly omnivorous; various adaptive hypotheses relating degrees of sociality to dietary causes would then be advanced for this comparative finding. The problem with this procedure, common to many comparative tests (Harvey & Mace 1982), is when most of the carnivorous species lie in one taxonomic family we do not have statistical independence in the comparative data (Ridley 1989). Statistical independence between taxonomic affiliation and behavioral/ecological (adaptive) association is necessary so that we can be certain of causal patterns in differences among taxa (Felsenstein 1985b).

Statistics aside, a more serious problem is faced when phylogenetic pattern may actually change the form of association between a trait and some behavioral/ecological association. For example, Huey (1987) considered in a hypothetical example the adaptive response of an array of six species classified among two genera (Figure 2) that live in sections of an environmental gradient. If we were interested in broad-scale evolutionary change we might perform a regression between the environmental feature and the species' adaptive response, as represented by the solid horizontal line in Figure 2b. However, accounting for the generic associations would clearly produce a different kind of adaptive response in the generic groupings, as in the dashed lines. This hypothetical example raises a fundamental problem that is involved in all comparative studies.

Sampling of species arrays must consider that "...species are part of a hierarchically structured phylogeny, and thus cannot be regarded for statistical purposes as if drawn independently from the same distribution" (Felsenstein 1985b: 1). In our hypothetical example, if we were to ignore the generic (phylogenetic) associations an incorrect conclusion would be drawn - the species response would appear to be unrelated to the environmental feature.

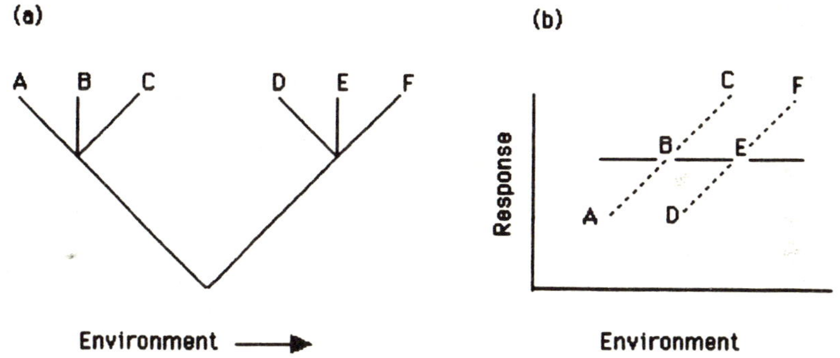

Figure 2. (a) A hypothetical phylogeny for two genera, each with three species, distributed along an environmental gradient. (b) Hypothetical adaptive responses versus environmental position for the six species in (a). The solid horizontal line represents the regression responses to the environment if phylogeny is ignored, where the two dashed lines represent regressions for the genera analyzed separately. Failure to consider phylogeny would lead to the incorrect conclusion that the adaptive response is unrelated to the adaptive feature (taken from Huey 1987, with permission).

PHYLOGENETIC APPROACHES TO BEHAVIORAL ADAPTATION: A NULL MODEL

Evolutionary explanations of behavior based on cross-taxonomic comparisons are often faced with the above problems. In recent years, a number of approaches have developed which aim to estimate or remove phylogenetic pattern from comparative tests of adaptation. These approaches may be generally classified into the following: Analysis of Variance (ANOVA) approach, which tests for taxonomic heterogeneity *prior to* analyzing comparative data (Harvey and Mace 1982; Clutton-Brock & Harvey 1984; Harvey & Clutton-Brock 1985; Gittleman 1986a, 1989b); the cladistic/outgroup approach, used primarily for categorical traits when direct phylogenetic information is available, is an outgroup technique for minimizing the number of evolutionary transitions necessary to account for the distribution of traits across taxa (Greene & Burghardt 1978; Gittleman 1981; Ridley 1983; Huey 1987; Huey & Bennett 1987; Lauder 1986; Coddington 1988); the contrasts approach, assuming that characters evolve by Brownian motion, uses an accepted phylogeny to define a series of contrasts (among traits) which should be statistically independent (Felsenstein 1985b, 1988). All of these approaches have their merit, depending on the amount and kind of comparative data (particularly of quantitative or categorical nature), available phylogenetic versus solely taxonomic information, and particular question at hand. Reviews of the analytical assumptions and goals of these methods may be found in Felsenstein (1985b, 1988), Huey (1987), Pagel & Harvey (1988), and Gittleman (1989a). In the following we present a new technique which holds benefits in comparison with some other methods. It has the flexibility of handling comparative data with and without phylogenetic information and, with some statistical tools, may extract as much phylogenetic effect in a comparative data set as possible. Thus, we may think of this as a "null model" approach.

Our approach is built upon that of Cheverud et al. (1985) who developed a model that estimates both a phylogenetic component and specific (adaptive) component of a trait for each species in a comparative data set. The phylogenetic component may be thought of as that part of a trait attributable to the ancestry of a species whereas the specific component represents a species' independent evolution that is the focus of interest in most comparative analyses. The phylogenetic component for a species is calculated by taking a weighted average of the trait for related species. The specific component (independent evolution) is thought of as a series of kicks or shocks to this essentially autoregressive process. Maximum-likelihood parameter estimation is used to obtain the proper mix of phylogeny and independent evolution that best fits the data. This autoregressive approach was used in an analysis of sexual dimorphism in body mass across primates, which showed that 50% of the variance among primates is phylogenetic, whereas 36% reflects overall size, and only 2% is due to habitat, diet, and mating system (Cheverud et al. 1985).

At the heart of this approach is the weighting assigned to each species in estimating other species' phylogenetic components. These weights reflect the relative phylogenetic distance of pairs of species. More accurately, they reflect the anticipated phenotypic similarity given these distances. Closely related species are assigned high weights, distant relatives are assigned low weights. Cheverud et al. arbitrarily let similarity fall off in a reciprocal manner with taxonomic distance. That is, irrespective of the trait, all species within the same genus are assigned a value of 1.0, all species within the same family a value of 1/2, and so on. Two problems immediately arise in using this scheme. First, for phylogenetically-linked traits, one might imagine that the rate at which similarity falls off varies with how conservative or plastic the trait is. Second, and more fundamentally, the empirical data may not actually warrant any sort of autoregressive model. These factors prompted Gittleman & Kot (in press) to extend and refine Cheverud et al.'s method.

To get an initial estimate on whether phylogeny may be important to trait variation, Gittleman & Kot (in press) introduced a descriptive (autocorrelation) statistic (Moran's I) which estimates correlation in a trait with taxonomic distance; a series of the statistics may be strung together into a *phylogenetic correlogram* to compare phylogenetic effects at different taxonomic ranks and to ascertain the appropriateness of an autoregressive model. Consider the example of clutch size in birds (data from Lack 1968), an extremely variable trait like many behavioral features and a variable which is correlated with numerous behavioral and ecological variables (see e.g. Lack 1968; Perrins 1977; Ricklefs 1977). A phylogenetic correlogram of the family Anatidae demonstrates strong phylogenetic effects (Figure 3, top left), with hierarchical patterns at different taxonomic levels. Congenerics (g) and noncongeners within the same tribe (t) are positively correlated; species that are merely in the same subfamily (sf) or family (f) are negatively correlated (standardized normal deviates are also presented, top right). It is clear that the correlations are statistically significant.

Nevertheless, as shown by the phylogenetic correlograms for clutch size, trait variation may fall off with phylogenetic distance in a nonobvious way. Thus Gittleman & Kot modified Cheverud et al.'s autoregressive model by using a maximum likelihood parameter to calculate distance among taxa, to provide greater flexibility in describing the relationship between phenotypic similarity and phylogenetic distance. The autoregressive model may then be used, for the example above, to decompose standardized clutch sizes across species (Figure 4, top) into a phylogenetic component (Figure 4, middle) and a specific (adaptive) component (Figure 4, bottom). In this case, the phylogenetic component accounts for 53% of the total variation in average clutch size. If the model indeed removes all of the phylogenetic component, as Gittleman & Kot predict, the specific component should be immune from correlation: As shown in Figure 3 (bottom), the calculated specific components are free of any significant phylogenetic correlation. An autoregressive approach is therefore based on the idea that it is this specific component that should be used in comparative analyses.

The autocorrelation approach implicitly assumes that phylogeny acts uniformly - an increment in genetic distance will cause the same loss of phenotypic similarity in every part of the tree. Nevertheless, this model may prove useful for distinguishing between phylogenetic and adaptive factors.

TOWARDS A REVAMPED COMPARATIVE PSYCHOLOGY

Griffin's writings (e.g. 1976) on cognitive ethology have helped to reawaken interest in the traditional goals of comparative psychology. He raised issues of animal mental states in a provocative manner accessible to a diverse scientific community and pulled together examples from insects to apes that were rarely discussed from the same perspective. Griffin's antipathy to laboratory psychology and his ignoring of the details of their concepts and methods have stimulated a reaction based on showing that the field was concerned all along with the issues Griffin raised or had dismissed (such as awareness) for good reason (Mason 1976). Nonetheless, it is clear that psychology did not adequately address either species diversity or biological relevance. On the

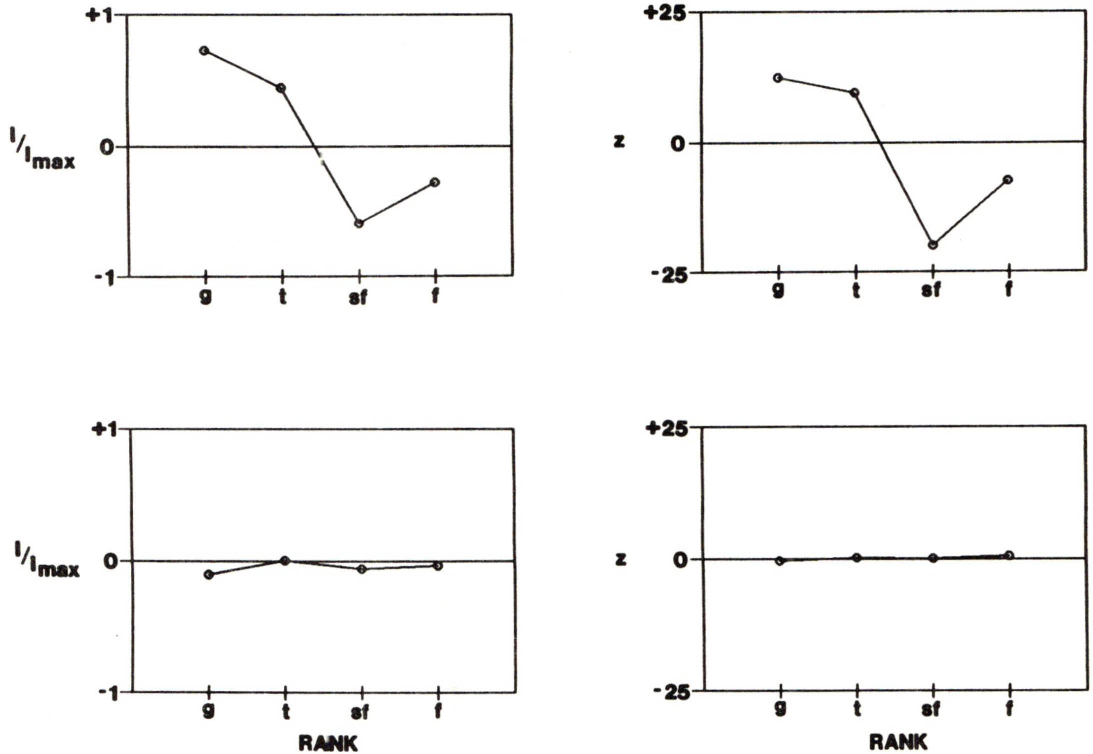

Figure 3. Phylogenetic correlograms of clutch size in birds. Moran's I and associated z values reflect phylogenetic correlation across taxonomic rank. See text for definitions and general discussion of analysis.

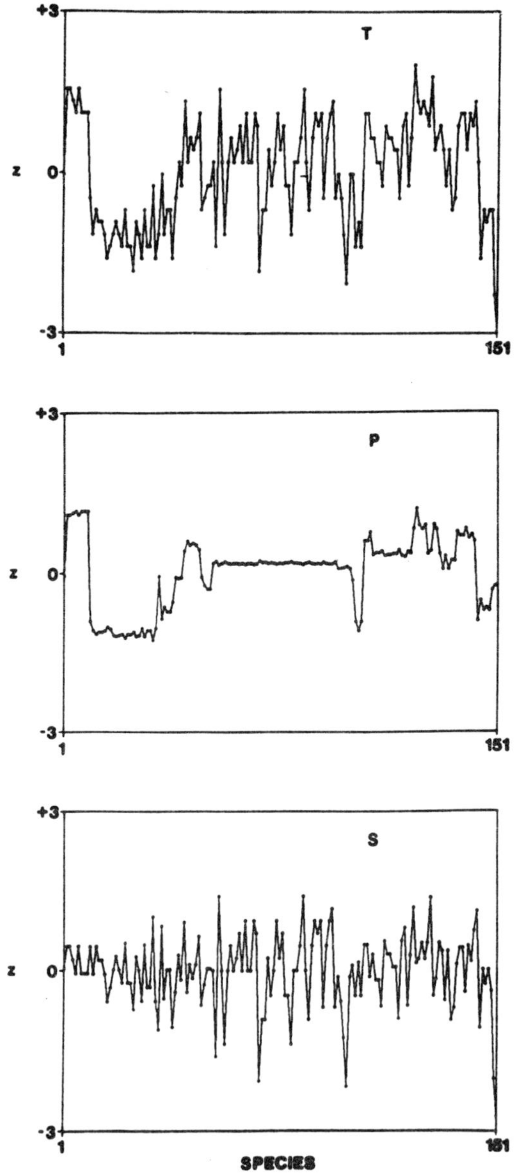

Figure 4. *z* values for clutch size following autocorrelational analyses breaking down values into: T = actual trait values; P = phylogenetic values; S = specific (adaptive) values. See text for detailed discussion.

other hand, Griffin did not provide a comparative method for analyzing the diverse mental abilities he called on scientists to study. Is there now a path to preserving and rejuvenating comparative psychology as a distinctive area of biology and psychology, bridging the links that many see when they argue that psychology and the social sciences are but a subset of biology; a biology defined as "The Science of Life" (Alexander 1987)? The methods outlined above may very well move us toward a definitive answer.

Comparative analyses in biology are now dealing with life history variables such as metabolic rate that are indirect reflections of other processes, such as fitness and response to diet. Although a comparative psychology of mental processes as envisioned by Romanes proved intractable, the issues involved endure, and rightly so. Today most comparative psychologists have adopted the framework of ethology (Dewsbury 1984). But is comparative psychology only ethology done by psychologists? Perhaps comparative psychology addresses ethological concerns but with somewhat more experimental, developmental, psychophysical, and learning theory sophistication. Or is there a joint subdiscipline of psychology and biology that can also address the classic issues of comparative psychology and the analysis of mental events? Are there behavioral, physiological, and neural characteristics that constrain the cognitive capacities of animals and behavioral organization, motivation, and experiential effects? The study of play may be an area where such considerations are important (Burghardt 1988).

The two issues here apply to 1) the use of cognitive and emotional characteristics in reconstructing a phylogeny using psychological characteristics in addition to or instead of morphological, life history, and movement patterns and 2) the attempt to partition adaptive and phylogenetic sources of variation in such psychological characteristics.

For an example of what this shift in perspective might mean consider comparative learning. Three approaches can be described: general process, ethological, and phyletic discontinuity (Burghardt 1973). The general process approach views learning as basically the same in all animals, involving only a few mechanisms. The approach has contributed sophisticated theory and research technology, and a wealth of information on selected species and problems, but until recently its advocates did not seriously value an understanding of basic natural history, phylogeny, or species diversity. However, the methods and concepts are now being exploited at a rapid rate in empirical work on foraging and in other areas.

In contrast, until recently the ethological approach focused on the importance of learning in particular contexts, especially the ecological and social settings in which the species, population, sex, or age class live (Johnston 1981 adds an important developmental perspective). The roles of genetic, sensory, and motor limitations were also addressed. Although homologies and phylogenetic factors were studied for behavior patterns, at the individual learning level there was a general restriction, as in Crook's social ethology, to adaptation to the current biological problems animals faced. The ethological and general process approaches can be integrated (e.g. Johnston 1981; Domjan & Galef 1983) if one realizes that current ecological adaptations are largely superficial mechanistic reorganizations of a few fundamental processes.

The phyletic discontinuity approach most clearly resembles that of the early comparative psychologists in that "levels of organization" put constraints on modes of behavior and behavioral capacities in a hierarchical progressive manner (Maier & Schneirla 1935). In this sense it is similar to the search for phylogenetic factors in the comparative methods outlined above. Currently, however, there is little evidence to show that these constraints hold within, but not across, large taxonomic groupings such as vertebrate classes or orders. Many (c.f. Macphail 1987) have been critical of research programs such as Bitterman's (1965) that attempted a phylogeny of learning types that differed across the vertebrates based on comparative work under carefully controlled conditions on only one or two species in each class. Perhaps the way such programs were executed aided their relative demise.

Macphail (1987) and commentators on his target article provide a stimulating discussion of the status of these three approaches in relation to comparative intelligence. Macphail strongly supported the general process approach by arguing that a) ecological learning specializations are not the issue, b) comparing a few arbitrarily selected and taxonomically diverse species is a good strategy, but c) all work purporting to show species differences between any two or more vertebrate species is flawed and thus "the null hypothesis should be accepted...that there are no differences, either quantitative or qualitative, among the intellects of nonhuman vertebrates" (p. 653). The requirements set forth by Macphail for demonstrating intelligence seem to involve demonstrating a general problem solving capacity not coupled with adaptation to a niche but, as he doesn't accept any definition of intelligence, his views seem untestable. Unfortunately, the lessons of recent comparative biology have been ignored in this and similar debates on species differences in psychological capacities.

One lesson is the importance of careful definitions of characters and their measurement. Is there a way in which behavioral abilities can be compared across taxa, using multiple well delineated characters so useful in phylogenetic approaches such as cladistics? Davis & Perusse (1988) have argued for the need for clear definitions in order to make progress in understanding comparative issues of numerical competence. Chevalier-Skolnikoff (1989) has tried to apply Piagetian principles to comparative studies of tool use. The vigorous commentaries on these *BBS* papers by Macphail, Davis & Perusse, and Chevalier-Skolnikoff raise issues that can only be resolved by incorporating rigorous phylogenetic analyses. Thus, in the case of problem solving, numerical competence, tool-use, and the underlying mechanisms of each, the general phylogenetic versus narrowly specific adaptation issue must be addressed. Certainly each capacity or mental character in a tree may not reflect a true biological character, but this is just as true of many of the reified anatomical, behavioral, developmental or physiological measures used in more traditional analyses. But the adoption of the perspective will force rigorous study of diverse species and the resulting patterns will help us in looking at parsimonious arrangements, primitive versus later specializations, and so forth. Medicus (1987) has presented a revised macroevolutionary phylogeny in the Romanes (1883) mold. Although crude, it does attempt to separate convergent versus homologous mental and behavioral traits that can be tested if sufficient data are carefully collected. Still, it is critical that closely related taxa be carefully studied and not just representative "types." For example, Pellis (1988) has shown that the targets of attack and defense in play fighting may have multiple origins in mammals that are obscured by superficial similarities.

Besides the topics listed above, indices of emotional expression, communication styles and repertoire, and social and foraging characters can be studied. Even issues of different cognitive abilities, self-awareness, consciousness, pain, and subjective states might be addressed with appropriate comparative methods if the dangers of anthropomorphism are addressed (Burghardt 1985a). A fear of anthropomorphism should not lead to an avoidance of these issues as it has so often in the past (see Fisher Volume I). Initially these data, as in the early phylogeny computer programs, will be mere presence - absence, but become progressively more quantitative as our database and measurement systems improve.

Will this proposal for addressing the classic concerns of comparative psychology be taken up? Perhaps the growing links between neuroscientists and psychologists working in cognition, motivation, and emotion, will finally force the issue. In any event cataloging and classifying have returned as essential endeavors because they are necessary to test any and all hypotheses of both pattern and process. The early ethologists proved the need for descriptive ethograms in order to make more valid comparative inferences. The call was not for rigidly constrained ethograms but objective ones necessary for the job at hand. We see the search for mental taxonomies in the same light.

CONCLUSIONS

Although comparative biology has a long tradition, progress over the last decade suggests that maverick ideas will continue for some time. We feel that application of the new comparative methods to classical questions in both ethology (McLennan et al. 1988) and comparative psychology (Medicus 1987) will enrich and give new and exciting insights to the comparative study of behavior at many levels (e.g. linguistics, Cavalli-Sforza et al. 1988). We see three lessons for future comparative studies in ethology and psychology.

First, methodologies of classification are filled with critical assumptions related to character weighting, parsimony, types of characters and so on that have dramatic effects on the classifications produced across taxa. Likewise, the evaluation of a specific comparative hypothesis of behavior will rest on the taxonomic classification used.

Second, the nagging problem of phylogeny and convergent adaptation will not go away. Undoubtedly, debate will continue because of historical precedent for questioning the concept of adaptation, dating back at least to Darwin, and problems pertaining to unobservable (i.e. phylogenetic, mental) causal factors will always face resistance (Taylor 1987; O'Hara 1988; Sober 1988). Part of the problem is that conclusions from comparative tests are mainly structured in an "adaptation or phylogeny" style, reminiscent of early debates artificially dichotomizing structural and functional explanations of morphology (Gould 1986) or even explanations of behavior as learned or innate (Lorenz 1965). The newer comparative approaches suggest a revised attitude concerning adaptation: Phenotypic traits should be explained in terms of total variation, with attempts to partition amounts of variation to both phylogeny and adaptation. Adaptive variation may occasionally be credited to phylogeny; phylogeny and adaptive variation often do covary. Even so, by first addressing phylogenetic constraints one may safely pursue comparative hypotheses of adaptation. The comparative approaches described here are likely to be the first of many to cast aside the dichotomy of phylogeny *or* adaptation while retaining both as valid concepts of analysis.

The third issue, not discussed here, but increasingly relevant to all discussions of evolution, is the possible role of development as a source of innovations for evolutionary change as well as for characters to be compared (Blass 1988). In fact, future progress in understanding the process of evolution may hinge specifically on our knowledge of development (Buss 1987; McKinney 1988), including behavior (Gottlieb 1987).

ACKNOWLEDGEMENTS

Preparation of this chapter was supported in part by NSF grant BNS-8709629 to GMB and the Science Alliance at the University of Tennessee. David Kenny generously helped with the phylogenetic analysis of Lorenz's data and Frank McKinney gave us helpful information. We also thank the editors, anonymous reviewers, and Harold A. Herzog, Jr. for a critical reading of an earlier version.

LITERATURE CITED

Alexander, R.D. 1987. *The Biology of Moral Systems.* New York: Aldine de Gruyter.

Arnold, S. 1981. The microevolution of feeding behavior. In: *Foraging Behavior: Ecological, Ethological and Psychological Approaches* (ed. by A. Kamil & T. Sargent), pp. 409-453. New York: Garland STPM.

Atz, J.W. 1970. The application of the idea of homology to behavior. In: *Development and Evolution of Behavior* (ed. by

L.R. Aronson, E. Tobach, D.S. Lehrman, & J.S. Rosenblatt), pp. 53-74. San Francisco, California: W.H. Freeman.

Barlow, G.W. 1977. Modal action patterns. In: *How Animals Communicate* (ed. by T.A. Sebeok), pp. 98-134. Bloomington, Indiana: Indiana University Press.

_____. 1989. Has sociobiology killed ethology or revitalized it? In: *Perspectives in Ethology*, Volume 8 (ed. by P. P. G. Bateson & P. H. Klopfer), pp. 1-45. New York: Plenum Press.

Beer, C. 1980. Perspectives in animal behavior comparisons. In: *Comparative Methods in Psychology* (ed. by M. C. Bornstein), pp. 17-64. Hillsdale, New Jersey: Lawrence Erlbaum Associates.

Bekoff, M. 1977. Quantitative studies of three areas of classical ethology: Social dominance, behavioral taxonomy, and behavioral variability. In: *Quantitative Methods in the Study of Animal Behavior* (ed. by B. A. Hazlett), pp. 1-46. New York: Academic Press.

_____. 1988. Motor training and physical fitness: Possible short- and long-term influences on the development of individual differences in behavior. *Developmental Psychobiology* 21, 601-612.

Bitterman, M.E. 1965. Phyletic differences in learning. *American Psychologist* 20, 396-410.

Blass, E.M. (ed.) 1988. *Handbook of Behavioral Neurobiology,* Volume 9 *Behavioral Ecology and Developmental Psychobiology.* New York: Plenum Press.

Boice, R. 1973. Domestication. *Psychological Bulletin* 80, 215-230.

Bornstein, M.C. 1980. On comparison in psychology. In: *Comparative Methods in Psychology* (ed. by M. C. Bornstein), pp. 1-7. Hillsdale, New Jersey: Lawrence Erlbaum Associates.

Burghardt, G.M. 1970. Intraspecific geographical variation in chemical food cue preferences of newborn garter snakes. *Behaviour* 36, 246-257.

_____. 1973. Instinct and innate behavior: Toward an ethological psychology. In: *The Study of Behavior* (ed. by J. A. Nevin & G. S. Reynolds), pp. 322-400. Glenview, Illinois: Scott, Foresman.

_____. 1975. Chemical prey preference polymorphism in newborn garter snakes Thamnophis sirtalis. *Behaviour* 52, 202-225.

_____. 1985a. Animal awareness: Current perceptions and historical perspective. *American Psychologist* 40, 905-919.

_____. (ed.) 1985b. *The Foundations of Comparative Ethology.* New York: Van Nostrand Reinhold.

_____. 1988. Precocity, play, and the ectotherm - endotherm transition: Profound reorganization or superficial adaptation. In: *Handbook of Behavioral Neurobiology.* Volume 9 (ed. by E. M. Blass), pp. 107-148. New York: Plenum Press.

Buss, L.W. 1987. *The Evolution of Individuality.* Princeton, New Jersey: Princeton University Press.

Cavalli-Sforza, L.L., Piazza, A., Menozzi, P. & Mountain, J. 1988. Reconstruction of human evolution: Bringing together genetic, archaeological, and linguistic data. *Proceedings of the National Academy of Sciences U.S.A.* 85, 6002-6006.

Chevalier-Skolnikoff, S. 1989. Spontaneous tool use and sensorimotor intelligence in Cebus compared with other monkeys and apes. *Behavioral and Brain Sciences* 12, 561-627.

Cheverud, J.M., Dow, M.M. & Leutenegger W. 1985. The quantitative assessment of phylogenetic constraints in comparative analyses: Sexual dimorphism of body weight among primates. *Evolution* 39, 1335-1351.

Chiszar, D. 1986. Motor patterns dedicated to sensory functions. In: *Chemical Signals in Vertebrates,* Volume 4 (ed. by D. Duvall, D. Müller-Schwarze, & R.M. Silverstein), pp. 37-44. New York: Plenum.

Clutton-Brock, T.H. & Harvey, P.H. 1984. Comparative approaches to investigating adaptation. In: *Behavioural Ecology* 2nd Edition (ed. by J. R. Krebs & N. B. Davies), pp. 7-29. Sunderland, Massachusetts: Sinauer.

Coddington, J.A. 1988. Cladistic tests of adaptational hypotheses. *Cladistics* 4, 3-22.

Coleman, W. 1980. Morphology in the evolutionary synthesis. In: *The Evolutionary Synthesis* (ed. by E. Mayr & W. B. Provine), pp. 174-180. Cambridge, Massachusetts: Harvard University Press.

Crawford, C.B. 1989. The theory of evolution: Of what value to psychology. *Journal of Comparative Psychology* 103, 4-22.

Crook, J.H. 1970. Social organization and the environment: Aspects of contemporary social ethology. *Animal Behaviour* 18, 197-209.

Dane, B. & van der Kloot, W.G. 1964. An analysis of the display of the goldeneye duck (*Bucephala clangula*). *Behaviour* 22, 282-328.

Darwin, C. 1859/1967. *On the Origin of Species.* New York: Atheneum.

_____. 1883. A posthumous essay on instinct. In: *Mental Evolution in Animals* (G.J. Romanes), pp. 353-384. London: Kegan, Paul, Trench & Co.

Davis, H. & Pérusse, R. 1988. Numerical competence in animals: Definitional issues, current evidence, and a new research agenda. *Behavioral and Brain Sciences* 11, 561-615.

Demarest, J. 1987. Two comparative psychologies. In: *Historical Perspectives and the International Status of Comparative Psychology* (ed. by E. Tobach), pp. 127-155. Hillsdale, New Jersey: Lawrence Erlbaum Associates.

Dethier, V.G. & Stellar, E. 1964. *Animal Behavior.* 2nd Edition. Englewood Cliffs, New Jersey: Prentice Hall.

Dewsbury, D.A. 1984. *Comparative Psychology in the Twentieth Century.* Stroudsburg, Pennsylvania: Hutchinson Ross.

_____. (ed.). 1985. *The Foundations of Comparative Psychology.* New York: Van Nostrand Reinhold.

_____. 1989. A brief history of the study of animal behavior in North America. In: *Perspectives in Ethology*, Volume 8 (ed. by P.P.G. Bateson & P.H. Klopfer), pp. 85-122. New York: Plenum Press.

Domjan, M. & Galef, B.G. 1983. Biological constraints on instrumental and classical conditioning: Retrospect and prospect. *Animal Learning & Behavior* 11, 151-161.

Drummond, H. 1981. The nature and description of behavior patterns. *Perspectives in Ethology* 4, 1-30.

Felsenstein, J. 1978. The number of evolutionary trees. *Systematic Zoology* 27, 27-33.

_____. 1979. Alternative methods of phylogenetic inference and their relationship. *Systematic Zoology* 28, 49-62.

_____. 1983. Parsimony in systematics: biological and statistical issues. *Annual Review of Ecology and Systematics* 14, 313-333.

_____. 1985a. Confidence limits on phylogenies: an approach using the bootstrap. *Evolution* 39, 783-791.

_____. 1985b. Phylogenies and the comparative method. *American Naturalist* 126, 1-25.

_____. 1988. Phylogenies and quantitative characters. *Annual Review of Ecology and Systematics* 19, 445-471.

Ferguson, G.W. 1971. Variation and evolution of the push-up displays of the side-blotched lizard genus *Uta* (Iguanidae). *Systematic Biology* 20, 79-101.

Freud, S. 1915/1959. Instinct and their vicissitudes. In: *The Collected Papers of Sigmund Freud.* Volume 4 (ed. by E. Jones), pp. 60-83. New York: Basic Books.

Ghiselin, M.T. 1980. The failure of morphology to assimilate Darwinism. In: *The Evolutionary Synthesis* (ed. by E. Mayr & W. B. Provine), pp. 180-193. Cambridge, Massachusetts: Harvard University Press.

Gittleman, J.L. 1981. The phylogeny of parental care in fishes. *Animal Behaviour* 29, 936-941.

_____. 1986a. Carnivore life history patterns: Allometric, phylogenetic, and ecological associations. *American Naturalist* 127, 744-771.

_____. 1989a. The comparative approach in ethology: Aims and limitations. In: *Perspectives in Ethology,* Volume 8 (ed. by P.P.G. Bateson & P.H. Klopfer), pp. 55-83. New York: Plenum Press.

_____. 1989b. Carnivore group living: Comparative trends. In: *Carnivore Behavior, Ecology, and Evolution* (ed. by J.L. Gittleman), pp. 183-207. Ithaca, New York: Cornell University Press.

Gittleman, J.L. & Kot, M. In Press. Adaptation: Statistics and a null model for estimating phylogenetic effects. *American Naturalist.*

Gottlieb, G. 1984. Evolutionary trends and evolutionary origins: Relevance to theory in comparative psychology. *Psychological Review* 91, 448-456.

_____. 1987. The developmental basis of evolutionary change. *Journal of Comparative Psychology* 101, 262-271.

Gould, S.J. 1986. Archetype and adaptation. *Natural History* Nov., 16-27.

Gould, S.J. & Lewontin, R.C. 1979. The spandrels of San Marco and the Panglossian paradigm: a critique of the adaptationist programme. *Proceedings of the Royal Society of London* B 205, 581-598.

Greene, H.W. & Burghardt, G.M. 1978. Behavior and phylogeny: Constriction in ancient and modern snakes. *Science* 200, 74-77.

Grier, J.W. 1984. *Biology of Animal Behavior.* St. Louis, Missouri: Times Mirror/Mosby College Publishing.

Griffin, D.R. 1976. *The Question of Animal Awareness.* New York: Rockefeller University Press.

Harvey, P.H. & Clutton-Brock, T.H. 1985. Life history variation in primates. *Evolution* 39, 559-581.

Harvey, P.H. & Mace, G.M. 1982. Comparisons between taxa and adaptive trends: Problems of methodology. In: *Current Problems in Sociobiology* (ed. by King's College Sociobiology Group), pp. 343-362. Cambridge: Cambridge University Press.

Heinroth, O. 1930/1985. Über bestimmte Bewegungsweisen der Wirbeltiere. *Gesellschaft naturforschender Freunde, Berlin, Sitzungsberichte*, 333- 342. Partial translation by C. J. Mellor & D. Gove In: *Foundations of Comparative Ethology* (ed. by G. M. Burghardt), pp. 339-342. New York: Van Nostrand Reinhold Company.

Hennig, W. 1966. *Phylogenetic Systematics.* Urbana, Illinois: University of Illinois Press.

Hinde, R.A. 1966. Animal Behaviour: A Synthesis of Ethology and Comparative Psychology. New York: McGraw-Hill.

Hodos, W. & Campbell, C.B. 1969. Scalae naturae: Why there is no theory in comparative psychology. *Psychological Review* 76, 337-350.

Huey, R.B. 1987. Phylogeny, history, and the comparative method. In: *New Directions in Ecological Physiology* (ed. by M.E. Feder, A.F. Bennett, W.W. Burggren, & R.B. Huey), pp. 76-98. Cambridge: Cambridge University Press.

Huey, R.B. & Bennett, A.F. 1987. Phylogenetic studies of coadaptation: Preferred temperatures versus optimal performance temperatures of lizards. *Evolution* 41, 1098-1115.

Hull, D.L. 1988. *Science as a Process.* Chicago, Illinois: University of Chicago Press.

James, W. 1890. *Principles of Psychology.* New York: Holt.

Jenssen, T A. 1971. Display analysis of *Anolis nebalosus* (Sauria, Iguanidae). Copeia 1971, 197-209.

Jerison, H. J. 1973. *Evolution of Brain and Intelligence.* New York: Academic Press.

Johnsgard, P.A. 1965. *Handbook of Waterfowl Behavior.* Ithaca, New York: Cornell University Press.

Johnston, T.D. 1981. Contrasting approaches to a theory of learning. *Behavioral and Brain Sciences* 4, 125-173.

Kluge, A.G. & Strauss, R.E. 1985. Ontogeny and systematics. *Annual Review of Ecology and Systematics* 16, 247-268.

Lack, D. 1968. *Ecological Adaptation for Breeding in Birds.* London: Methuen.

Lauder, G.V. 1986. Homology, analogy, and the evolution of behavior. In: *Evolution of Animal Behavior: Paleontological and Field Approaches* (ed. by M. H. Nitecki & J. A. Kitchell), pp. 9-40. New York: Oxford University Press.

Livezey. B.C. 1986. A phylogenetic analysis of recent anseriform genera using morphological characters. *Auk* 103, 737-754.

Lockard, R.B. 1971. Reflections on the fall of comparative psychology: Is there a message for us all? *American Psychologist* 26, 168-179.

Lorenz, K.Z. 1941/1970. Vergliechende Bewegungsstudien an Anatinen. *Journal für Ornithologie* 79 (supplement), 194-294. Translated by R. Martin in: *Studies in Animal and Human Behaviour,* Volume 2 (K. Lorenz), pp. 14-114. Cambridge, Massachusetts: Harvard University Press.

_____. 1950. The comparative method in the study of innate behaviour patterns. *Symposia of the Society for Experimental Biology* 4, 221-268.

_____. 1965. *Evolution and Modification of Behavior.* Chicago, Illinois: University of Chicago Press.

_____. 1966. *On Aggression.* New York: Harcourt Brace Jovanovich.

Macphail, E.M. 1987. The comparative psychology of intelligence. *Behavioral and Brain Sciences* 10, 645-695.

Maddison, W.P., Donoghue, M.J. & Maddison, D.R. 1984. Outgroup analysis and parsimony. *Systematic Zoology* 33, 83-103.

Maier, N.R. & Schneirla, T.C. 1935. *Principles of Animal Psychology.* New York: McGraw-Hill.

Mason, W. 1976. Review of *The Question of Animal Awareness. Science* 194, 903-931.

Mayr, E. 1969. *Principles of Systematic Zoology.* New York: McGraw-Hill.

_____. 1982. *The Growth of Biological Thought.* Cambridge, Massachusetts: Harvard University Press.

McDougall, W. 1908. *An Introduction to Social Psychology.* London: Methuen.

McKinney, F. 1975. The evolution of duck displays. In: *Function and Evolution of Behaviour* (ed. by G. Baerends, C. Beer, & A. Manning), pp. 331-357. Oxford: Clarendon.Press.

_____. 1978. Comparative approaches to social behavior in closely related species of birds. *Advances in the Study of Behavior* 8, 1-38.
McKinney, M. (ed.) 1988. *Heterochrony in Evolution.* New York: Plenum Press.
McLennan, D.A., Brooks, D.R., & McPhail, J.D. 1988. The benefits of communication between comparative ethology and phylogenetic systematics: A case study using gasterosteid fishes. *Canadian Journal of Zoology* 66, 2177-2190.
Medicus, G. 1987. Toward an etho-psychology: A phylogenetic tree of behavioral capabilities proposed as a common basis for communication between current theories in psychology and psychiatry. *Ethology & Sociobiology* 8 (Suppl. 3), 131S-150S.
Morgan, C.L. 1894. *An Introduction to Comparative Psychology.* London: Edwin Arnold.
Morris, D. 1967. *The Naked Ape.* New York: McGraw-Hill.
O'Hara, R.J. 1988. Homage to Clio, or, toward an historical philosophy for biology. *Systematic Zoology* 37, 142-155.
Pagel, M.D. & Harvey, P.H. 1988. Recent developments in the analysis of comparative data. *Quarterly Review of Biology* 63, 413-440.
Pellis, S.M. 1988. Agonistic versus amicable targets of attack and defense: Consequences for the origin, function, and descriptive classification of play-fighting. *Aggressive Behavior* 14, 85-104.
Perrins, C.M. 1977. The role of predation in the evolution of clutch size. In: *Evolutionary Ecology* (ed. by B. Stonehouse & C. Perrins), pp. 181-191. Baltimore, Maryland: University Park Press.
Ricklefs, R.E. 1977. A note on the evolution of clutch size in altricial birds. In: *Evolutionary Ecology* (ed. by B. Stonehouse & C. Perrins), pp. 193-214. Baltimore, Maryland: University Park Press.
Ridley, M. 1983. *The Explanation of Organic Diversity: The Comparative Method and Adaptations for Mating.* Oxford: Clarendon Press.
_____. 1986. *Evolution and Classification.* London: Longman.
_____. 1989. Why not to use species in comparative tests. *Journal of Theoretical Biology* 136, 361-364.
Romanes, G. J. 1882. *Animal Intelligence.* London: Kegan, Paul, Trench, & Co.

____. 1883. *Mental Evolution in Animals.* London: Kegan, Paul, Trench, & Co.

Schleidt, W.M., Yakalis, M.D. & McGarry, J. 1984. A proposal for a standard ethogram, exemplified by an ethogram of the bluebreasted quail *Coturnix chinensis. Zeitschrift für Tierpsychologie* 64, 193-220.

Simpson, G.G. 1961. *Principles of Animal Taxonomy.* New York: Columbia University Press.

Sober, E. 1988. *Reconstructing the Past.* Cambridge, Massachusetts: MIT Press.

Sokal, R.R. & Sneath, P.H.A. 1963. *Principles of Numerical Taxonomy.* San Francisco, California: W.H. Freeman.

Stamps, J.A. & Barlow, G.W. 1973. Variation and stereotype in the displays of *Anolis aeneus* (Sauria: Iguanidae). *Behaviour* 42, 67-94.

Taylor, P.J. 1987. Historical versus selectionist explanations in evolutionary biology. *Cladistics* 3, 1-13.

Thornhill, R. 1984. Scientific methodology in entomology. *Florida Entomologist* 67, 74-96.

Tinbergen, N. 1952. "Derived" activities: Their causation, biological significance, origin, and emancipation during evolution. *Quarterly Review of Biology* 27, 1-32.

Wickler, W. 1961. Ökologie und Stammesgeschichte von Verhaltensweisen. *Fortschriffte Zoologie* 13, 303-365.

____. 1973. Ethological analysis of convergent adaptation. *Annals of the New York Academy of Sciences* 223, 65-69.

Wilson, E.O. 1975. *Sociobiology: The Modern Synthesis.* Cambridge, Massachusetts: Harvard University Press.

Yerkes, R. 1907. *The Dancing Mouse.* New York: Macmillan.

9. How the Mismatch between the Experimental Design and the Intended Hypothesis Limits Confidence in Knowledge, as Illustrated by an Example from Bird-song Dialects

Donald E. Kroodsma

INTRODUCTION

As students of animal behavior, we observe, describe, ask questions, develop hypotheses, and then perform experiments, usually in that order. Each of these stages is crucial in the scientific process, but it is the experiment on which our decisions and plans often pivot. Alternative outcomes to that experiment may strengthen or weaken our faith in certain ideas, and then we may repeat part or all of the observe-to-experiment cycle again and again as we come progressively closer to a more complete understanding of the issues. Well-designed and well-executed experiments for our explicitly stated hypotheses are therefore critical for taking us ever closer to understanding the essence of the world around us.

An experiment that is inadequately designed is severely limited in what it can tell us about the world. Even worse, however, is that strong conclusions drawn from such an experiment may confuse and mislead the experimenter as to the next appropriate steps in the research program. Science progresses by successive refinements of the question being pursued, but without conclusive answers to a given question, one cannot be sure of what the next question should be. As a consequence, few refinements in the questions might be attempted, perhaps because of a lack of confidence in the results, or unwarranted confidence may launch an investigator into ill-chosen follow-up questions.

It is important, therefore, for researchers in any subdiscipline to evaluate critically 1) what is being learned about the natural world itself, 2) what is being learned about the methods of inquiry that have been used, and 3) what is being learned about how those methods limit confidence in developing knowledge. To challenge one's assumptions, to question "ruling theories" and their accompanying "theory tenacity," to dispel "confirmation bias" (Loehle 1987), in sum, to rock the comfortable *status quo*, are all necessary, though sometimes painful, processes that are part of maturing disciplines.

With these convictions and concerns, I first discuss a general experimental design that is useful for sampling "populations" of both subjects and behavioral stimuli. I illustrate how failure to state the hypothesis in explicit terms may lead to inappropriate experimental designs and statistical tests for the intended hypothesis (i.e. pseudoreplication; Hurlbert 1984) and to actual tests of unintended and (usually) biologically uninteresting hypotheses. I then demonstrate the consequences of this problem with a survey of sampling designs actually used to test if songbirds (oscines, suborder Passeres, Order Passeriformes) perceive differences between home and non-home song dialects. My conclusion is, as my title states, that inappropriate sampling designs prevent confidence in our conclusions.

SAMPLING RELEVANT STIMULI FOR AN INTENDED HYPOTHESIS

Of primary concern in experimental testing is the "typology trap," in which the one stimulus is assumed to be representative or typical of all other stimuli from the same class. In the most extreme form of this problem, multiple responses from many subjects to one stimulus are used in statistical testing as if each response were actually to a different stimulus from the same class. Because a particular class of stimuli is represented by only one exemplar, however, the experimenter's conclusions are limited to that particular stimulus and cannot safely be generalized to other stimuli in that class. To generalize (usually done unknowingly) to the entire class constitutes pseudoreplication, because the experimenter mistakenly believes that the collected data will enable a test of the distribution of responses to that entire class of stimuli, not just a test of the distribution of responses by different animals to one particular stimulus from the two larger, composite distributions.

To illustrate this particular problem, consider testing a particular hypothesis that "Female rose-breasted grosbeaks (*Pheucticus ludovicianus*) when choosing a mate during the breeding season are not more responsive to brightly colored males in second-year plumage than they are to duller colored males in first-year plumage." For this hypothesis, it is inappropriate to assess the responses of females to only one stuffed individual, mounted in some appropriate life-like pose, from each age class of males. The two mounted individuals will differ in many unintended ways (such as plumage quality, fine nuances of pose, proportions of final stuffed specimen, etc.) in addition to the one feature of male plumage on which the experimenter is focusing. With only one representative specimen from each class, one cannot be certain that plumage color is the factor to which females are responding.

If only one stuffed male is used to represent each class of stimuli, the researcher can learn little about the composite distribution of responses to the population of specimens that represent each plumage class (see Figure 1). If the response to A1 and to B1 (a single individual from each plumage class) had not been significantly different, for example, one could state with some degree of confidence that the females responded no differently only to the two particular specimens that had been used in the experiment. Even if the response to B1 had been greater than to A1, the results still do not falsify any of the following three statements: the *mean* response intensity to dull-colored 1-year-olds is less than, equal to, or greater than the *mean* response intensity to more brightly colored 2-year-olds. Without an appropriate sample of specimens from each age class, no confident statement can be made about the mean response for each class. To assume the one mounted stimulus is representative of, or identical to, all other stimuli from the class ignores the nature of variability in the real world.

TESTS OF DIALECT VARIATION IN BIRD SONG

Geographic Variation

Songs of a given songbird species vary on both a macro- and a micro-geographic scale (Mundinger 1982). Because male songbirds typically learn their songs, variation in songs on both geographic scales often reflects only cultural and not genetic differences in populations (e.g. Kroodsma & Canady 1985). Here, however, I am less concerned with the variation that occurs over great distances (even though it, too, may be cultural) than I am with the specific kinds of local variation that a young songbird might encounter in dispersing from its natal territory (the area where it was hatched) to a breeding territory. If a songbird (e.g. of the species listed and discussed later in Table 2) perceives its auditory world in a fashion at all similar to the way we perceive and analyze it, the bird will identify micro-geographic areas in which conspecifics sing relatively similar songs.

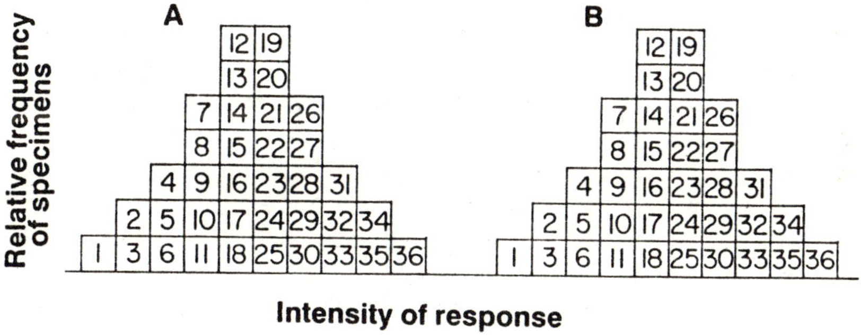

Figure 1. Graphical illustration of the "typology trap," in which one stimulus is assumed to be representative of, or identical to, other stimuli from a given class. Illustrated are the response distributions to two classes of stimuli, such as the response of female rose-breasted grosbeaks to males with dull-colored first-year plumages (A) and brightly-colored second-year plumages (B). Each response distribution is constructed by using the mean (or median) response to a sample of many different stuffed specimens (1, 2, 3,...36) that comprise that class. With increasing overlap of distributions A and B, using the response to a single specimen to represent the response to all specimens of the class produces increasing confusion. With only partial overlap, for example, using specimens A36 and B1 could give the impression of no significant difference between the two distributions. If the two distributions had the same mean and variance (i.e. coincided perfectly), one could "demonstrate" by using a single stuffed specimen to represent each class (i.e. by using pseudoreplication; Hurlbert 1984) that the mean response to one-year-olds is less than (e.g. A1 vs. B36), equal to (A10 versus. B10), or greater than (A36 versus. B1) the mean response to two-year-olds.

For purposes of discussion, I here refer to these micro-geographic areas as "dialects," regardless of whether or not boundaries between those areas have been identified. These song dialects have been described for a number of species (see Kroodsma & Baylis 1982). Typically, songs for a given dialect are identified by one or more components in the song. Songs of Nuttall's white-crowned sparrows (scientific names listed later in Table 2), for example, often consist of four components: an introductory phrase, complex syllables, simple syllables, and a terminal phrase. Birds in a given area often share three or four of these song components with one another, and birds in neighboring areas may differ consistently in one or more components (Baker & Cunningham 1985). In some species abrupt boundaries exist between different dialect areas (e.g. Baptista 1975; Baker & Thompson 1985), but in other species such boundaries may not have been sought and identified (e.g. Kroodsma et al. 1984) or may not exist if songs change more gradually over distance. Thus, among many songbirds, neighbors or groups of neighbors typically share more features of their songs with each other than they do with more distant individuals.

The second important aspect of song variation on the local scale is that interacting males sing *relatively* similar songs, but no two males, even within the same dialect, sing identical songs. Songs of males within the same dialect are, by definition, more similar to each other than they are to songs of males in other dialects, but considerable variability among males still exists. No "typical" or "representative" song can actually be chosen to represent all songs of the dialect. "Subdialects" of males may be identifiable, for example, or males on one end of a dialect area may differ consistently in some relatively minor way from those on another end of the dialect (Baker et al. 1984; Baker & Cunningham 1985). One of my basic premises in this essay is that neglecting the existence of this type of variation in stimuli can severely compromise interpretation of experimental work.

Song Variation and Hypothesis Testing with Playback Studies

An examination of the hierarchical nature of song variation (Table 1) will help visualize, I believe, the kinds of data one could gather to test specific hypotheses. First, there are over 4000 species of songbirds (Bock & Farrand 1980); micro-geographic variation probably exists in most species, but it has been studied

Table 1. Song variability: sampling "populations" of species (1-4000+), populations of dialects for each species (A, B, C, . . .), populations of song variants within each dialect (A1, A2, A3, . . .), and populations of responding subjects (each "x" a different subject) for testing hypotheses about song dialects.

<u>Species 1</u>   <u>Species 2</u>   <u>Species 3</u>   <u>Species 4</u>   . . . .   <u>Species 4000+</u>

Species 1

Dialect A			Dialect B			Dialect C			
A1	A2	A3 . . . An	B1	B2	B3 . . . Bn	C1	C2	C3 . . . Cn	. . . . .
x	x	x x	x	x	x x	x	x	x x	. . . . .
x	x	x x	x	x	x x	x	x	x x	. . . . .
x	x	x x	x	x	x x	x	x	x x	. . . . .
x	x	x x	x	x	x x	x	x	x x	. . . . .
x	x	x x	x	x	x x	x	x	x x	. . . . .
x	x	x x	x	x	x x	x	x	x x	. . . . .
x	x	x x	x	x	x x	x	x	x x	. . . . .
x	x	x x	x	x	x x	x	x	x x	. . . . .
x	x	x x	x	x	x x	x	x	x x	. . . . .
x	x	x x	x	x	x x	x	x	x x	. . . . .

in only a few. For each one of those species (1, 2, 3,...4000), a number of dialect populations (A, B, C...) may exist, and within each dialect population, individuals sing variants (e.g. A1, A2, A3 ...) of the identified local dialect (for heuristic purposes, assume that each individual sings only one song form).

Some specific hypotheses may be formulated to address this variation in songs. These hypotheses can be stated as either null hypotheses or as denials (see Popper 1959; Platt 1964); the types of experimental designs one would choose to test the hypotheses should not differ as a consequence. I begin with questions about (1) variation in response to different songs from one particular dialect, and then proceed to more general questions about (2) two particular dialects, (3) representative dialects from a given species, and finally (4) dialects among different songbird species. I use these four sample hypotheses, with crucial words italicized, to illustrate the kinds of data one could gather to address specific questions.

H_1: Birds of a *given* dialect (in a given species) respond no differently to different songs from the *same local* dialect. An appreciation for the variability in response to different songs from the same dialect is crucial for an understanding of how birds perceive the songs in their immediate environment. Surprisingly few tests of this hypothesis have actually been attempted, but ample evidence now exists to indicate that different recorded songs from a single dialect (identified by human standards) are not treated identically by birds. For Nuttall's white-crowned sparrow, for example, Tomback et al. (1983) specifically tested responses of males in a given dialect (say, A) to similar pairs of songs from four dialects (A1 and A2, B1 and B2, C1 and C2, D1 and D2); they found statistically significant differences in the singing responses by A males to A1 and A2, two songs of the home dialect, and to B1 and B2, two songs of a nearby dialect. In a special case of song discrimination within a dialect, Baker et al. (1981c) also demonstrated that males responded differently to songs of neighbors and strangers within the same dialect (see Falls 1982 for additional examples of neighbor-stranger discrimination). In another study, two song variants from the same dialect were treated differently by responding males (Baker et al. 1984). One of four songs of the blue-winged warbler seemed to evoke especially strong responses from males (Kroodsma et al. 1984). Finally, Searcy et al. (1982) used four different local song types and found that female swamp sparrows

(*Melospiza georgiana*) responded, on average, 2.5 times more strongly to the most potent song than they did to the leastpotent song.

H_2: Birds of a *given* dialect respond no differently to songs of their home dialect than they do to songs of a *specified* neighboring dialect. In a test of this second hypothesis, the variation in responsiveness to different exemplars of songs from each dialect must be taken into account. Most importantly, the responses to the population of A songs (not one specific Ai) must be compared to the responses to the population of B songs (not one specific Bi). Given probable response differences (whether statistically significant at p = 0.05 or not) by A males to A1 and A2, for example, one cannot compare responses by A males to song A1 and to song B1 and claim that any difference in response is because the songs were from different dialects. Response differences equally as great might have been obtained had two songs from the same dialect been used.

Neither can one inflate the actual sample of playback songs by using multiple responses to each of several playback songs from a class as if each response were to a different playback song. A common practice is to use several songs from each dialect area (e.g. A1, A2, A3 and B1, B2, B3) and to test several different subjects with each playback song. Responses to all A songs and to all B songs are then pooled, and the two data sets are compared with a simple two-sample statistical test. Such pooling is not appropriate even if no significant differences occur among the responses to different A songs (A1 versus A2 vs A3) and among the responses to different B songs (B1 versus B2 versus B3). One wants to know whether any response difference that occurs between Ai and Bi songs is over and above the response difference that occurs among the Ai and among the Bi. In this example only three songs from each of two distributions (responses to A songs and to B songs) have been used, and inflating the sample size of playback songs by using multiple responses to each song is statistically equivalent to the pooling fallacy described by Machlis et al. (1985).

Although including several song exemplars from each dialect region on the same stimulus tape may be an improvement, this procedure does not provide the rigorous test that is needed. With several songs, a researcher cannot be sure that he is closer to estimating the mean of the response distribution to A or B songs or whether he is estimating the mean response to the most (or even least) potent song in the sample. The mean response and response variability to each A exemplar and to each B exemplar need to be preserved for an ANOVA-like (analysis of variance) statistical test. Only with such a test can it be determined whether or not birds of dialect A respond differently to songs of their own A dialect and those of the neighboring B dialect.

H_3: Birds of a given species (perhaps in a restricted geographic area) respond no differently to songs of the home dialect than they do to songs of a *neighboring* dialect (other dialect categories could also be tested, such as home versus nonneighboring dialect, neighboring vs. nonneighboring dialect, etc.). This third hypothesis differs from the second in one significant way. It is more general because it does not identify the particular dialect in which the tests are to be done. If one tests and rejects this third hypothesis, one would like to accept (or state that the data are consistent with) an alternative, namely, that birds of this particular species distinguish home from neighboring dialects, not just that birds of a specific dialect, say A, distinguish songs of dialect A from songs of dialect B.

Ironically, although a question about two particular dialect areas (hypothesis 2 above) requires multiple sample songs from each of those two dialects, the more general and perhaps biologically more interesting hypothesis about dialects in general does not *require* multiple exemplars from each dialect area. The more important sample units here are the number of home dialects that have been tested with at least one home exemplar and one neighboring exemplar. At the very least, it seems, one could (though n = 5 would not be recommended) visit five dialect areas and test one male (or several males) in each with one home exemplar and one neighboring exemplar. Additional song exemplars for each dialect would be desirable, of course, but they are not necessary for a test of this particular hypothesis. Especially powerful here is the use of reciprocal playbacks (e.g. Weary et al. 1987), in which the home song for one test becomes the neighboring song for the next test; results are convincing if males respond consistently to their respective home and neighboring dialects. If the phenomenon is robust and males consistently respond more strongly to either the home or the neighboring dialect, and if one had predicted *a priori* the direction of response, then a one-tailed test with only n = 5 dialects tested and only one playback in each dialect would yield a statistically significant difference (p = 0.03).

H_4: *Songbirds*, or a subgroup of songbird species, respond no differently to songs of the *home* and a *neighboring* dialect. This fourth hypothesis is even more general than the third hypothesis, and the important samples are now the number of species in which one has sampled the responses of males (or a male) to at least one exemplar of a home and one of a neighboring dialect. A confident statement about each particular species can not be made if one uses only one pair of dialects for each species, just as information about each particular dialect is lacking in a simplified test of the third hypothesis. By using all available playback studies in the literature (i.e. a meta-analysis), one can perhaps actually test whether songbirds in general respond more strongly to home than to nonhome song dialects.

Hypotheses actually Tested: A Literature Survey

In my survey of the literature (Table 2), I found 22 published studies that had tested response differences of birds to songs of the local dialect and to songs of another (often neighboring) dialect. These studies differed from one another in many ways. Over a dozen species were tested. Some studies focused on males, others on females, and some on both. Some tests occurred in the laboratory, but many were done in the field. Some studies used reciprocal designs, in which birds in each of two (or more) dialects were tested with each others' songs, but most did not.

A thorough review of these studies would evaluate the merits of each study and the validity of the conclusions drawn. In such scrutiny, I would focus on how many individuals were tested, whether sample sizes were inflated by using multiple responses from the same individual or multiple responses to the same song exemplar, whether data were collected blindly, whether individuals heard stimuli in random (or semi-random) order, and so on (see Kroodsma 1989a,b). My goal here is simpler, however, and I wish merely to examine how each study sampled the populations of relevant stimuli and subjects, because those aspects of the design determine the potential hypotheses that the particular study could have tested.

The column in Table 2 that is most informative indicates the number of song stimuli that were used to represent each song dialect. One study (14) specifically tested more than one song stimulus from each of four dialects (the home and three foreign dialects); tests were pairwise comparisons of responses to each stimulus separately, and a number of differences in response to the two exemplars of the same dialect did occur. Another study (4) tested for differences in response to three songs from a given dialect, but when no differences in response were found all data were pooled (inappropriately, I would argue) for later analysis. Similarly, in my study (18a) I used four different pairs of song stimuli, with one stimulus of each pair from one of two song populations; in the statistical test, however, I pooled the multiple responses to each pair of song stimuli for one two-sample test. Several other studies used the same approach (see footnote e in Table 2). This pooling approach is inappropriate because the multiple responses to each of several stimuli are not independent data points (see Machlis et al. 1985). Several other studies used one stimulus tape but used up to three different songs on the same stimulus tape. But unless one obtains the mean and variance of responses to the separate stimuli, one cannot obtain an appropriate estimate of the mean for the response distribution to the population of stimuli from a particular class.

These entries suggest that many investigators have been aware that using a single stimulus to represent a class of stimuli is inappropriate, but that their concern was not translated into appropriate experimental designs. In the statistical test for a "dialect effect" in most of these cited studies, the investigators used only one stimulus or several stimuli as if they were one, and the multiple responses to that "one" stimulus were then used as if they were the mean responses to multiple stimuli from a given class. Investigators undoubtedly believed, as I did (study 18), that they were testing hypotheses about dialects (hypothesis 2 or 3 discussed above), when in actuality they were testing a hypothesis of reduced interest about particular songs. Conclusions from these studies are therefore compromised because the studies do not determine the variability in responsiveness to the separate stimuli that comprise the larger class of stimuli.

Relatively simple changes in these experimental designs can change the actual hypotheses tested. In most studies the number of subjects tested was large, but the effective number of song stimuli used to represent each dialect was inappropriately small (usually one). In a nonreciprocal test of two dialects, by simply testing more stimuli and fewer subjects per stimulus, and by expending relatively little (if any) extra energy, one can test more interesting hypotheses, such as hypotheses 1 and 2 outlined above. By distributing effort among different dialects (or species), and

Table 2. A survey of sampling designs used for testing hypotheses about song dialects.

Study[a]	Species[b]	Sex[c]	Dialects Tested[d]			Exemplars per Dialect[e]	Birds Tested[f]
			Home	Foreign	R?		
1	Northern Cardinal	M	2	2		1	28
2	Indian Hill Mynah	M	2	2		1 (4)	2
3	White-crowned Sparrow	M,F	2 (3)	3	R	1	>100
4	Song Sparrow	M	3	3	R	1 (1-3)	46
5	Wren	M	2	2	R	1	58
6	White-crowned Sparrow	F	1	1		1 (3)	13
7	White-crowned Sparrow	M	1	2		1	75
8	White-crowned Sparrow	M,F	1	3		1	21
9	European Starling	M	1	1		1 (3)	6
10	White-crowned Sparrow	F	1	3		1 (2)	20
11	Corn Bunting	M	1	1		1	10
12	Corn Bunting	M	2	2	(R)	1	30
13	White-crowned Sparrow	F	1	1		1 (3)	20
14	White-crowned Sparrow	M	1	3		2	141
15	White-crowned Sparrow	M,F	3	3	R	1 (2)	187
16	Redwing	M	1 (3)	1		1	13

17	Yellowhammer	M	1	1		1 (2)	20
18a	Blue-winged Warbler (I)	M	1	1		1 (4)	25
18b	Blue-winged Warbler (II)	M	2	2	R	1 (1-2)	32
19	Darwin's finches	M	7	7		1	79
20	White-crowned Sparrow	F	1	1		1	32
21	Yellowhammers	F	1	1		1	13
22	Great Tit	F	1	1		1 (3)	10

a. References are listed in chronological order. 1, Lemon 1967; 2, Bertram 1970; 3, Milligan & Verner 1971; 4, Harris & Lemon 1974; 5, Kreutzer 1974; 6, Baker et al. 1981a, 1982; 7, Baker et al. 1981b; 8, Petrinovich & Patterson, 1981; 9, Adret-Hausberger 1982; 10, Baker 1983; 11, McGregor 1983; 12, Pellerin 1983; 13, Spitler-Nabors & Baker 1983; 14, Tomback et al. 1983; 15, Baker et al. 1984; 16, Bjerke 1984; 17, Hansen 1984; 18ab, Kroodsma et al. 1984 (tests on the Type I & Type II song, respectively, of this warbler species); 19, Ratcliffe & Grant 1985; 20, Baker et al. 1987a; 21, Baker et al. 1987b; 22, Baker et al. 1988.

b. Scientific names as follows: northern cardinal, *Cardinalis cardinalis*; Indian hill mynah, *Gracula religiosa*; white-crowned sparrow, *Zonotrichia leucophrys*; song sparrow, *Melospiza melodia*; wren, *Troglodytes troglodytes*; European starling, *Sturnus vulgaris*; corn bunting, *Emberiza calandra*; redwing, *Turdus iliacus*; yellowhammer, *Emberiza citrinella*; blue-winged warbler, *Vermivora pinus*; Darwin's finches, *Geospiza* spp.; great tit, *Parus major*.

c. Sex of responding individuals. M, Males; F, Females.

d. The number of home and the number of foreign (i.e., nonhome) dialects tested in each study, and whether or not the playbacks were done reciprocally in different dialects (R). In study 3, pilot work was begun in a third dialect. In study 16, data from 3 dialects were pooled in a single statistical analysis. Study 12 was done reciprocally, but data for the two dialects were pooled, thus preventing any finer analysis of how birds in each of the two dialects responded.

e. The number of song exemplars used to represent each dialect. In parentheses are the number of actual song stimuli used in the experiments. In some studies (6, 10, 13, 22) the different song stimuli were on the same tape, and in other studies (2, 4, 9, 15, 17, 18ab) different tapes or isolated stimuli were used but all data were pooled for a single analysis; in both types of analyses, then, the mean and variance due to each stimulus are lost in the statistical analysis, and the data are treated as if they were from a single stimulus (which I represent outside of the parentheses as a single exemplar used to represent the dialect).

f. The total number of individuals or pairs or territories tested. For study 3, entry is probably >100, but exact number cannot be determined because some birds were used more than once.

by expending less effort within each dialect (or species), one can test the more general hypotheses 3 and 4 discussed above.

I am convinced that some phenomena will be better understood and confusion will be reduced with improved experimental designs. In some studies, for example, birds have responded more strongly to a song from the neighboring dialect than to the home dialect, but in most studies just the reverse has occurred. With overlapping distributions in responses to stimuli from two classes (see Figure 1), one would expect such flip-flops in results at times. With more stimuli in the experimental design to represent each class we will gain confidence in answers to some of these questions.

My Goals and Nongoals Made Explicit

My primary goal in this survey is *only* to illustrate that the hypotheses actually tested have frequently not been the ones that have been intended. Sampling designs now in use often preclude the asking of biologically interesting questions. I hope that an increased awareness of these issues will allow researchers to refine experimental questions and hence improve our interpretations of how animals perceive stimuli in their environment.

Several nongoals need to be made explicit, the first of which is that I do not mean to imply that the studies I cite in Table 2 have no or little merit. The aspect of the study on which I focus (i.e. sampling design) may be only of secondary importance in a larger study addressing other issues. Or, in some studies, the data can be interpreted from a different perspective and considerable confidence gained from the efforts expended. For example, Ratcliffe & Grant (1985) tested populations of Darwin's finches on seven different islands and found that six of the seven populations discriminated a local song from a foreign song. A metaanalysis on those data (actually a test of hypothesis 4 above) allows a general statement to be made about the ability of *Geospiza* finches, but not particular species, to discriminate local from foreign songs. Other studies have increased the power of their tests by using reciprocal designs, thus demonstrating that idiosyncratic stimuli would very likely not have produced such consistent results. Many analyses provide important data on variability in responsiveness to the chosen song stimulus from a given class. They have also provided other important refinements in how we go about asking birds to reveal a discrimination among stimuli. Each study also represents an important data point for a test of hypothesis 4, whether songbirds in general respond differently to home and foreign song dialects. If such a meta-analysis were to be done, one would want to realize that one particular family of songbirds, the Emberizidae, is over-represented in the sample. To actually generalize to the oscines, one would need to obtain a representative sample of different lineages within the suborder.

Second, I do not want to leave the impression that I actually advocate testing the hypotheses that I have listed and discussed in this essay. I used those hypotheses only as a heuristic tool, to illustrate how one would sample the world of song variation given a specifically stated question. One of the key problems is that identifying "dialect" areas is not a simple and objective task upon which different researchers can agree. More importantly, however, is that I think biologically more interesting questions can now be asked. Science often progresses through a series of experiments, with earlier experiments in a series only establishing a line of inquiry. With continued progress in a subdiscipline, refinements in the questions are constantly made, and early experiments (e.g. Lemon 1967; Milligan & Verner 1971; and others listed in Table 2) are recognized for what they truly are, as pioneering efforts rather than as the final word. As each publication appears, it becomes the "latest" word, the most recent refinement, only to become a "pioneering effort" with time. Given the results of papers cited in Table 2, for example, I would now be surprised if a study reported no difference in response to familiar signals (home dialect) and to unfamiliar signals (non-home dialect), which is the essence of hypotheses 2-4 in this essay. I would perhaps first suspect the methods used by the experimenter. If those methods proved satisfactory, I would begin to ask biologically relevant questions about how different were the familiar and unfamiliar signals actually used in the playbacks, how great was the distance between the two locations tested, how rapidly signals change over distance, how likely it is that the birds may actually be familiar with songs from distant locations, what kinds of movements do birds of (supposedly local) populations undergo, and so on. Authors listed in Table 2 deserve credit, in proportion to their number of citations, for both their pioneering efforts and for the refinements in questions they have attempted. With continued refinements in both the questions asked and in the sampling designs that are used to address those questions, researchers have the potential to understand far more fully the evolution and perception of animal signals.

Third, some reviewers of my efforts (Kroodsma 1989a,b) are concerned that my suggestions make it impossible to do experimental bioacoustics and therefore difficult to publish such work. Our ultimate goal is not, of course, to publish papers, but to understand the evolution and perception of animal signals. When important advances in understanding are made, publication will naturally follow. To argue (implicitly, of course) that inappropriate experimental designs and inflation of sample sizes are acceptable because doing "better science" would be more difficult confuses, in my opinion, our collective interests in gaining knowledge about animal behavior. My goal is simply to achieve a certain candor about what we have done and about what we can conclude, because I believe such candor is healthy for a growing discipline. If certain controls in an experiment are too difficult to achieve, let the assumptions and qualifications be candidly stated. Perhaps there should be, in addition to the standard Introduction, Methods, Results, and Discussion sections, a final section, entitled Reservations, in which all of one's assumptions and qualifications could be explicitly stated. With a preestablished reservoir for those thoughts, perhaps a reader could assess more easily the relative significance of our achievings.

CONCLUSIONS

For experimental research in which the responses of animals are tested with various types of stimuli, sampling designs are often inappropriate to test intended hypotheses. The primary difficulty is that experimenters use too few stimuli to represent an entire class of stimuli. At its worst, multiple responses by different subjects to only one stimulus are used in statistical tests as if those responses were the mean responses to many different stimuli from the stimulus class. The actual hypothesis tested, however, is about a specific exemplar of a class, not about the classes of stimuli themselves.

I believe that using experimental designs that match our intended hypotheses will boost drastically the confidence that we can have in interpreting and explaining how animals respond to stimuli in their environment. Having an estimate of both the mean and variance in responsiveness to a *population* of stimuli, not just to *one* particular example from the population, is the crucial factor (see Figure 1 and Table 1). When the population of stimuli is not adequately sampled, no confident statement can be made about the class of stimuli from which a particular example was taken. The statistics that are used in such studies generally involve pseudoreplication and generate only a "veneer of rigor" (Hurlbert 1984).

Confidence in our work will be increased, but will our conclusions change? I don't know. For robust phenomena, in which response distributions are widely separated (see Figure 1), our conclusions will most likely not change. For finer discriminations, individual studies using too few stimuli may produce misleading conclusions, thereby confusing both the experimenter and reader. But which are the coarse and which the fine discriminations that we are asking animals to make? In what kind of study must we use many stimuli and in which study can we use fewer stimuli? We cannot be sure until we have some appreciation for the variability in responsiveness to stimuli from a particular class. There is no escape from the importance of understanding variability in stimuli and variability in responsiveness to those stimuli.

ACKNOWLEDGEMENTS

I thank the NSF for financial support (BNS-8812084). I also gratefully acknowledge colleagues who have offered valuable advice: T. Armstrong, M. Bekoff, B. Byers, T. Class, T. Highsmith, D. Jamieson, A. Kamil, I. Pepperberg, D. Spector, and M. Sutherland.

LITERATURE CITED

Adret-Hausberger, M. 1982. Social influences on the whistled songs of starlings. *Behavioral Ecology and Sociobiology* 11, 241-246.

Baker, M.C. 1983. The behavioral response of female Nuttall's White-crowned Sparrows to male song of natal and alien dialects. *Behavioral Ecology and Sociobiology* 12, 309-315.

Baker, M.C., Bjerke, T. K., Lampe, H.U. & Espmark, Y.O. 1987b. Sexual response of female yellowhammers to differences in regional song dialects and repertoire sizes. *Animal Behaviour* 35, 395-401.

Baker, M.C. & Cunningham, M.A. 1985. The biology of bird-song dialects. *The Behavioral and Brain Sciences* 8, 85-133.

Baker, M.C., McGregor, P.K. & Krebs, J.R. 1988. Sexual responses of female great tits to local and distant songs. *Ornis Scandinavica* 18, 186-188.

Baker, M.C., Spitler-Nabors, K.J. & Bradley, D.C. 1981a. Early experience determines song dialect responsiveness of female sparrows. *Science* 214, 819-821.

____. 1982. The response of female mountain White-crowned Sparrows to song from their natal dialect and an alien dialect. *Behavioral Ecology and Sociobiology* 10, 175-179.

Baker, M.C., Spitler-Nabors, K.J., Thompson, A.D., Jr. & Cunningham, M.A. 1987a. Reproductive behaviour of female white-crowned sparrows: effect of dialects and synthetic hybrid songs. *Animal Behaviour* 35, 1766-1774.

Baker, M.C. & Thompson, D.B. 1985. Song dialects of White-crowned Sparrows: historical processes inferred from patterns of geographic variation. *Condor* 87, 127-141.

Baker, M.C., Thompson, D.B. & Sherman, G.L. 1981c. Neighbor/stranger discrimination in white-crowned sparrows. *Condor* 83, 265-267.

Baker, M.C., Thompson, D.B., Sherman, G.L. & Cunningham, M.A. 1981b. The role of male vs. male interactions in maintaining population dialect structure. *Behavioral Ecology and Sociobiology* 8, 65-69.

Baker, M.C., Tomback, D.F., Thompson, D.B., Theimer, T.C. & Bradley, D.C. 1984. Behavioral consequences of song learning: Discrimination of song types by male white-crowned sparrows. *Learning and Motivation* 15, 428-440.

Baptista, L.F. 1975. Song dialects and demes in sedentary populations of the White-crowned Sparrow (*Zonotrichia leucophrys nuttalli*). *University of California, Berkeley, Publications in Zoology* 105, 1-52.

Bertram, B. 1970. The vocal behaviour of the Indian Hill mynah, *Gracula religiosa. Animal Behaviour Monograph* 3, 81-192.

Bjerke, T.K. 1984. The response of male Redwings *Turdus iliacus* to playback of different dialects. *Cinclus* 7, 24-27.

Bock, W.J. & Farrand, J., Jr. 1980. The number of species and genera of recent birds: A contribution to comparative systematics. *American Museum Novitates* No. 2703, 1-29.

Falls, J.B. 1982. Individual recognition by sounds in birds. In: *Acoustic Communication in Birds*, Volume 2 (ed. by D. E. Kroodsma and E. H. Miller), pp. 237-278. New York: Academic Press.

Hansen, P. 1984. Neighbor-stranger song discrimination in territorial yellowhammer *Emberiza citrinella* males, and a comparison with responses to own and alien song dialects. *Ornis Scandinavica* 15, 240-247.

Harris, M.A. & Lemon, R.E. 1974. Songs of song sparrows: Reactions of males to songs of different localities. *Condor* 76, 33-44.

Hurlbert, S.H. 1984. Pseudoreplication and the design of ecological field experiments. *Ecological Monographs* 54, 187-211.

Kreutzer, M. 1974. Réponses comportementales des males *Troglodytes* (Passeriformes) a des chants spécifiques de dialectes différents. *Revue du Comportement Animal* 8, 287-295.

Kroodsma, D.E. 1989a. Suggested experimental designs for song playbacks. *Animal Behaviour* 37, 600-609.

_____. 1989b. Inappropriate experimental designs impede progress in bioacoustic research: A reply. *Animal Behaviour* 37, 717-719.

Kroodsma, D.E. & Baylis, J.R. 1982. Appendix: A world survey of evidence for vocal learning in birds. In: *Acoustic Communication in Birds*, Volume 2 (ed. by D.E. Kroodsma & E.H. Miller), pp. 311-337. New York: Academic Press.

Kroodsma, D.E. & Canady, R.A. 1985. Differences in repertoire size, singing behavior, and associated neuroanatomy among marsh wren populations have a genetic basis. *Auk* 102, 439-446.

Kroodsma, D.E., Meservey, W.R., Whitlock, A.L. & Vanderhaegen, W.M. 1984. Blue-winged Warblers (*Vermivora pinus*) "recognize" dialects in type II but not type I songs. *Behavioral Ecology and Sociobiology* 15, 127-132.

Lemon, R.E. 1967. The response of cardinals to songs of different dialects. *Animal Behaviour* 15, 538-545.

Loehle, C. 1987. Hypothesis testing in ecology: Psychological aspects and the importance of theory maturation. *Quarterly Review of Biology* 62, 397-409.

Machlis, L., Dodd, P.W.D. & Fentress, J.C. 1985. The pooling fallacy: Problems arising when individuals contribute more than one observation to the data set. *Zeitschrift für Tierpsychologie* 68, 201-214.

McGregor, P.K. 1983. The response of corn buntings to playback of dialects. *Zeitschrift für Tierpsychologie* 62, 256-260.

Milligan, M.M. & Verner, J. 1971. Inter-population song dialect discrimination in the White-crowned Sparrow. *Condor* 73, 208-213.

Mundinger, P. 1982. Microgeographic and macrogeographic variation in the acquired vocalizations of birds. In: *Acoustic Communication in Birds*, Volume 2 (ed. by D. E. Kroodsma & E. H. Miller), pp. 147-208. New York: Academic Press.

Pellerin, M. 1983. Variability of response of the corn bunting, *Emberiza calandra*, to songs of different dialects. *Behavioural Processes* 8, 157-164.

Petrinovich, L. & Patterson, T.L. 1981. The responses of White-crowned Sparrows to songs of different dialects and subspecies. *Zeitschrift für Tierpsychologie* 57, 1-14.

Platt, J.R. 1964. Strong inference. *Science* 146, 347-353.

Popper, K.R. 1959. *The Logic of Scientific Discovery*. New York: Basic Books.

Ratcliffe, L.M. & Grant, P.R. 1985. Species recognition in Darwin's finches (*Geospiza*): 3. Male responses to playback of different song types, dialects and heterospecific songs. *Animal Behaviour* 33, 290-307.

Searcy, W.A., Searcy, M.H. & Marler, P. 1982. The response of swamp sparrows to acoustically distinct song types. *Behaviour* 80, 70-83.

Spitler-Nabors, K.J. & Baker, M.C. 1983. Reproductive behavior by a female songbird: differential stimulation by natal and alien song dialects. *Condor* 85, 491-494.

Tomback, D.F., Thompson, D.B. & Baker, M.C. 1983. Dialect discrimination by White-crowned Sparrows: Reactions to near and distant dialects. *Auk* 100, 452-460.

Weary, D.M., Lemon, R.E. & Date, E.M. 1987. Neighbour-stranger discrimination by song in the veery, a species with song repertoires. *Canadian Journal of Zoology* 65, 1206-1209.

10. The Influence of Models on the Interpretation of Vigilance

Steven L. Lima

Foraging animals are usually simultaneously both predator and prey. This simple truism dictates that many animals face a profound conflict between predator avoidance and efficient food intake, because behavioral decisions minimizing the risk of predation are often antithetical to efficient food intake. Such a conflict may exist at several levels of decision-making, from broad-scale habitat selection to diet selection (Lima & Dill 1990). Here, I will focus on a conflict faced by many higher vertebrates that arises from the simple act of ingesting food itself.

The nature of this conflict is illustrated in Figure 1. An animal with its head up and eyes scanning for predators is relatively safe, but it cannot feed. An animal with its head down can feed, but its ability to detect predators is severely compromised because its attention is narrowly focused on food items. Thus an animal faced with this conflict must decide how to trade-off vigilance against feeding in order to maximize its Darwinian fitness (i.e. survival for present purposes).

This conflict and ensuing tradeoff have been the subject of several behavioral studies over the last 20 years (Lima & Dill 1990). My goal here is to critically examine these studies and the conclusions drawn from them; a major theme throughout is that our present interpretation of antipredatory vigilance is perhaps more a function of human intuition than of a critical examination of the processes thought to underlie observed behavior. I will focus on four main areas in the study of antipredatory vigilance: the effect of group size, cooperative vigilance, the scanning process itself, and the role of mathematical modelling. I begin with the effect of group size, the early studies of which set the stage for all subsequent work.

THE EFFECT OF GROUP SIZE

Many animals forage in groups, and group size may have a strong influence on how individuals tradeoff vigilance against

Figure 1. Representative forager in a nonvigilant feeding position (head down) and vigilant, nonfeeding position (head up). (Figure kindly drawn by K.L. Wiebe.)

feeding. The reason is simple. As group size increases, there are more eyes to scan for approaching predators. Thus, each individual in the group can devote less time to vigilance and more time to feeding as group size increases, without seriously affecting the group's ability to detect predators (assuming that predator detection is somehow rapidly transmitted to all group members). This effect was mentioned long ago (Allee 1938) but was not examined in detail until the 1970s. At this point in time, there was much discussion of the costs and benefits of sociality, and several researchers independently converged on the vigilance benefits of sociality. For instance, Powell (1974) showed that a group of starlings (*Sturnus vulgaris*) detected predators sooner than solitary individuals. This result held despite the fact that individuals decreased their vigilance as group size increased. Dimond & Lazarus (1974) independently confirmed the "group size effect" on individual vigilance in flocks of geese. Since this early work, the group size effect has been demonstrated in many studies, mostly on birds (see Lima & Dill 1990). Some representative examples are shown in Figure 2a-c.

Perhaps the most influential work on antipredatory vigilance was a brief model of group predator detection developed by Pulliam (1973). This seminal model provided a simple and elegant statement of the many-eyes hypothesis and served as a focus for the great majority of studies on the group size effect.

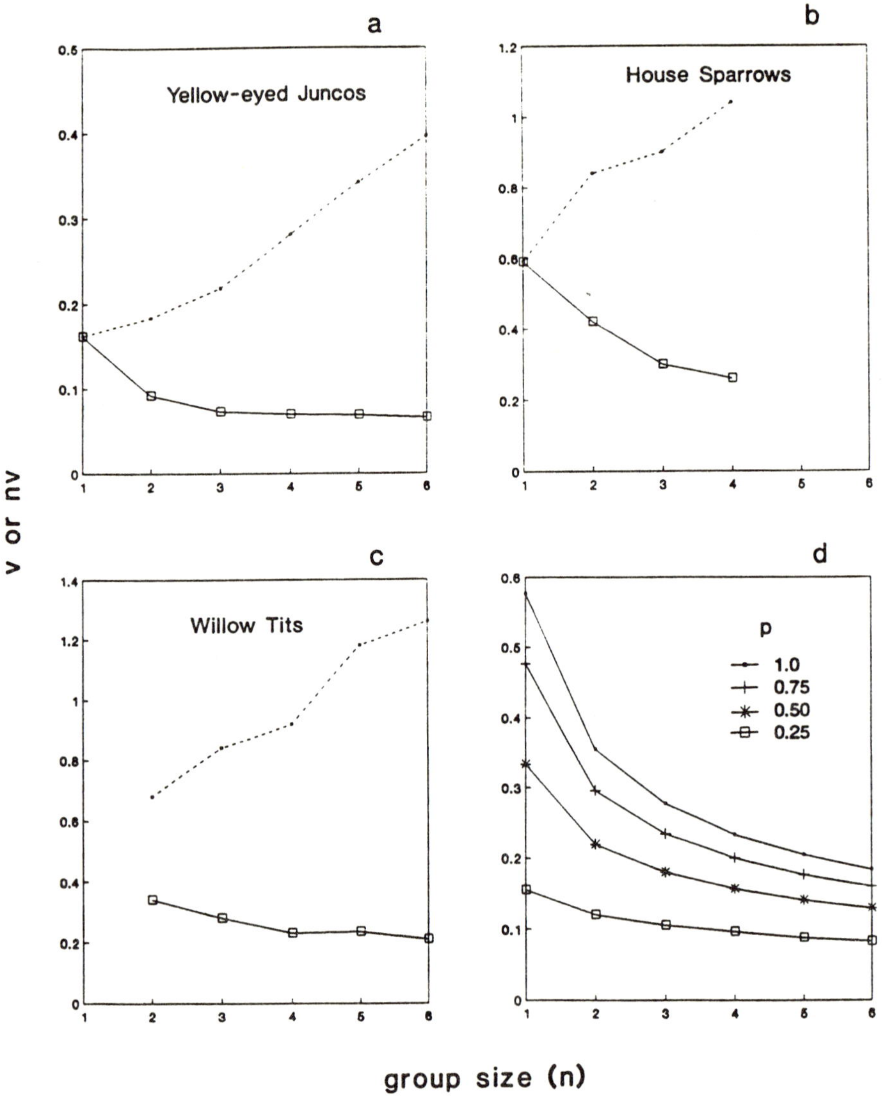

Figure 2. (a) - (c) Representative examples of individual vigilance as a function of group size in yellow-eyed juncos (Pulliam et al. 1982), willow tits (Ekman 1987), and house sparrows (Lima 1987a). Shown are v (——), the average proportion of time spent vigilant by individuals, and "total group vigilance" given by the quantity nv (---). (d) Individual vigilance (v) predicted by Eq. (2) for the indicated probabilities of predatory attack.

For purposes of discussion, I will develop a simple model of vigilance which expresses the essence of the ideas in Pulliam (1973).

A MODEL OF VIGILANCE BEHAVIOR

Assume that vigilance and feeding are mutually exclusive activities. Following Parker & Hammerstein (1985), let V(n) be the probability that at least one member of a group of size n is vigilant when an attack occurs. V(n) is essentially the probability that every group member escapes from the predator, thus 1-V(n) is the probability that the predator is successful. The probability of escape during a successful attack is (n-1)/n, assuming only one death. Thus a given animal's probability of surviving an attack is [1-V(n)](n-1)/n+V(n), which can be rearranged to (n-1)/n + V(n)/n.

Note that V(n) incorporates the many-eyes hypothesis. Following the assumptions in Pulliam (1973), $V(n) = 1-(1-v)^n$ where v is the proportion of time spent vigilant by a given group member. This form of V(n) assumes independent scanning by the n group members. Let p be the probability of the group being attacked by a predator, and let S be an individual's probability of avoiding starvation. A group member thus has the following probability of surviving the time period in question:

$$P(\text{survival}) = (1-p)S + p\{(n-1)/n + V(n)/n\}S. \qquad (1)$$

Assuming that S has the reasonable form of $S = 1-v^2$, and substituting the above relationship for V(n) into (1), we have (after simplification):

$$P(\text{survival}) = [1-p(1-v)^n/n][1-v^2]. \qquad (2)$$

The optimal level of vigilance (v*) is that which maximizes the probability of survival.

Figure 2d shows the relationship between group size (n) and v*. Regardless of the probability of attack (p), v* decreases at a decelerating rate as n increases. When compared to actual vigilance behavior (Figure 2a-c), the heuristic power and appeal of this model is apparent: It readily explains the "classical" response of vigilance to group size. Several variants of this simple model that explicitly consider scanning rates, etc. (e.g.

Pulliam et al. 1982; Lima 1987b), produce similarly striking correspondences between theory and observation.

There have been no major challenges to the basic view of predation and vigilance embodied in Eq. (2), nor is it really my goal here to do so. I will, however, raise some concerns stemming from a simple fact: Models of vigilance behavior have rarely been tested quantitatively. While it is possible to test certain assumptions of the model (e.g. independent scanning in group members), the difficulty in testing the model's numerical predictions stems from the difficulty in measuring many parameters, such as the probability of being attacked, and a host of other parameters (regarding the process of attack and escape, etc.) in more realistic models (e.g. Pulliam et al. 1982; Lima 1987b). With this in mind, I will probe a bit more deeply into the group size effect. A main theme is that simple models such as that developed above are such powerful heuristics that they have "channelled" our interpretation of vigilance away from some potentially important matters.

Simple Models and Reality

Perception of group size. One major reason for the strong appeal of Eq. (2) is its emphasis on group size (n), the parameter most easily measured. However, even the nature of n is not clear. For instance, Metcalfe (1984) and others report that vigilance responds to forager density rather than group size *per se*. Furthermore, Elgar et al. (1984) found that a house sparrow separated from others by as little as 1.2 m (but in visual contact) scans as if it were alone. The bottom line here is that we have remarkably little understanding of the way in which group size is perceived by various animals. More troubling is the possibility that the effects of forager density on vigilance may imply a process fundamentally different from the many-eyes hypothesis in Eq. (2), perhaps one involving antipredatory tactics other than group vigilance.

On the awareness of vigilance. Since it is not clear how various animals perceive n, it may come as no surprise that it is not clear whether social foragers have any regard for the vigilance of their group mates. In fact, I can cite no studies which directly examine this question. Information about the vigilance of group mates is presumably acquired visually in most animals (especially birds). However, Sullivan's (1984) observation that downy woodpeckers' (*Picoides pubescens*) lower vigilance in response to titmice (*Parus* spp.) that they can only hear and not see suggests that these woodpeckers reduce vigilance in situations where they cannot possibly be aware of the vigilance of flock mates.

This point concerning the awareness of vigilance means that there is no solid evidence for the $(1-v)^n$ factor in Eq. (2), which is the crux of the many-eyes hypothesis. Just for fun, let's say that the forager is aware of n but only its own level of vigilance. Therefore the exponent n in Eq. (2) is set equal to 1, and Eq. (2) reduces to

$$P(\text{survival}) = [1-p(1-v)/n][1-v^2]. \qquad (3)$$

The n remaining in (3) represents the dilution of risk with increasing group size (again, assuming one death in a successful attack). Taking the derivative of (3) with respect to v and setting it equal to zero shows that optimal vigilance is given by

$$v^* = [\sqrt{z^2+3y^2} - z]/3y$$

where $y = p/n$ and $z = 1 - y$. As seen in Figure 3, Eq. (3) produces optimal behavior very similar in form to Eq. (2), which incorporates the many-eyes hypothesis. An overall greater level of vigilance is predicted by the many-eyes model, but this difference cannot be exploited empirically unless confidence can be placed in quantitative predictions and the models themselves. In any case, the point of this exercise is clear: the dilution of risk with increasing group size may account for much of the group size effect, even if a forager *is* aware of the vigilance of others (see also Packer & Abrams 1990).

Group size and attack. A further point on interpreting the group size effect is the possibility that the probability of being attacked (p) is group-size-dependent (e.g. Caraco 1979a). For instance, if larger flocks detect predators with greater certainty, then perhaps predators avoid attacking larger groups. Such an effect could add greatly to the decrease in vigilance with group size. There is not much evidence from field work that p is a function of n (but see Lindström 1989), which partly reflects the lack of our ability to determine p itself.

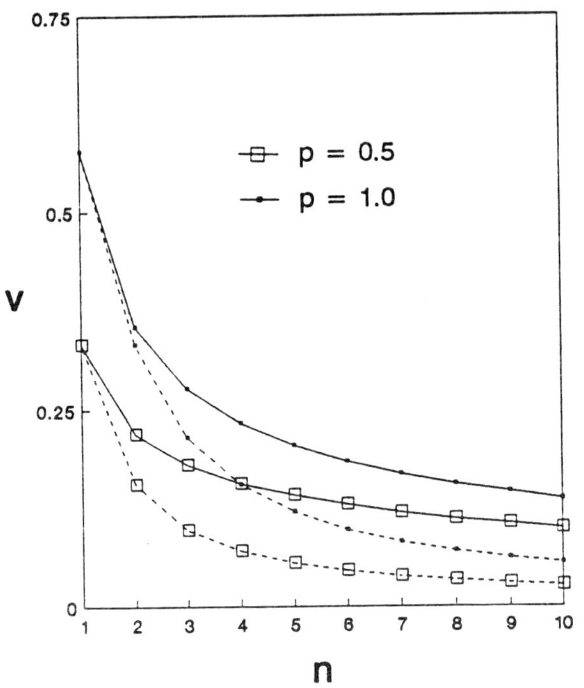

Figure 3. Individual vigilance predicted by Eq. (2) with the many-eyes hypothesis included (——) and excluded (---). These relationships are shown for two values of p.

<u>Competition for limited food resources</u>. Virtually all models of vigilance implicitly assume unlimited food resources. This essentially ensures that the group size parameter n appears only in the many-eyes factor (e.g. $(1\text{-}v)^n$ in Eq. [2]) and the dilution factor (e.g. 1/n in Eq. [2]) of a given model. With limited food resources, however, n may strongly influence vigilance not only by decreasing the risk of predation, but also by increasing competition for food.

One way to approach this "competition effect" is to view antipredatory vigilance as a component of food handling time. If n animals are competing for a limited amount of food, then there is a premium on consuming food as quickly as possible (c.f. Barnard et al. 1983). One obvious way to achieve this end is to decrease vigilance. Since competition will increase with increasing group size, it may account for a major portion of the group size effect traditionally ascribed to antipredatory effects.

Very few studies have seriously considered competition as a possible factor in vigilance (but see Bertram 1980). Furthermore, no studies to date have examined the problem directly through experimentation. Thus, it is currently impossible to assess the extent to which the group size effect is dominated by competition for food, although many studies effectively side-stepped the problem by eliminating (short-term) competition through the use of superabundant food (e.g. Lima 1987a). Surprisingly, studies suggesting that competition is not necessarily an overriding factor in the group size effect deal with sleeping animals; the observation that animals interrupt their sleep (to scan their environment) less frequently with increasing group size (Lendrem 1984b; da Silva & Terhune 1988) cannot easily be construed as evidence of competition for sleep.

Object of vigilance. So far, I have accepted the premise that vigilance is antipredatory in nature. What is the evidence for such a premise? A few studies have demonstrated that recent sightings of predators, etc., actually lead to higher vigilance in birds (Caraco et al. 1980a; Lendrem 1984a; Sullivan 1984; Glück 1987). Predation is also strongly implicated by observations that foragers adjust vigilance in response to the distance to safe refuge or the presence of visual obstructions (see Lima 1987b). However, it is probably fair to say that the group size effect itself is the main support for this predation premise. Clearly, the logical problems not withstanding, a forager with its head down (e.g. Figure 1) has compromised not only its ability to detect predators, but also its ability to detect a host of lesser threats (Dimond & Lazarus 1974). How might such considerations alter our interpretation of the ubiquitous group size effect?

A few recent studies have begun to address the possibility of nonantipredatory vigilance. For instance, Thompson & Lendrem (1985; see also Barnard & Thompson 1985) found that vigilance in certain shorebirds may be directed towards food-robbing gulls. A few studies also suggest that vigilance may be directed toward other, potentially aggressive group members in social birds (Waite 1987a,b; Withiam et al. ms; Knight & Knight 1986) and mammals (Caine & Marra 1988; Roberts 1988).

The extent to which our present interpretation of the group size effect should be influenced by nonpredatory factors is not clear. Of particular concern is the role of intragroup aggression, which can be considerable (e.g. Caraco 1979b). Overall, however, it is unlikely that antiaggression vigilance is of overriding importance in the group size effect for two main reasons. First, experimental studies (Waite 1987a,b; Withiam et al. 1990) induced unusually high rates of aggression by forcing birds to feed in highly constrained situations uncharacteristic of most natural situations. Second, since aggression often increases with group size (e.g. Caraco 1979b), one might generally expect vigilance to increase with group size (Bertram 1980). However, only Knight & Knight (1986) report such an effect.

In summary, while the foregoing discussion is largely critical of studies on vigilance and the group size effect, my main point is really that the interpretation of vigilance in social animals may reflect more the elegance of simple heuristic models than experimentation examining key behavioral postulates. I suspect, however, that the simple many-eyes hypothesis embodied in Eq. (2) will survive as an important heuristic tool even in the face of more critical experimentation. I can cite two main reasons for this optimism. First, an increase in predation risk has demonstrable effects on vigilance (see above and Lima 1987b). Second, larger groups do in fact detect predators sooner than smaller ones (Powell 1974; Kenward 1978; Lazarus 1979), and I find it difficult to believe that natural selection has not molded animals to take this into account and be less vigilant with increasing group size.

COOPERATION IN VIGILANCE

All models of social vigilance assume (implicitly or explicitly) that natural selection has molded animals to take into account the vigilance of other group members when deciding upon their own level of vigilance. This assumption, however, actually presents a major problem for the vigilance model developed above. The reason is basic to the evolutionary process: natural selection should produce inherently selfish animals. In terms of vigilance, what will stop a selfish individual from "parasitizing" the vigilance of its group mates by decreasing its vigilance (and thereby increasing its time spent feeding)? In the absence of human conventions such as legally-binding contracts between individuals, nothing will stop attempts at "cheating". Because the simple model developed above does not consider the possibility of such cheating, it implicitly assumes a level of intragroup cooperation that may not be evolutionarily stable to cheating (i.e. cheaters should enjoy a relative advantage over cooperators and thus be favored by natural selection). A game-theoretical modelling approach is necessary to determine evolutionarily stable optimal behavior in such situations (see Maynard Smith 1982; Parker 1984).

Pulliam et al. (1982) recognized the possibility of cheating as a major problem for the many-eyes hypothesis, and developed a game-theoretical model of vigilance behavior to compare and contrast selfish and cooperative vigilance. Their model, based on scanning rates as per Pulliam (1973), showed that scanning rates in selfish groups are lower than those in cooperative groups. Furthermore, they found that vigilance observed in socially foraging yellow-eyed juncos (*Junco phaeonotus*) was close to that predicted for cooperative groups (see Table 1). To explain the existence of this presumably unstable vigilance pattern, Pulliam et al. tentatively proposed that juncos employ a "judge" strategy, where each flock member remains cooperative as long as others do so. This strategy is functionally equivalent to the Tit-for-Tat strategy proposed for stable cooperation in other behavioral contexts (Axelrod & Hamilton 1981).

TABLE 1. Observed Scanning Rates of Yellow-eyed Juncos vs. Predicted Cooperative and Selfish Scanning Optima (from Pulliam et al. 1982)

	Scanning Rate (per min.)		
Flock Size	Observed	Cooperative	Selfish
1	13.9	15.9	15.9
2	7.85	6.2	0.0
3	6.22	5.9	0.0
4	6.02	5.5	0.0
5	5.87	5.2	0.0
6	5.66	4.9	0.0
7	5.58	4.7	0.0
8	5.59	4.5	0.0
9	4.88	4.4	0.0
10	4.65	4.0	0.0

Further theoretical work by Parker & Hammerstein (1985) confirmed these basic results on social vigilance using Eq. (4), which is essentially a game-theoretical version of Eq. (2).

$$P(\text{survival of } v \text{ in group using } \hat{v}) = [1-p(1-v)(1-\hat{v})^{n-1}/n][1-v^2] \quad (4)$$

Here, v represents the vigilance of a potential cheater, and $\hat{v}$ is the (evolutionarily stable) vigilance of the remaining n-1 individuals. The selfish optimum is determined by taking the derivative of (4) with respect to v, setting it equal to zero, and solving for v after setting $\hat{v} = v$; setting $\hat{v} = v$ in (4) before performing this mathematical procedure yields the cooperative optimum. Assuming $p = 1$, Parker & Hammerstein also found that selfish groups are less vigilant than cooperative ones (Figure 4). In addition, because vigilance decreases with group size regardless of selfish or cooperative strategies, Parker & Hammerstein proposed that the two might be distinguished by the quantity nv. As seen in Figure 4, nv increases with group size for cooperative groups, and decreases for selfish groups. A survey of empirical results often suggests cooperation in vigilance; some real-world examples of increasing nv are illustrated in Figure 2a-c.

These results are remarkable. They suggest widespread cooperative vigilance among largely nonrelated animals (at least in birds), even though such cooperation has been difficult to demonstrate elsewhere (see Packer 1986).

Before accepting cooperation as fact, however, I must stress some important points concerning the interpretation of vigilance. First, all of the above-mentioned problems in interpreting the basic group size effect apply to cooperation. In particular, the complete lack of direct evidence that social animals have any regard for the vigilance of others (beyond the group size effect itself) leaves the "judge" strategy in question.

Furthermore, consider the implications of previous work concerning the perception of group size (see above). This work suggests that animals perceive only a fraction of the n animals in a group. For simplicity, assume that the perceived group size (m) is given by $m = 1+k(n-1)$ where k indicates the fraction of the group perceived by the animal in question; $m = n$ if $k = 1$. Substituting m for n in Eq. (4), and solving for selfish optima,

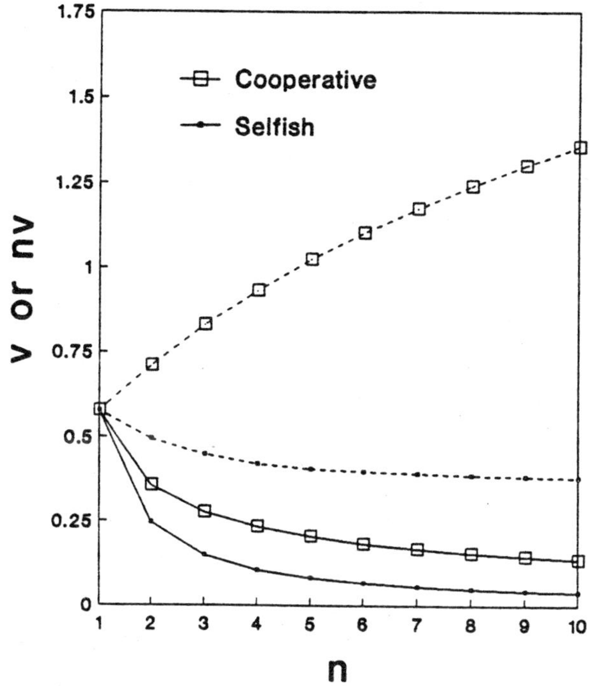

Figure 4. Selfish and cooperative vigilance levels predicted by Eq. (4). Individual vigilance v (——) and nv (---) are shown. The parameter p is set at 1.0.

significantly alters the above results regarding cooperation. As seen in Figure 5a, the basic group size effect holds over several values of k. However, as k decreases, the quantity nv may increase with n (but not m), even though these are selfish optima (Figure 5b). Does this explain the real-world cases of increasing nv (Figure 2a-c)? It is presently impossible to say given our lack of understanding of the perception of group size.

Recent theoretical work shows further how conclusions regarding cooperation depend critically on the way in which the overall predator-prey interaction is modelled. For instance, Packer & Abrams (1990) point out that Parker & Hammerstein's model (the "PH model") applies only when a successful predator kills randomly within the group of n foragers. By realistically relaxing several implicit assumptions in the PH model, Packer &

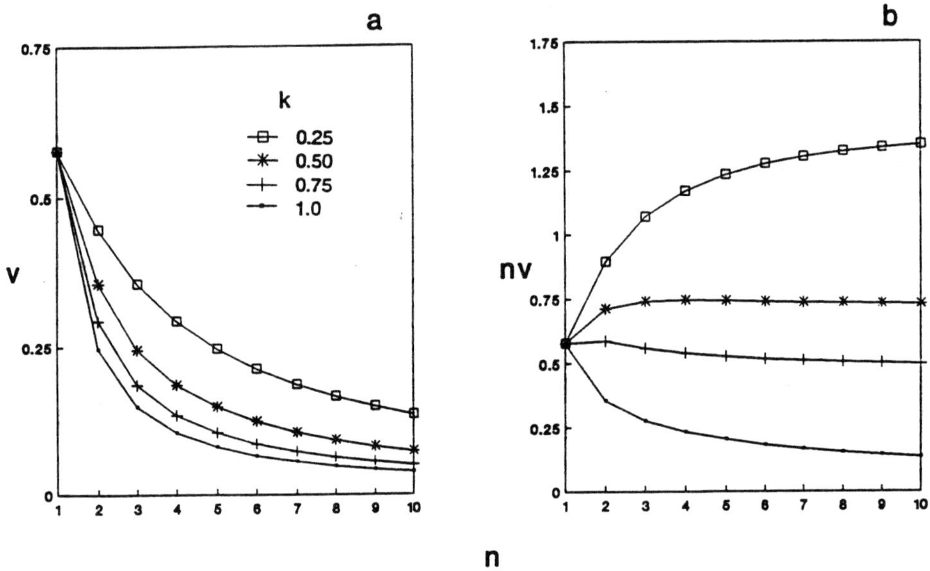

Figure 5. Selfish vigilance and the "incomplete" perception of group size. (a) Individual vigilance as a function of actual group size (n) for the indicated values of k. (b) The quantity nv for the values of k as per (a).

Abrams identified several situations where cooperative groups should be *less* vigilant than selfish groups. For instance, this result may hold if predators preferentially attack the least vigilant members of foraging groups, as suggested by FitzGibbon (1989). Selfish vigilance may also exceed cooperative vigilance if the chances of escape depend upon the vigilance of other group members at the time of attack. Such an effect of vigilance on escape is suggested by Elgar (1986), who found that nonvigilant house sparrows react more slowly to attack than vigilant sparrows.

There are other reasons for expecting similarities in selfish and cooperative behavior. All of the models of selfish/cooperative vigilance mentioned above assume that the group is concerned with surviving only one attack or one short period of time (e.g. one day). Clearly, most animals will be concerned with multiple attacks or longer periods of time. I have shown (Lima 1989) that apparent cooperation may be evolutionary stable in such situations; essentially, cheating only leads to the death of groupmates, who may be essential to long-term survival. Kaitala et al. (1989) independently came to a similar conclusion in re-analyzing Pulliam et al.'s (1982) model. In particular, they found that the very low levels of selfish vigilance predicted by Pulliam et al. (see Table 1) are not to be expected when considering behavior over the long-term.

Overall, despite a promising start based upon simple, heuristically elegant models, I must conclude that there is little unambiguous empirical support for cooperative vigilance in social foragers. In fact, the many "selfish" ways of achieving apparent cooperation may make selfish and cooperative strategies virtually impossible to distinguish empirically without great confidence in quantitative tests of various models. Perhaps future studies should focus on simple, well-defined situations where unambiguous qualitative tests may be possible.

I conclude this section by noting that the study of cooperation may have been side-tracked somewhat by the historical development of the study of vigilance. The assumption of many-eyes independently vigilant for predators has been carried through all empirical and theoretical studies since Pulliam (1973) and earlier, including those concerned with cooperative vigilance. However, perhaps truly cooperative vigilance involves sentinels; i.e. highly vigilant, nonfeeding individuals which stand guard over the remaining n-1 feeding individuals (e.g. Rasa 1987). It is easily shown that n group members alternately sharing 1/n of the total vigilance burden achieve a higher level of survival than n independently scanning individuals (Lima, in prep.) In fact, independent scanning may itself represent a lack of cooperation. The obvious problem with cheating in sentinel vigilance is that there is no guarantee that other group members will take their watch after a given sentinel wishes to feed. It is perhaps no coincidence that sentinels are often observed in stable, family groups (e.g. Ferguson 1987; Rasa 1987) which are presumably favorable situations for the evolution of cooperation (Hamilton 1964; Trivers 1971).

PATTERNS IN INDIVIDUAL VIGILANCE

The process of vigilance involves a repeating sequence of head lifts, scans of the environment, and nonvigilant inter-scan intervals where actual feeding takes place. As with other subjects, the early models of vigilance played a major role in shaping the study of patterns in the vigilance of individuals.

Scan Initiation

Pulliam (1973) assumed that individual foragers initiate scans of constant length according to an exponential probability distribution re^{-rt}, where r is the overall rate of scan initiation, and t is the time since terminating the last scan. This distribution implies that a scan is initiated with constant probability in any small time interval, regardless of t. Hence, the logical basis for this "exponential" assumption is that a predator cannot easily time an attack against such a forager.

This assumption is intuitively reasonable, but there are overriding mathematical needs for it as well. First, it implies that the forager has control over a single variable r, which completely specifies the probability distribution. The exponential assumption also yields very simple probabilities concerning predator detection. For instance, let "a" represent the time a forager has to detect an attack once the predator is no longer concealed by vegetation, etc. The probability that the forager fails to detect the attack is therefore e^{-ra}. With n independently scanning individuals, this probability is e^{-nra}. Most other probability distributions would yield much more "messy" probabilities (c.f. Hart & Lendrem 1984) that may involve more than one parameter under a forager's control; such multi-parameter optimization models are mathematically more difficult.

This exponential assumption has evoked several tests, all with birds. Bertram (1980), Caraco (1982), and Studd et al. (1983) showed that the lengths of interscan intervals fit nicely to exponential distributions. Lendrem (1983) found interscan intervals in blue tits (*Parus caeruelus*) became more exponentially distributed as the risk of predation increased. The exponential assumption was largely unsupported by Elcavage & Caraco (1983), Sullivan (1985), and Pöysä (1987); Hart & Lendrem (1984) re-analyzed Bertram's (1980) data in detail and also found a lack of support. Lendrem et al. (1986) suggested that a slightly more complicated exponential distribution, where r is a time-dependent function, can help explain some of these discrepancies. More recently, Desportes et al. (1989) used spectral analysis to suggest that scanning patterns are random in appearance, but underlaid by regular patterns nonetheless.

These studies represent one of the few areas where the behavioral processes underlying vigilance have been examined critically. They also, however, demonstrate the power of the simple many-eyes hypothesis to channel research into perhaps overly-narrow areas of discourse. With respect to understanding patterns in vigilance behavior, perhaps a more fruitful approach would also have addressed the simple question of what pattern should we expect to observe. In this regard, Desportes et al. (1989) correctly suggest that when predators attack without regard to prey behavior, regular (i.e. constant) interscan intervals are superior to exponential scanning in detecting predators. I suggest that even if predators time their attacks relative to prey vigilance (c.f. Hart & Lendrem 1984), an exponential pattern might not be expected, particularly if the parameter "a" is known with certainty. Overall, because so much work has focused on a mathematically convenient assumption, relatively little is understood about this aspect of the scanning process.

Scan Length

Two further assumptions in simple scanning-rate models of vigilance (e.g. Lima 1987b) are worthy of mention: (i) scan length, s, is appreciable and constant; and (ii) predator detection is instantaneous. These assumptions are, in fact, somewhat contradictory, for why would scans be of appreciable length if predator detection is instantaneous? It turns out that assumption (i) is necessary because assumption (ii) dictates that the optimal scanning rate approaches infinity as s becomes smaller. Clearly, a forager has control over s, and the fact that scans are appreciable in length probably means that there are differing consequences for one s over another.

Relatively little work has examined scan lengths, perhaps because the above two assumptions seem rather innocuous. The assumption of constant s has some support (Pulliam et al. 1982), while other studies show that s decreases markedly with group size (Metcalfe 1984). Glück (1987) found that s increases in solitary birds which have recently sighted a predator, while Pöysä (1987) found foraging constraints may affect scan length. McVean & Haddlesey (1980) and Elcavage & Caraco (1983) provide evidence for variable and constant s, respectively, in the same species (house sparrows).

The adaptive value of altering vigilance via scan length (in addition to scanning rate) is not clear. It seems reasonable that longer scans yield more information about the immediate risk of predation. Perhaps this is why solitary individuals may exhibit longer scans than grouped foragers (Metcalfe 1984). However, longer scans may actually prevent attack (e.g. FitzGibbon 1989). Furthermore, does a decrease in scan length with increasing group size mean each individual requires less information, or does it reflect competition for food?

I must once again caution that in interpreting the fine-scale patterns in vigilance, care should be taken not to equate vigilance solely with predator detection (which is the strong tendency). A particular problem to avoid is the violation of the assumption of mutually exclusive food ingestion and vigilance; in some cases, scan lengths may strongly reflect food handling times if scanning and handling are not mutually exclusive (e.g. Lima 1988). In other cases, scanning may be influenced by multiple objects of vigilance (e.g. predators, aggressors, food-robbers, etc.).

MODELS AND THE STUDY OF VIGILANCE BEHAVIOR

I have argued that simple heuristic models have played a major role in the study of vigilance behavior. At the risk of promoting the abandonment of such modeling, I have focused mainly on its more negative effects on the interpretation of vigilance behavior. In this section, however, I wish to elaborate more on the role of modeling, and suggest that it has played, and will continue to play, an important and positive role.

It is something of a theoretician's cliché to state that models act to clarify the important issues at hand, and thus stimulate meaningful research. However, I believe this is precisely what the early models have done. Pulliam (1973), in particular, stimulated a great deal of relatively well-focused research that has identified a genuine phenomenon (the group size effect) in many taxa (Lima & Dill 1990). Perhaps a more important need for models relates to a simple fact: Human intuition has not always been the best guide to interpreting vigilance.

Consider first the idea of cheating and cooperation in vigilance. It is highly intuitive that selfish animals should be less vigilant than cooperative ones (Pulliam et al. 1982; Parker & Hammerstein 1985). However, modeling that exposed the hidden assumptions in the early models (Packer & Abrams, 1990; Kaitala et al. 1989) showed that intuition is less than adequate in many reasonable situations (see above). This theoretical discourse on the problem of cheating/cooperation in vigilance, at the very least, should give the empiricist reason to tread carefully in this matter.

The same may be said for intuition and the influence of environmental factors on vigilance. For instance, sparrows may be more vigilant the farther they must feed from the safety of dense brushy cover (e.g. Barnard 1980). This result is intuitively clear: the increase in risk with increasing distance to the refuge leads to an increase in vigilance. However, a simple model (Lima 1987b) indicates that the opposite result is also a reasonable expectation (see also Lima 1987a). This model makes clear two nonintuitive main points regarding the interpretation of vigilance: (i) that vigilance need not always increase with increasing risk, and (ii) a behavioral response to one factor may actually reflect a response to one or more correlated factors (e.g. an increase in the likelihood of attack with increasing distance to cover).

Overall, I believe that models will be essential to the proper interpretation of vigilance behavior. However, I again must stress a few points concerning the use of such models. First, most models are quantitatively untestable given our current inability to measure many parameters. Second, little is known about the perceptions of the animals being studied, thus our models reflect mainly the perceptions of the modelers themselves. Finally, all models are gross caricatures of reality. For instance, consider Eq. (2). There is no explicit consideration of scan lengths or scanning rates, nor is there any mention of the relationship between the latter and time exposed to attack. There is no consideration of perceived group size, competition for food, aggression, or the fact that groups size is rarely constant over time. Thus, the quantitative predictions derived from such models may be meaningless and even misleading. Models of vigilance behavior are merely heuristic devices to be used as guides to research. As such, their use should be accompanied by a healthy dose of skepticism.

CONCLUSION

A great many studies have established the generality of the group size effect in the vigilance of several taxa (see Lima & Dill 1990). In fact, the many-eyes hypothesis interpretation of the group size effect may be approaching the status of dogma. However, I have argued here that our current interpretation of this effect is related more to the power of simple models to mimic observed behavior rather than experimentation examining critical behavioral processes. The same can be said for other aspects of vigilance behavior.

Future empirical and theoretical work must be more focused and critical if further progress is to be made in the study of vigilance behavior; in my view, the halcyon "Golden Age" is over. I have outlined several topics in need of further research, many of which can be approached experimentally. In pursuing these and other topics, researchers should take care to: (i) allow for nonpredatory interpretations of vigilance; (ii) explicitly consider the perceptions of the animals themselves; (iii) stay within the realm of ecological reason; and (iv) understand the influence, importance, and limitations of models in the study of vigilance behavior.

LITERATURE CITED

Allee, W.C. 1938. *The Social Life of Animals.* New York: Norton and Company.

Axelrod, R. & Hamilton, W.D. 1981. The evolution of cooperation. *Science* 211, 1390-1396.

Barnard, C.J. 1980. Flock feeding and time budgets in the house sparrow (*Passer domesticus* L.). *Animal Behaviour* 28, 295-309.

Barnard, C.J. & Thompson, D.B.A. 1985. *Gulls and Plovers.* New York: Columbia University Press.

Barnard, C.J., Brown, C.A.J. & Gray-Wallis, J. 1983. Time and energy budgets and competition in the common shrew (*Sorex araneus* L.). *Behavioral Ecology and Sociobiology* 13, 13-18.

Bertram, B.C.R. 1980. Vigilance and group size in ostriches. *Animal Behaviour* 28, 278-286.

Caine, N.G. & Marra, S.L. 1988. Vigilance and social organization in two species of primates. *Animal Behaviour* 36, 897-904.

Caraco, T. 1979a. Time budgeting and group size: a theory. *Ecology* 60, 611-617.

_____. 1979b. Time budgeting and group size: a test of theory. *Ecology* 60, 618-627.

_____. 1982. Flock size and the organization of behavioral sequences in juncos. *Condor* 84, 101-105.

Caraco, T., Martindale, S. & Pulliam, H.R. 1980. Avian time budgets and distance to cover. *Auk* 97, 872-875.

_____. 1982. Avian flocking in the presence of a predator. *Nature* 285, 400-401.

da Silva, J. & Terhune, J.M. 1988. Harbour seal grouping as an antipredator strategy. *Animal Behaviour* 36, 1309-1316.

Desportes, J.-P., Metcalfe, N.B., Cezilly, F., Lauvergeon, G. & Kervella, C. 1989. Tests on the sequential randomness of vigilant behaviour using spectral analysis. *Animal Behaviour* 38, 771-777.

Dimond, S. & Lazarus, J. 1974. The problem of vigilance in animal life. *Brain, Behavior, and Evolution* 9, 60-79.

Ekman, J. 1987. Exposure and time use in willow tit flocks: The cost of subordination. *Animal Behaviour* 35, 445-452.

Elcavage, P. & Caraco, T. 1983. Vigilance behaviour in house sparrow flocks. *Animal Behaviour* 31, 303-304.

Elgar, M.A. 1986. Scanning, pecking, and alarm fights in house sparrows. *Animal Behaviour* 34, 1892-1894.

Elgar, M.A., Burren, P.J. & Posen, M. 1984. Vigilance and perception of flock size in foraging house sparrows (*Passer domesticus* L.). *Behaviour* 90, 215-223.

Ferguson, J.W.H. 1987. Vigilance behavior in white-browed sparrow-weavers, *Plocepasser mahali*. *Ethology* 76, 223-235.

FitzGibbon, C.D. 1989. A cost to individuals with reduced vigilance in groups of Thomson's gazelles hunted by cheetahs. *Animal Behaviour* 37, 508-510.

Glück, E. 1987. An experimental study of feeding, vigilance, and predator avoidance in a single bird. *Oecologia* 71, 268-272.

Hamilton, W.D. 1964. The genetical evolution of social behavior. *Journal of Theoretical Biology* 7, 1-52.

Hart, A. & Lendrem, D.W. 1984. Vigilance and scanning patterns in birds. *Animal Behaviour* 32, 1216-1224.

Kaitala, V., Lindström, K. & Ranta, E. 1989. Foraging, vigilance, and risk of predation in birds - a dynamic game study of ESS. *Journal of Theoretical Biology* 138, 329-345.

Kenward, R.E. 1978. Hawks and doves: Attack success and selection in goshawk flights at wood-pigeons. *Journal of Animal Ecology* 47, 449-460.

Knight, S.K. & Knight, R.L. 1986. Vigilance patterns in bald eagles feeding in groups. *Auk* 103, 263-272.

Lazarus, J. 1979. The early warning function of flocking in birds: An experimental study with captive quela. *Animal Behaviour* 27, 855-865.

Lendrem, D.W. 1983. Predation risk and vigilance in the blue tit (*Parus caeruleus*). *Behavioral Ecology and Sociobiology* 14, 9-13.

_____. 1984a. Flocking, feeding and predation risk: Absolute and instantaneous feeding rates. *Animal Behaviour* 32, 298-299.

_____. 1984b. Sleeping and vigilance in birds. II. An experimental study of the Barbary dove (*Streptopelia risoria*). *Animal Behaviour* 32, 243-248.

Lendrem, D.W., Stretch, D., Metcalfe, N. & Jones, P. 1986. Scanning for predators in the purple sandpiper: A time-dependent or time-independent process? *Animal Behaviour* 34, 1577-1578.

Lima, S.L. 1987a. Distance to cover, visual obstructions, and vigilance in house sparrows. *Behaviour* 102, 231-238.

_____. 1987b. Vigilance while feeding and its relation to the risk of predation. *Journal of Theoretical Biology* 124, 303-316.

_____. 1988. Vigilance and diet selection: a simple example in the dark-eyed junco. *Canadian Journal of Zoology* 66, 593-596.

_____. 1989. Iterated prisoner's dilemma: An approach to evolutionarily stable cooperation. *American Naturalist* 134, 828-838.

Lima, S.L. & Dill, L.M. 1990. Behavioral decisions made under the risk of predation: A review and prospectus. *Canadian Journal of Zoology* 68.

Lindström, A. 1989. Finch flock size and risk of predation at a migratory stopover site. *Auk* 106, 225-232.

Maynard Smith, J. 1982. *Evolution and the Theory of Games.* New York: Cambridge University Press.

McVean, A. & Haddlesey, P. 1980. Vigilance schedules among house sparrows *Passer domesticus*. *Ibis* 122, 533-536.

Metcalfe, N.B. 1984. The effects of mixed-species flocking on the vigilance of shorebirds. Who do they trust? *Animal Behaviour* 32, 986-993.

Packer, C. 1986. Whatever happened to reciprocal altruism? *Trends in Ecology and Evolution* 1, 142-143.

Packer, C. & Abrams, P. 1990. Should cooperative groups be more vigilant than selfish groups? *Journal of Theoretical Biology*, in press.

Parker, G.A. 1984. Evolutionarily stable strategies. In: *Behavioral Ecology: An Evolutionary Approach,* 2nd Edition. (ed. by J.R. Krebs & N.B. Davies), pp. 30-61. London: Blackwell.

Parker, G.A. & Hammerstein, P. 1985. Game theory and animal behaviour. In: *Evolution: Essays in Honour of John Maynard*

Smith. (ed. by P.J. Greenwood, P.H. Harvey, & M. Slatkin), pp. 73-94. New York: Cambridge University Press.

Powell, G.V.N. 1974. Experimental analysis of the social value of flocking by starlings (*Sturnus vulgaris*) in relation to predation and foraging. *Animal Behaviour* 22, 501-505.

Pöysä, H. 1987. Feeding-vigilance trade-off in the teal (*Anas crecca*): Effects of feeding method and predation risk. *Behaviour* 103, 108-122.

Pulliam, H.R. 1973. On the advantages of flocking. *Journal of Theoretical Biology* 38, 419-422.

Pulliam, H.R., Pyke, G.H. & Caraco, T. 1982. The scanning behavior of juncos: a game-theoretical approach. *Journal of Theoretical Biology* 95, 89-103.

Rasa, O.A.E. 1987. Vigilance behavior in dwarf mongooses: selfish or altruistic? *South African Journal of Science* 83, 587-590.

Roberts, S.C. 1988. Social influences on vigilance in rabbits. *Animal Behaviour* 36, 905-913.

Studd, M., Montgomerie, R.D., & Robertson, R.J. 1983. Group size and predator surveillance in foraging house sparrows (*Passer domesticus*). *Canadian Journal of Zoology* 61, 226-231.

Sullivan, K.A. 1984. Information exploitation by downy woodpeckers in mixed-species flocks. *Behaviour* 91, 294-311.

_____. 1985. Vigilance patterns in downy woodpeckers. *Animal Behaviour* 33, 328-330.

Thompson, D.B.A. & Lendrem, D.W. 1985. Gulls and plovers: Host vigilance, kleptoparasite success and a model of kleptoparasite detection. *Animal Behaviour* 33, 1318-1324.

Trivers, R.L. 1971. The evolution of reciprocal altruism. *Quarterly Review of Biology* 46, 35-57.

Waite, T.A. 1987a. Dominance-specific vigilance in the tufted titmouse: Effects of social context. *Condor* 89, 932-935.

_____. 1987b. Vigilance in the white-breasted nuthatch: Effects of dominance and sociality. *Auk* 104, 429-434.

Withiam, M.L., Lemmon D. & Barkan, C.P.L. 1990. Scanning and social dominance in black-capped chickadees (*Parus atricapillus*). Unpublished manuscript.

11. Levels of Analysis and the Functional Significance of Helping Behavior

Walter D. Koenig and Ronald L. Mumme

INTRODUCTION

The field of behavioral ecology includes a diverse range of problems having to do with the developmental, physiological, genetical, and ecological bases of behavior. Partly as a result of this diversity, it is not unusual for controversies to arise from a failure to place hypotheses within their appropriate context. When this happens, advances can sometimes best be made by combining a critical evaluation of the available data with a more philosophical analysis of the questions involved.

In this chapter we pursue such an approach to the question of the evolution of helping behavior in cooperative (or communal) breeders: species in which individuals other than a male-female pair cooperate in rearing offspring. Because it involves apparently altruistic behaviors in the absence of parent-offspring relatedness, cooperative breeding is of interest both to behavioral ecologists and evolutionary biologists. Recent reviews include Brown (1987) and Stacey & Koenig (1990) for birds, Gittleman (1985) for mammals, Emlen (1984) for birds and mammals, and Taborsky & Limberger (1981) for fishes.

Much of the literature on cooperative breeding is devoted to the search for adaptive explanations of helping behavior. Recently, however, Jamieson (1986, 1989a) and Jamieson & Craig (1987) have presented a critique of functional explanations of helping behavior in birds, suggesting instead that this phenomenon might be an unselected consequence of the delayed dispersal and group-living that characterizes cooperative breeders. Our goal here is to provide a reconsideration of this issue. First, we define what helpers are and the kinds of aid they appear to offer. Second, we discuss levels of analysis (Sherman 1988, 1989; Jamieson 1989b) as it applies to the question of why some individuals care for offspring that are not their own. Third, we briefly clarify how Jamieson & Craig's (1987) alternative explanation for the evolution of helping behavior combines different levels of analysis, and therefore obfuscates the issue of whether helping behavior has current adaptive utility. Finally, we focus on the kinds of data relevant to hypotheses that helping behavior is selectively advantageous and assess the current status of our ability to reject the alternative hypothesis that it is not. Parallel papers by Ligon & Stacey (1989) and Emlen et al. (1990) discuss some of these same issues and reach conclusions similar to ours.

WHAT ARE HELPERS AND WHAT DO THEY DO?

A helper, as originally defined by Skutch (1961: 198), is an individual that "assists in the nesting of an individual other than its mate, or feeds or otherwise attends a bird of whatever age which is neither its mate nor its dependent offspring." A more recent definition by Brown (1987: 300-301) is similar: "An individual that performs parent-like behavior toward young that are not genetically its own offspring...Helpers may be altruistic, cooperative, or selfish. Note that breeding status and conferral of benefit or harm to recipient or helper are irrelevant to the definition."

Provisioning of dependent young is the most conspicuous form of helping behavior, but helpers may also perform a variety of other alloparental duties, including building the nest (e.g. white-winged choughs *Corcorax melanorhamphos*; Rowley 1978), incubating eggs and brooding young (e.g. acorn woodpeckers *Melanerpes formicivorus*; Koenig et al. 1983), defending young against predators (e.g. bicolored wrens *Campylorhynchus griseus*; Austad & Rabenold 1985), and cleaning or grooming young (e.g. *Lamprologus brichardi*; Taborsky 1984).

Although Brown's (1987) definition indicates that helpers can provide alloparental care without necessarily conveying fitness benefits to recipients, it is equally important to note that the converse is also true: helpers can convey important fitness benefits to recipients in ways that are not directly attributable to the alloparental care they provide. For example, in addition to caring for nondescendant young, helpers in a number of species also contribute to the defense of group territories (e.g. Hunter 1985; Mumme & de Queiroz 1985), which can reduce the amount of time breeders must spend in territorial defense (e.g. Kinnaird

& Grant 1982; Taborsky 1984). Helpers can also improve anti-predator defenses in ways unrelated to alloparental care. Family groups of Florida scrub jays (*Aphelocoma coerulescens*), which typically contain one or more nonbreeding helpers, have a coordinated sentinel system that serves primarily to detect aerial predators during the fall and winter (McGowan & Woolfenden 1989). The improved sentinel performance in larger groups may be responsible for the higher annual survivorship observed among breeders assisted by helpers in this species (Woolfenden & Fitzpatrick 1984).

Similarly, the presence of nonbreeders in the acorn woodpecker significantly increases annual survivorship of male breeders, but this increase in survivorship does not appear to be attributable to the alloparental care that nonbreeders provide (Koenig & Mumme 1987). Thus, not all the effects of helpers on fitness is necessarily attributable to alloparental behavior *per se*. This issue is discussed in further detail below.

HELPING BEHAVIOR AND LEVELS OF ANALYSIS

Following Tinbergen (1963), Sherman (1988, 1989) proposed the existence of four different levels of analysis in behavioral research: evolutionary origins, functional consequences, ontogenetic process, and mechanisms, the latter including both cognitive and physiological processes. Any behavior can be explained at each of these levels (e.g. Holekamp & Sherman 1989), but only hypotheses within a particular level are legitimate alternatives. Although we recognize the existence of several alternative philosophical frameworks by which behavioral hypotheses may be categorized (c.f. Dwyer 1984; Horn 1990), we believe that Sherman's levels provide a relatively straightforward and heuristically useful scheme for organizing many ideas in this field. Below we discuss some of the hypotheses at each level that address the question: Why do some individuals feed offspring not their own?

Evolutionary Origins

Of interest here is the context in which helping behavior evolved. One plausible hypothesis is that helping arose as a byproduct of selection in some other context, such as on parents to feed begging offspring (e.g. Williams 1966: 208; Jamieson 1989a). Second, helping might be present in a particular species as a consequence of phylogenetic inertia (Edwards & Naeem 1990). Note, however, that this explanation only shifts the question of the ultimate evolutionary origin of the trait from the species of interest to the one from which it phylogenetically "inherited" the trait. Alternatively, helping behavior could have arisen directly as an adaptive response to natural selection reflecting the particular ecological and demographic conditions of the population or species under consideration.

Discriminating among hypotheses at this level is notoriously difficult. It is usually impossible to obtain a clear glimpse of the evolutionary origin of a trait much less to examine those origins experimentally. However, we can envision two approaches that are potentially productive. The first is to search for probable evolutionary origins by comparing the behavior and phylogeny of closely related species. An example is the recent cladistic analysis of the distribution of cooperative breeding in passerine birds by Edwards & Naeem (1990). By providing evidence that phylogeny plays a role in determining the presence of cooperative breeding, these authors demonstrate that this phenomenon may not be an independently evolved adaptation in some species (see also Russell 1989). Their analyses, however, also demonstrate that helping behavior is frequently lost from a clade, contrary to the suggestion that, once present, helping is difficult to eliminate. The resulting patchwork pattern in the presence of this trait is consistent with the hypothesis that selection is acting within at least some lineages to maintain or eliminate this behavior, depending on the ecological conditions faced by each species.

A second approach that addresses the issue of whether helping could have originated as a byproduct of selection in some other context is to determine the correlation between helping behavior and other behaviors hypothesized to be the direct agents of selection. For example, one might examine the correlation between the quantity and quality of alloparental care given by individuals when they are helpers and the parental care they provide as parents. No correlation (or an inverse correlation) would effectively reject Jamieson & Craig's (1987) hypothesis that helping behavior arose as a byproduct of selection on parents to feed begging offspring, while a positive correlation would be consistent with this hypothesis (as well as others, including that nonbreeders gain experience by helping that makes them better parents later in life).

This latter approach does not directly address evolutionary origins: Knowing the correlation between two traits does not tell us how closely they were linked when they arose. Nonetheless, such analyses, using both interspecific and intraspecific data, may provide the least ambiguous test possible of Jamieson and Craig's unselected hypothesis for the origin of helping behavior.

Functional Consequences

At this level, the question of interest is whether helping confers selective advantages to the individuals involved or whether it is selectively neutral, or even maladaptive. Answering this question involves determining not only whether helping behavior correlates with fitness, but whether it is the direct target of selection rather than a byproduct of selection for a correlated trait (e.g. Lande & Arnold 1983).

There are three general ways in which the fitness benefits of helping behavior can accrue to helpers: (1) kinship or indirect fitness benefits (*sensu* Brown & Brown 1981), (2) any of several direct (or Darwinian) fitness advantages to the helper, such as increased experience, and (3) reciprocal benefits. If helping has no direct fitness benefit, it could still be present for at least three reasons: (1) selection on a correlated character, (2) phylogenetic inertia, or (3) drift. Evidence for the functional significance of helping behavior is discussed and critically evaluated in detail below.

Ontogenetic Processes

Hypotheses at this level concern the development of helping behavior in individuals, particularly in reference to age, sex, social environment, and previous experience. Virtually all research on the ontogeny of helping behavior has been descriptive in nature (e.g. Stallcup & Woolfenden 1978; Lawton & Guindon 1981; Jamieson & Craig 1987; Jamieson 1988); there has been little general theory concerning the ontogeny of helping behavior and virtually all hypotheses have been *ad hoc* and species-specific.

Physiological Processes

The fourth level of analysis is that of mechanisms, including both cognitive and physiological processes (Sherman 1988). There is little discussion in the cooperative breeding literature of the former. Two plausible physiological explanations of helping behavior are that it is (1) an expression of the same stimulus-response that produces feeding of young by parents (e.g. Jamieson & Craig 1987; Jamieson 1989; Ligon & Stacey 1989) and (2) a byproduct of seasonal hormonal changes in nonbreeders.

LEVELS OF ANALYSIS: JAMIESON & CRAIG'S CRITIQUE

Jamieson & Craig (1987) and Jamieson (1986, 1989a) offer an alternative to the hypothesis that helping behavior is a direct product of natural selection. They propose that helping behavior originated and is currently maintained nonadaptively as a result of its tight linkage with the clearly adaptive behaviors associated with normal parental care. Viewed in this context, helping behavior is an unselected consequence of delayed dispersal, group living, and the physiological processes that normally lead individuals to feed begging offspring.

Examined in the framework of levels of analysis, Jamieson & Craig's hypothesis actually consists of sub-hypotheses on all four of the levels discussed above. First, at the level of evolutionary origins, helping behavior originated as an unselected consequence of group living. Second, at the level of functional consequences, helping behavior presently has no selective value but is maintained indirectly by virtue of its genetic correlation with provisioning behavior of parents. Third, on the level of ontogenetic processes, helping behavior develops as a neotenic shift in the timing of the expression of provisioning behavior. Fourth, at the level of physiological processes, provisioning behavior is elicited from potential helpers by begging juveniles via a stimulus-response mechanism.

Knowledge at any one level of analysis may provide important clues, but cannot supercede, hypotheses at other levels. For example, consider the possibility that there is a phylogenetic component to helping behavior in some taxa (Edwards & Naeem 1990). This hypothesis specifically addresses evolutionary origins. However, consider species A in taxon B, and imagine that we know, from an analysis using appropriate null models, that helping behavior occurs more frequently in taxon B than expected by chance. This knowledge of the possible importance of phylogenetic inertia in taxon B would then make it more plausible that helping behavior might be of no current adaptive utility in species A, at least compared to a species in a taxon in which helping behavior is not found more frequently than expected.

In this example, information at the level of evolutionary origins provides clues about possible functional consequences or the lack thereof. However, helping behavior could still be of current functional significance regardless of its evolutionary origin or its developmental and physiological bases (Greene 1986; Sherman 1988).

Jamieson & Craig's failure to distinguish clearly between hypotheses at different levels of analysis leads to considerable confusion. For example, Jamieson & Craig (1987: 80) suggest as an alternative to "increased fitness" type arguments (level of functional consequences) that feeding of nestlings in cooperative breeders is "maintained by the same stimulus-response mechanism that results in parents feeding their own young or host species feeding parasitic young" (level of physiological processes). In an earlier paper, Jamieson (1986: 203) suggests that if helping behavior is an elicited response to the presence of nestlings, then "the functional question 'What is the selective advantage of helping behavior' could be replaced by a developmental question, 'How does parenting behavior develop in nonbreeders of cooperative species?' "

In fact, neither the stimulus-response subhypothesis nor the ontogeny of helping behavior in cooperative species can supersede explanations at other levels of analysis. The hypothesis that helping behavior is explained at the physiological level by a stimulus-response mechanism is entirely consistent with functional hypotheses proposing that helping behavior is maintained as a direct result of natural selection. Similarly, helping behavior may confer a selective advantage regardless of its ontogeny.

This mixing levels of analysis is particularly dangerous when evaluating alternative, testable hypotheses for helping behavior. For example, Jamieson (1989a: 403) proposes that if helping behavior has been the direct product of selection, then there should be a greater predisposition of naive juveniles to provision nestlings in cooperative species than in closely related, noncooperative species. Results of such a test, if positive, would indeed reject Jamieson's subhypothesis that helping behavior is the result of the same stimulus-response mechanism responsible for parental behavior. However, the test does not reject alternatives consistent with the hypothesis that helping behavior is an unselected trait; for example, the increased predisposition for provisioning could still be a side effect of the hormonal state of individuals which, purely as a side effect of group living, is different in naive juveniles of cooperative compared to noncooperative breeders. Jamieson's (1989) proposed test therefore cannot reject hypotheses concerning current utility of helping behavior; such behavior may or may not be of adaptive utility regardless of the outcome.

To avoid the fruitless debate and semantic confusion that arise when levels of analysis are mixed (Sherman 1988), we will devote the rest of this chapter to a detailed consideration of the functional consequences of helping behavior. First we critically evaluate evidence suggesting that helping behavior results in selective benefits for helpers and/or recipients. We then briefly discuss some of the evidence for the hypothesis that helping behavior does not have current selective utility.

FITNESS CONSEQUENCES OF HELPING BEHAVIOR

The Evidence

For a trait to have current adaptive utility, it must have an overall positive effect on fitness. Here we evaluate the empirical evidence that individuals increase their fitness by acting as helpers. A summary is provided in Table 1. Another classification of the potential fitness effects of helpers with a somewhat different philosophical orientation is presented by Emlen & Wrege (1989).

Helpers increase the reproductive success of related breeders. Groups with helpers have been shown to produce significantly more young than groups without helpers in at least five species of mammals (Gittleman 1985) and ten species of birds (Brown 1987). In a few of these, such as the white-fronted bee-eaters (*Merops bullockoides*) studied by Emlen & Wrege (1988, 1989) and the bicolored wrens studied by Austad & Rabenold (1985), the effect is so pronounced that it is unlikely to be due to confounding variables. Helping in these species yields significant fitness benefits both to the individuals helped, who are able to raise more offspring, and to their closely related helpers, who gain indirect fitness benefits.

However, in many species the apparent helper effect, if present at all, is modest and could occur because helpers are associated with groups which are reproductively superior for other reasons (Lack 1968). For example, because helpers are usually the offspring of breeders from prior years, breeders with helpers are almost always more experienced and often on territories of higher quality than breeders without helpers.

Hence, increased reproductive success of groups with helpers could be due to their experience or better territories rather than helpers *per se*.

There are at least three approaches toward resolving this difficulty. The first is to restrict analyses to sets of birds with comparable prior histories and living on similar territories. For example, Gibbons (1987), studying helping by juvenile moorhens (*Gallinula chlorops*), restricted analyses to experienced pairs, to pairs attempting renests or second nests during different parts of the season, to pairs with territories of a particular size class, and pairs breeding in each of the three years of his study in order to control for the potentially confounding effects of experience, season, territory quality, and yearly variation, respectively. Based on these analyses, Gibbons (1987) concluded that helpers resulted in an increase in the number of young fledged.

With longer-term data it is even possible to control for individual variation, rendering this approach as robust as the best experimental studies. An excellent example is the analysis of the Florida scrub jay by Woolfenden & Fitzpatrick (1984). They restricted their analysis to resident breeders that had attempted reproduction on the same territories at least twice with and without helpers and found a significant increase in reproductive success due to helpers. When further restricting the analysis to pairs (either previously successful or not) breeding on the same territory with and without helpers, a difference still emerged, but was no longer significant.

A second approach is to control statistically for the effects of experience and territory quality. This approach is particularly useful in species whose group composition is complex. For example, Koenig & Mumme (1987) performed an analysis of variance of reproductive success in the acorn woodpecker including four variables: (1) whether or not a turnover in breeders had occurred in the prior year, (2) number of breeder males, (3) number of breeder females, and (4) number of nonbreeding helpers. They found no significant effect of nonbreeding helpers on group reproductive success when the other three variables were controlled, contrary to the results of univariate analyses. Similarly, Zack & Ligon (1985) found no influence of group size on reproductive success of gray-backed fiscal shrikes (*Lanius excubitorius*) controlling for cover and Nias (1986) found no effect of number of helpers on reproduction in superb fairy-wrens (*Malurus cyaneus*) controlling for a suite of vegetational characters. Such studies reinforce the hypothesis

Table 1. Evidence for the Hypothesis that Helping Behavior has Current Adaptive Utility

Observation	Species	References
A. Fitness consequences of helping behavior		
1. Helpers increase the reproductive success of related breeders (current indirect fitness benefits)	Many (descriptive)	Gittleman 1985; Brown 1987
	Grey-crowned babbler	Brown et al. 1982
	Florida scrub jay	R. L. Mumme unpubl. data
	Lamprologus brichardi	Taborsky 1984
2. Helpers increase the survival of related breeders (future indirect fitness benefits)	Groove-billed anis (male)	Vehrencamp 1978
	Florida scrub jay (male and female)	Woolfenden & Fitzpatrick 1984
	Pied kingfisher (female)	Reyer 1984
	Bicolored wren (male and female)	Austad & Rabenold 1986
	Acorn woodpecker (male)	Koenig & Mumme 1987
	Splendid fairy-wren (female)	Russell & Rowley 1988
3. Helpers lighten the load of breeders	Dwarf mongoose	Rood 1978
	Grey-crowned babbler	Brown et al. 1978
	Green woodhoopoe	Ligon & Ligon 1978b
	White-browed sparrow weaver	Lewis 1982
	Galápagos mockingbird	Kinnaird & Grant 1982
	Stripe-backed wren	Rabenold 1984
	Pied kingfisher	Reyer 1984
	Bicolored wren	Austad & Rabenold 1985
	Beechey jay	Raitt et al. 1984
	Moorhen	Eden 1987; Gibbons 1987
4. Helpers gain direct fitness benefits		
a. Increased access to breeding space	Florida scrub jay	Woolfenden & Fitzpatrick 1984
b. Increased experience	Brown jay	Lawton & Guindon 1981
	White-winged chough	Heinsohn et al. 1988
	Florida scrub jay	Woolfenden & Fitzpatrick 1984
	Spendid fairy-wren	Rowley & Russell 1989
	Purple gallinule	Hunter 1987
c. Forming associations leading to increased future reproductive success	Arabian babbler	Carlisle & Zahavi 1986
	Green woodhoopoe	Ligon & Ligon 1978
	Bell miner	Clarke 1989
	Galápagos mockingbird	Curry 1988a
	Brown hyena	Owens & Owens 1984
	Gray-breasted jay	Brown & Brown 1980; Caraco & Brown 1986
	Pied kingfisher	Reyer 1984
d. Gaining access to group resources	*Lamprologus brichardi*	Taborsky 1985
B. Functional patterns of helping behavior		
1. Differential feeding of relatives	Bell miner	Clarke 1984
	White-fronted bee-eater	Emlen & Wrege 1988
	Galápagos mockingbird	Curry 1988a
	Brown hyena	Owens & Owens 1984
	Pied kingfisher	Reyer 1984, 1986

Table 1 (continued)

Observation	Species	References
2. Helping behavior is influenced by specific ecologic conditions	White-fronted bee-eater Galápagos mockingbird Pied kingfisher	Emlen 1981, 1984 Curry 1988b Reyer & Westerterp 1985
3. Sex ratio of nestlings influenced by helpers	Red-cockaded woodpecker	Gowaty & Lennartz 1985
4. Relationship between helping and reproductive opportunities	Dunnock Acorn woodpecker Moorhen	Davies 1985; Burke et al. 1989 Stacey 1979; Koenig in prep. Gibbons 1986

that apparent helper effects in many species may be due to confounding variables.

A third approach to this problem is experimental. We are aware of four such studies. First, Brown et al. (1982) compared the number of young fledged in second nests by groups of grey-crowned babblers (*Pomatostomus temporalis*) with multiple helpers to groups that had been reduced to trios (presumably the breeding pair and one helper), and found that subsequent reproductive success of the experimental groups was one-third times that of the controls. Second, Taborsky (1984), studied the cichlid fish *Lamprologus brichardi* in the laboratory under controlled conditions. Although helpers in this species do not feed offspring, they do engage in a variety of parental duties including cleaning of eggs, larvae, and fry. He found that pairs with helpers produced larger clutches and that egg survival was possibly enhanced when helpers were present. Third, Mumme (in preparation) removed helper Florida scrub jays and was able to demonstrate that helpers help by reducing nest predation, as suggested by Woolfenden & Fitzpatrick (1984).

In contrast to the results of these three studies, which support the hypothesis that helping behavior increases the reproductive success of breeders, is the work of Leonard et al. (1989) who removed moorhen juvenile helpers and observed no effect on subsequent survival or reproductive success of the breeding pair. This suggests that the enhanced reproductive success observed in groups with helpers by Gibbons (1987) was due to interactions with confounding variables, despite his attempt to control for such interactions by restricting his analyses to comparable subsets of groups.

Unfortunately, the experimental approach is not without its shortcomings. There are at least three potential difficulties. First, experimental removals of potential helpers from social groups usually will affect both group size and alloparental behavior, thereby confounding estimates of the fitness effects of alloparental behavior *per se*. This point is discussed further below. Second, the social disruption of removing helpers from experimental groups is not easily controlled. For example, most experimental groups in Brown et al.'s (1982) study simply stopped breeding after three to five nonbreeding group members (potential helpers) were removed, while most unmanipulated control groups attempted to raise additional broods. Although the analysis of Brown & Brown (1981) suggests that the cessation of reproductive activities in experimental groups was due to the loss of potential helpers, their data are also consistent with the alternative hypothesis that the reproductive failure of experimental groups was caused by the social disruption resulting from the removal of a substantial fraction of the social unit. The only solution to this difficulty may be a combination of experimental studies and exhaustive nonmanipulative work.

Third, it is possible that helpers help and yet result in no increase in reproductive success (or survivorship) compared to groups without helpers. Group living incurs several automatic disadvantages and entails no automatic advantage (Alexander 1974; Hoogland & Sherman 1976). Consequently, it is entirely plausible that the aid provided by helpers (the "alloparental effect") may in some species simply counter the otherwise negative effect of larger group size (the "group-living effect") on reproductive success (Figure 1). This would be the case if, for example, larger groups depleted their resources more quickly or thoroughly than smaller groups and thus had less food for reproduction. In such a system, neither detailed empirical work nor experimental studies would reveal an increase in reproductive success in groups with helpers, yet helpers are significantly enhancing reproductive success compared to groups of similar size in which helpers do not help. Indeed, groups in such a system might even experience *lower* reproductive success than pairs without helpers, despite the aid-giving behavior of helpers. Separating the alloparental and group-living effects could only be accomplished by careful study of the relationship between group size, reproductive success, and variation in the degree of help provided by individual helpers, or by manipulating helping

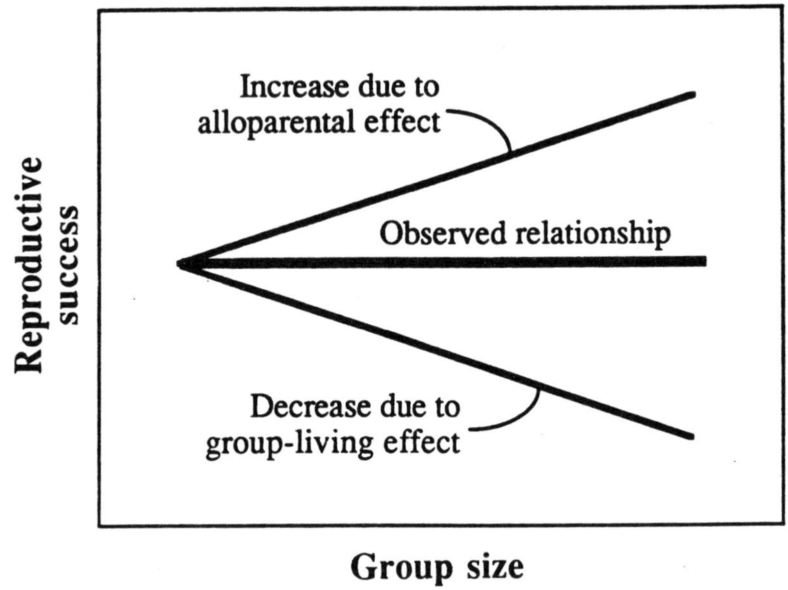

Figure 1. Potential opposing relationship between alloparental and group-living effects in cooperative breeders. In this hypothetical example, helpers have a significantly positive effect on fitness (the alloparental effect) which is cancelled by the adverse effect of group size; the resulting relationship between group size and group reproductive success is flat. It is equally possible that the alloparental effect could be negative, but the relationship between group size and group reproductive success could be positive as a consequence of a beneficial group-living effect.

behavior without altering group size. Such analyses have yet to be performed. We will discuss this problem in further detail below.

Helpers increase the survival of related breeders. The majority of workers have focused on the effects of helpers on current reproduction. However, helpers may also influence the survivorship of breeders. When helpers are related to the breeders, as is usually the case, increased survival of breeders augments the fitness of both parties. Increased survivorship of breeders as a consequence of helpers has been documented in at least six species (Table 1).

The effect of helpers on breeder survivorship may seem slight yet still contribute substantially to the indirect fitness benefits of helping behavior (Mumme et al. 1989). For example, annual survivorship of Florida scrub jay breeders with helpers is only 8% greater than those without helpers (Woolfenden & Fitzpatrick 1984). Yet this difference is sufficient to provide 46% of the estimated indirect fitness benefits of helping. In four other species for which data are available, future indirect fitness benefits arising from increased survivorship are estimated to provide between 29% and 49% of the total gain in indirect fitness accruing to helpers (Mumme et al. 1989).

These results indicate that increased survivorship of breeders may yield important fitness benefits to helpers by increasing their future indirect fitness. However, as with the relationship between helpers and reproductive success, the apparent increased survivorship of breeders when helpers are present could be due largely or in part to confounding variables such as territory quality. Unfortunately, the experimental studies necessary to reject this alternative have yet to be attempted. Thus, direct experimental evidence that helping behavior increases breeder survivorship is thus far lacking.

Helpers lighten the load of breeders. In at least ten species, breeders reduce their feeding rates when assisted by helpers (Table 1). This "lightening of the load" of breeders will have fitness consequences only if it increases breeder survival, reproductive success, or both. For example, lightening of the load by helpers correlates with a higher incidence of renesting (increasing annual reproductive success) in several species (e.g. white-browed sparrow weavers *Plocepasser mahali* [Lewis 1982], grey-crowned babblers [Brown & Brown 1981], stripe-backed wrens *Campylorhynchus nuchalis* [Rabenold 1984], and bicolored wrens [Austad & Rabenold 1985]) and with increased breeder survival in the pied kingfisher *Ceryle rudis* (Reyer 1984).

Helpers gain direct fitness benefits. By increasing the reproductive success or survivorship of a breeder, helpers increase the breeder's direct fitness and their own indirect fitness. There is also considerable evidence to suggest that helpers may increase their own direct fitness; that is, improve their own survivorship or future reproductive success. A variety of different mechanisms for this have been suggested. These are discussed below.

Helpers gain access to breeding space. Woolfenden & Fitzpatrick (1984) have shown that helpers in the Florida scrub jay, by augmenting reproductive success and thus the size of their group, increase the size of their group's territory. This in turn enables male helpers to "bud off" and inherit a portion of the territory for their own reproduction.

Helpers gain experience. By helping, helpers may acquire experience which increases their direct fitness by allowing them to be more successful breeders later in life. First suggested by Skutch (1961), this hypothesis implies that, for whatever reason, young birds in cooperative breeders are unable to acquire the skills necessary to successfully raise young, and was thus dubbed the "skill hypothesis" by Brown (1987). Brown (1987) also discusses several lines of circumstantial evidence for this hypothesis, including reduced foraging efficiency of birds with recognizable immature plumage, a character observed in many helpers, and smaller body mass in birds of normal helper age in many cooperative breeders.

Better evidence for this hypothesis comes from the brown jay (*Psilorhinus morio*) and the purple gallinule (*Porphyrula martinica*) in which the effectiveness of helping behavior increases with helper age and with experience, even within a season (Lawton & Guindon 1981; Hunter 1987), and the white-winged chough, in which foraging success and feeding rates increase with age (Heinsohn et al. 1988). Brown jays and white-winged choughs, but not purple gallinules, have immature plumages and may require several years to attain full adult characters.

The definitive test for this effect is to compare the reproductive success of birds as a function of the help they provided as auxiliaries. Such a test has yet to be performed. However, several workers have tested for an effect of the number of years that individuals spent as auxiliaries on their subsequent reproductive success. The results are equivocal. No difference between the subsequent reproductive success of helpers and nonhelpers was found in white-fronted bee-eaters (Emlen & Wrege 1989) or in acorn woodpeckers (Koenig & Mumme 1987). Male Florida scrub jays with three or more years of helping experience were reported to reproduce more successfully than those with one or two years of experience (Woolfenden & Fitzpatrick 1984), but additional data have failed to support this trend (G.E. Woolfenden personal communication). Female splendid fairy-wrens (*Malurus splendens*) with prior helping experience tended to reproduce slightly better than those without experience, but the difference was not significant (Rowley & Russell 1990).

Helpers establish social relationships. By caring for dependent offspring, helpers may form associations that improve their own reproductive success later in life. A variety of different mechanisms has been proposed, hypothesizing the critical association to be (1) among the helpers, (2) between the helpers and the nestlings being fed, and (3) between the helpers and the breeders being helped.

The first of these mechanisms was suggested for Arabian babblers (*Turdoides squamiceps*) by Carlisle & Zahavi (1986). These authors detailed various interactions and interference between helpers which they interpreted as efforts to establish their status vis-à-vis other helpers. They suggested, with little supporting documentation, that the probability of a helper obtaining the collaboration of other helpers in establishing and defending territories later in life increases with the enhanced dominance that comes from helping.

Better supported is the hypothesis that helping forges reciprocal associations between the helpers and the recipient nestlings. Ligon & Ligon (1978a, 1983), for example, proposed that recipients of aid in green woodhoopoes (*Phoeniculus purpureus*) may later cooperate with donors (helpers) to compete for vacant breeding territories and even, through a process of delayed reciprocity or generational mutualism (Brown 1983), serve as helpers during a donor's own breeding attempts. Wiley & Rabenold (1984) proposed that such delayed reciprocity might be important in any cooperative breeder in which helpers queue for succession to breeding status within groups. Their suggestion that this occurs regularly in stripe-backed wrens has not been confirmed (Rabenold 1985), but reciprocation of help to donors occurs regularly in Galápagos mockingbirds *Nesomimus parvulus* (Curry 1988a), brown hyenas *Hyaena brunnea* (Owens & Owens 1984), and bell miners *Manorina melanophrys* (Clarke 1989). Clarke (1989) further shows reciprocation to be independent of genetic relatedness, thereby strengthening the hypothesis that some direct fitness benefit is derived from helping behavior in this species.

A similar "helper-resource" hypothesis was suggested as a possible explanation for the apparently indiscriminate feeding of young within a social unit observed in gray-breasted jays (*Aphelocoma ultramarina*) by Brown & Brown (1980). The proposed benefit of reciprocal feeding in this species is to decrease the variance in time between feedings for nestlings (Caraco & Brown 1986) rather than to increase the probability of helpers gaining breeding status, as in the green woodhoopoe.

The third social relationship that may be established by helping behavior is that between the helper and the breeders whose offspring are fed. An excellent example of this effect has been documented in the pied kingfisher by Reyer (1980, 1984, 1986). This species is unique in having two categories of male helpers: "primary" helpers, which are closely related to the breeders, and "secondary" helpers, which are not. Reyer (1984) showed that secondary helpers were more likely to be mated in the subsequent year, usually with the female they had helped, than were birds that did not choose to become secondary helpers. Reyer (1986) also showed that by bringing fish to the nest, secondary helpers reduced the probability of being attacked by the breeding pair and that secondary, but not primary, helpers tend to concentrate on feeding the breeding female rather than the nestlings. These observations support the interpretation that secondary helpers assist the breeding female in order to increase their chances of mating with her in subsequent years (Reyer 1980, 1986).

Reyer's (1986) observations are also consistent with the hypothesis that feeding by secondary helpers is "payment" for being accepted as a group member. This possibility, first proposed as an explanation of helping by Gaston (1978), is also supported by Taborsky's (1985) work on a fish, *Lamprologus brichardi*, discussed next.

Helpers gain access to group resources. Numerous studies have proposed that auxiliaries, by remaining on their natal territory, gain access to critical group resources which afford them increased survival. As in the secondary helpers of pied kingfishers, helping by such auxiliaries might be payment for being allowed access to these resources (Gaston 1978).

Only one study thus far has demonstrated a direct relationship between helping behavior and increased survival of helpers. In the territorial fish *Lamprologus brichardi*, Taborsky (1984, 1985) experimentally demonstrated that large helpers, generally expelled by breeders, are tolerated when a pair is exposed to predators or conditions of severe inter- or intra-specific competition for space, conditions under which helpers are especially important to the survival and success of a pair. Helpers in this species benefit significantly by having access to shelter sites, and thus the interpretation that helping is payment for access to these sites, which have a strong influence on survival, appears reasonable.

A Caveat: Distinguishing between Group-living and Alloparental Effects

As should now be evident, helpers appear to have a variety of effects on their own fitness and on the fitnesses of other individuals. It is thus tempting to conclude that helping behavior is generally not a neutral trait. Unfortunately, virtually all of the studies summarized above are vulnerable to an important criticism: Many (or even all) of the purported effects of helpers on fitness that have been documented may merely be a consequence of living in groups ("group-living" effects) rather than helping behavior ("alloparental" effects) *per se* (Figure 1). This criticism is not trivial. If we wish to demonstrate an adaptive function to helping behavior independent of the benefits of living in groups, we must be certain that the alloparental behavior performed by helpers has a direct effect on fitness and that this effect is not due simply to the presence of additional individuals within the social unit.

The main problem in separating these two effects is that in most cooperatively breeding species, all helpers at a particular nest are members of a cohesive social unit and virtually all nonbreeding members of that social unit act as helpers. As a result of this close linkage, it is extremely difficult to disentangle the effects of the two phenomena. This is the major criticism leveled by Jamieson (1989a) of studies purporting to demonstrate that helping behavior results in significant fitness benefits. As a simple example, consider a monogamous, permanently territorial species of cooperative breeder in which each breeding pair shares its territory with one or more mature, nonbreeding offspring from previous seasons. Nonbreeders assist their parents by participating in territory defense and by feeding dependent nestlings and fledglings. By helping to defend the territory, nonbreeders lighten the load on the breeders, thereby allowing them to devote more time to finding food for their dependent young. The extra food provided to dependent young by breeders and nonbreeding helpers results in reduced nestling starvation and increased reproductive success.

The nonbreeding helpers in this example influence food delivery to dependent young, and hence reproductive success, in two ways: directly, by feeding dependent young themselves, and indirectly, by helping to defend the group territory. Only the former, alloparental effect can be marshalled as evidence for the adaptive significance of helping behavior, while the latter, group-living effect cannot.

The problem of distinguishing between group-living and alloparental effects deserves considerably more attention than it has thus far received. Removal experiments do not circumvent this difficulty: The removal of potential helpers confounds group-living and alloparental effects just as surely as do nonmanipulative descriptive studies. Similarly, nonbreeders could conceivably lighten the load of related breeders, gain access to group resources and breeding sites, gain experience, and form reproductively valuable relationships simply by living and interacting within their social unit without performing any alloparental behavior *per se*.

At least four types of evidence potentially address this problem. First are data from species, such as the pied kingfisher (Reyer 1984), bell miner (Clarke 1984), white-fronted bee-eater (Emlen & Wrege 1988), and Galápagos mockingbird (Curry 1988b), where group size and helping behavior are not closely linked. In these species, positive effects of helpers on fitness are much more likely to be due to alloparental behavior rather than benefits associated with living in groups (Reyer 1984; Emlen & Wrege 1988). A second approach would be to perform removal experiments in species where young birds remain on their natal territories as nonbreeding territorial auxiliaries but do not act as helpers, such as in the green jay *Cyanocorax yncas* (Gayou 1986), or only rarely act as helpers, as in the northwestern crow *Corvus caurinus* (Verbeek & Butler 1981).

A third approach would be to undertake detailed analyses of the mechanisms by which helpers may enhance fitness (Brown & Brown 1981; Mumme in preparation). If such studies are sufficiently detailed, they should facilitate efforts to disentangle group-living effects from alloparental effects on fitness. Finally, one could manipulate helping behavior without simultaneously altering group size, possibly by careful use of hormonal implants. Such experiments would be very difficult to perform, but would nonetheless be the most powerful and unambiguous means of examining the effects of alloparental behavior on fitness.

FUNCTIONAL PATTERNS OF HELPING BEHAVIOR

Patterns of helping behavior that appear to have been modified or "fine-tuned" by selection suggest adaptation by design and thus that helping is a direct product of natural selection (Emlen et al. 1990). Here we discuss four ways in which helpers appear to alter their behavior in adaptive ways; Emlen et al. (1990) discuss several additional patterns.

Differential Feeding of Relatives

Bell miners, white-fronted bee-eaters, and Galápagos mockingbirds all have flexible breeding systems in which birds may act as helpers, breeders, both, or neither simultaneously within a season. Further, individuals often have the opportunity to help at more than one concurrent nest. These characteristics allow for tests of kin discrimination by helpers which would be difficult or impossible in most cooperative breeders.

Evidence for differential feeding of kin in bell miners (Clarke 1984) is suggestive but equivocal on account of small sample size and the inclusion of breeders provisioning their own offspring (Payne et al. 1985; Emlen & Wrege 1988). More conclusive are data from from the white-fronted bee-eater, where Emlen & Wrege (1988) found that kinship between potential donors and recipients was a highly significant predictor of whether or not birds acted as helpers and that helpers preferentially chose to aid their closest genetic relatives within their social unit 94% of the time. Equally good is the evidence from Galápagos mockingbirds, where Curry (1988a) showed that auxiliaries are more likely to become helpers when they are able to aid close relatives compared to distant or unrelated nestlings and that, when a choice was available, helpers invariably fed nestlings to which they were more closely related.

Two other studies yield data relevant to differential feeding of relatives by helpers. In the brown hyena, Owens & Owens (1984) found that potential male helpers fed infant half-siblings but not more distantly related cubs. Finally, in the pied kingfisher, the higher feeding rates of primary compared to secondary helpers correlates with their closer genetic relatedness to recipients (Reyer 1984, 1986).

Evidence that helping behavior is influenced by genetic relatedness provides strong support for the hypothesis that helping behavior has current adaptive function. These data are also relevant to at least two other hypotheses of general interest in behavioral ecology. First, they support the hypothesis that indirect fitness benefits have had a strong influence on the evolution of helping behavior in these species (e.g. Emlen & Wrege 1989). Second, they address the issue of how donors should dispense aid among potential recipients. Despite the importance of kinship to the decision of whether or not to help shown by white-fronted bee-eaters and Galápagos mockingbirds, degree of genetic relatedness was not found to influence the amount of aid offered in either these species or two others in which the possibility of such a correlation was tested (stripe-backed wrens [Rabenold 1985] and splendid fairy-wrens [Payne et al. 1985]). As discussed by Emlen & Wrege (1988), these findings are generally in accord with a "diminishing returns model," which predicts that helping should be an all-or-none response.

Helping Behavior is Influenced by Specific Ecological Conditions

The considerable data indicating that cooperative breeding occurs under specific ecological conditions (Koenig & Pitelka 1981; Emlen 1982, Stacey & Ligon 1987; Ford et al. 1988) cannot be used as evidence for an adaptive basis of helping behavior because of the high concordance between group living and the expression of helping behavior by auxiliaries (Jamieson 1989a). However, at least three studies indicate that the manifestation of helping behavior by auxiliaries is influenced by ecological conditions, and thus provide evidence for an adaptive basis of helping behavior.

Emlen (1984) found a correlation between the degree of environmental harshness, as measured by rainfall, and the proportion of the population of white-fronted bee-eaters that act as nonbreeding helpers. Although indirect, these data suggest that individual bee-eaters may assess the ecological potential for independent breeding and act as helpers when conditions are poor (Emlen 1981). Evidence for a more drastic interaction providing less ambiguous support for an adaptive value of helping was described by Curry (1988b), who found evidence that in dry years dominant breeders may recruit helpers by interfering with, and ultimately causing the failure of, the breeding efforts of competing subordinates within their group.

Direct evidence of a relationship between energetic stress and helping behavior is available from the pied kingfisher, where the use of doubly-labeled water revealed that energetically stressed male breeders were more likely to accept unrelated helpers than nonstressed birds (Reyer & Westerterp 1985). Their work thus provides excellent evidence that helping behavior in this species confers fitness benefits to breeders.

Sex Ratio of Nestlings Influenced by Helpers

In the red-cockaded woodpecker (*Picoides borealis*), as in several species of cooperative breeders, helpers are almost exclusively young males aiding their parents. Gowaty & Lennartz (1985) presented evidence that the nestling sex ratio in this species is biased toward males and that "nontenured" females (those not having bred previously in the study area) were more likely to produce sons than daughters.

This finding has led to considerable discussion concerning the possibility of local resource enhancement (Clark 1978) influencing the sex ratio of species with helpers (Gowaty & Lennartz 1985; Emlen et al. 1986; Lessells & Avery 1987). Emlen et al. (1986), for example, hypothesized that, by helping, nonbreeders can be thought of as repaying breeders, thereby rendering their own production less costly than that of the nonhelping sex. In the case of red-cockaded woodpeckers, this means that a son is cheaper to produce than a daughter, and therefore, because parents should invest equally in the production of sons and daughters (Fisher 1930), the sex ratio should favor sons.

The repayment model yields specific predictions concerning the sex ratios of species with helpers: in particular, in species in which helpers are males, the sex ratio bias should be proportional to the fitness enhancement (e.g. increase in reproductive success and survivorship) accruing to breeders as a consequence of having a helper, while in species in which both males and females act as helpers, there should be no such correlation. These and other predictions have yet to be tested. However, the apparently biased sex ratio in red-cockaded woodpeckers provides evidence that helping behavior may be an important evolutionary force. Certainly no interpretable relationship between helping behavior and nestling sex ratio is predicted if helping behavior has no functional basis.

Relationship between Helping and Reproductive Opportunities

In most of the species discussed thus far helpers are nonbreeders. In such species, provisioning of young is decoupled from the reproductive opportunities of helpers at that nest, which are small or nonexistent. In some species, however, more than a single individual of one sex may contribute genetically to the offspring in a communal nest. Because breeders later cooperate in raising the communal offspring, only some of which may be their own, individuals in such plural breeding species are "breeding" helpers (Brown 1987).

The variations on this theme are remarkably diverse. In the cooperatively polyandrous Galápagos hawk *Buteo galapagoensis* and Harris' hawk *Parabuteo unicinctus* (Mader 1979; Faaborg & Bednardz 1990) up to five males may consort and possibly mate with a single female. In the groove-billed ani *Crotophaga sulcirostris*, up to four pairs of birds may nest jointly in a communal nest (Vehrencamp 1978). In the dunnock *Prunella modularis*, two males may share a single female or more than one female, but the females nest solitarily (Davies 1985). The lack of a traditional pair bond and the mate-sharing by both sexes results in a mating system known as polygynandry. Finally, in the acorn woodpecker (Koenig et al. 1984) and pukeko *Porphyrio porphyrio* (Craig 1980), several cobreeding males may share several joint-nesting females. These mating systems are among the most complex known.

The benefits of provisioning offspring for a breeder are dependent on a variety of factors (Winkler 1987). The variable most relevant here is an individual's probability of having parented offspring (opportunity of parentage). Several studies have focused specifically on the relationship between provisioning rate and opportunity of parentage in species in which more than one male share a female and/or more than one female nest jointly.

The most complete of these is recent work on the dunnock by Burke et al. (1989). These authors were able to correlate the degree of reproductive access mate-sharing males enjoyed with a particular female with not only the degree to which the males subsequently fed nestlings but also, by use of DNA fingerprinting, with the actual number of offspring fathered by the male in the nest. Their results indicated a high correlation among these three variables; that is, males that had relatively great access to a female during her fertile period parented a high proportion of offspring in her nest and fed relatively often, while those with little or no access to a female neither parented offspring nor fed at her nest.

Less definitive data are available for the acorn woodpecker. In New Mexico, Stacey (1979) showed that males joining groups prior to egg laying subsequently helped provision offspring, while those joining after egg-laying did not. In California, Koenig (1990) performed experiments whose results suggest that males experimentally denied opportunity of parentage may either destroy the nest of their cobreeders or help provision the offspring depending, in part, on their dominance status vis-à-vis their cobreeders: Dominants, with a presumably high probability of parenting offspring in a renest, destroy the nest, while subordinates do not and instead assist in raising the nondescendant, but related, offspring in the nest.

A third relevant study is that of Gibbons (1986) on the moorhen, in which females may either nest jointly or parasitize the nests of other females. Joint-nesting females, which were usually close relatives, cooperated in parental care, while parasites, which were unrelated, did not. Similar results have been reported for the white-fronted bee-eater (Emlen & Wrege 1986). The coexistence of these behaviors indicates that helping is not simply a byproduct of laying eggs in a nest, but is part of an overall reproductive strategy fundamentally different from intraspecific nest parasitism.

These studies all support the hypothesis that provisioning behavior in mate-sharing and joint-nesting species has functional significance. This conclusion is most strongly supported in the dunnock, where helping behavior by breeders is correlated both with opportunity of parentage and actual genetic contribution to a particular nesting attempt.

NONADAPTIVE HELPING BEHAVIOR?

In the previous section we discuss cases in which helping may confer fitness advantages to helpers and recipients. In other cases, however, helping behavior may have no effect on fitness or even be maladaptive. A widely-cited example of helping behavior proposed to not be adaptive is Price et al.'s (1983) report of helpers in the cactus finch (*Geospiza scandens*) and the medium ground-finch (*G. fortis*) in the Galápagos. Helpers in these species occurred on only one island and were found only following a particularly dry year. All were unpaired and usually held the territory adjacent to the offspring they fed. Helpers were apparently unrelated to the offspring they fed, in many cases had bred in prior years (hence reducing the probability that essential experience was gained), and did not appear to be likely to mate with the breeding female whose offspring they fed in the following season. Having rejected these potential advantages to helping, Price et al. (1983) concluded that helping was misdirected parental care and not adaptive.

There are, however, reasons to be cautious concerning this conclusion. Helpers contributed a substantial proportion (up to 24.9%) of regurgitations at some nests; it seems reasonable that this amount might either increase reproductive success or lighten the load of breeders. Benefits to the helpers might also include increasing their probability of acquiring a mate, as found in the pied kingfisher (see above). Although Price et al. (1983) found no significant difference in the probability that helpers and nonhelpers obtained mates within a year, the difference (91% of helpers obtained mates compared to 69% of nonhelpers) is suggestive. Only a great deal of additional data could conclusively reject this possibility. Given the rarity of helping in these populations, such data are unlikely to be forthcoming.

The equivocal nature of Price et al.'s (1983) conclusion reflects an important asymmetry between the data necessary to accept the hypothesis that helping behavior has a positive effect on fitness compared to that necessary to reject this hypothesis. One cannot directly test the hypothesis that helping behavior is selectively neutral. To accomplish this goal it would be necessary to eliminate all of the ways in which helping may confer fitness benefits to the donor or recipients. This is clearly a difficult undertaking.

Thus, just as one should not automatically assume that a behavior has an adaptive explanation (Gould & Lewontin 1979), one must be equally circumspect in concluding that no adaptive explanation exists. This is well illustrated by brood parasitism (Payne 1977) and interspecific helping behavior (Shy 1982). Both involve parent-like behavior expressed in contexts where it may appear that any benefit to the helper is exceedingly unlikely. However, to always assume that this is true would run the risk of missing rare but biologically interesting and instructive cases in which these phenomena may confer fitness advantages to helpers

(Smith 1968; McKaye 1977). Right or wrong, adaptive hypotheses promote valuable empirical testing and should not be dismissed out of hand (see also Alcock 1987).

CONCLUSION

A characteristic feature of cooperatively breeding species is helping behavior: parent-like behavior directed toward young other than one's own genetic offspring. In the past two decades, a large literature has developed focusing on the search for adaptive explanations of helping behavior. A variety of mechanisms by which helping may increase fitness have been proposed, all leading to increased lifetime inclusive fitness of individuals, usually nonbreeders, that help compared to those that do not help. The data supporting this contention are widespread. Unfortunately, many of the relevant studies fail to control for the effects of potentially confounding variables. The most important of these is that enhanced fitness of helpers may in many cases result from living in social groups rather than from performing alloparental care *per se*. These studies thus do not provide unambiguous support for the hypothesis that helping behavior is of current adaptive utility.

More convincing evidence for this hypothesis comes through "argument by design": examples in which helping behavior has apparently been "fine-tuned" by selection (Williams 1966; Emlen et al. 1990). Four such functional patterns of helping behavior are discussed, including differential feeding of relatives, cases in which the propensity to help is apparently influenced by specific ecological conditions, cases in which the sex ratio of nestlings is apparently influenced by helpers, and cases in which there is a clear, interpretable relationship between helping and reproductive opportunities. These examples provide sound evidence for the current selective utility of helping behavior in a variety of species.

Jamieson (1986, 1988, 1989a) and Jamieson & Craig (1987) have recently challenged this conclusion. Their critique of helping behavior consists of a series of subhypotheses addressing the significance of helping behavior on four different levels of analysis (Tinbergen 1963; Sherman 1988): evolutionary origins, functional consequences, ontogeny, and physiological processes. They propose that helping behavior arose "epigenetically" in communal-living species as a byproduct of selection for normal parental behavior, and suggest a physiological mechanism (the normal expression of a stimulus-response link between begging offspring and provisioning behavior) by which helping behavior could be maintained in the absence of any current adaptive utility.

We agree with several aspects of Jamieson & Craig's (1987) arguments. It is quite possible that helping behavior has originated, in at least some cases, as a byproduct of selection in some other context. Brown & Brown (1980), for example, suggested that cooperative breeding, by allowing birds to coexist within social units, enables and ultimately capitalizes on the "mistake" of feeding unrelated offspring. Helping behavior might very well persist only in species where the initial "mistake" can be turned into a selective benefit for the helper, the individuals helped, or both. Jamieson & Craig (1987) are also correct to suggest that phylogenetic factors may have played a role in determining the potential for some species to exhibit cooperative breeding (Edwards & Naeem 1990). Furthermore, the physiological mechanism which they propose as leading to the expression of helping behavior may be accurate, at least for some species.

Where we differ from Jamieson & Craig is in how we view the relevance of these hypotheses to the current functional utility of helping behavior. Jamieson (1989a) and Jamieson & Craig (1987), stress that the evolutionary origin, ontogeny, and physiological basis of helping behavior are critical pieces of information needed to discern its functional utility. Such information, however, explains helping behavior at different levels of analysis and thus may complement, but does not compete with, adaptive hypotheses for the functional utility of helping behavior. Even if it were possible to know that helping behavior in a particular species was present due to phylogenetic constraints (that is, was "inherited" from an ancestral population rather than originated *de novo*) and that helping behavior was expressed through a stimulus-response mechanism, it would still be necessary to test hypotheses for the current functional utility of helping behavior before accepting the hypothesis that it is an "unselected" trait.

Where do we go from here? First, we recommend that researchers keep in mind that problems can be profitably addressed at more than one level, no one of which is inherently better than the others. Second, researchers must be careful to recognize which level they are addressing. In our view, the substantive criticisms of "functional explanations" for helping behavior presented by Jamieson (1986, 1989a) and Jamieson &

Craig (1987) run the risk of being overlooked or dismissed out of hand as a result of Jamieson's (1989b) refusal to accept that hypotheses at one level do not compete with or exclude hypotheses at other levels (Sherman 1988, 1989).

Finally, there is considerable need for rigorous work addressing the significance of helping behavior at all levels of analysis. Very little is known about the hormonal and physiological mechanisms of helping behavior and the importance of phylogeny has only recently been rigorously considered. At the level of functional consequences, the evidence for helping behavior being the direct product of selection is strong in some species but in general is not sufficiently unimpeachable that workers in this area can afford to blithely ignore the alternatives proposed by Jamieson (1989a) and Jamieson & Craig (1987). Only by careful analyses at all levels will we eventually understand the evolutionary bases of this intriguing biological phenomenon.

ACKNOWLEDGEMENTS

We thank Marc Bekoff, Nick Davies, Janis Dickinson, Steve Emlen, Andy Horn, Dale Jamieson, Ian Jamieson, Marty Leonard, Frank Pitelka, Paul Sherman, and Mark Stanback for comments on the manuscript, and Scott Edwards, Steve Emlen, Andy Horn, Ian Jamieson, Marty Leonard, Dave Ligon, Paul Sherman, and Pete Stacey for access to unpublished manuscripts. Our work on cooperative breeding has been supported by the NSF, most recently through grants DEB87-04992 to WDK and BSR86-00174 to RLM.

LITERATURE CITED

Alcock, J. 1987. Ardent adaptationism. *Natural History* 96, 4.

Alexander, R.D. 1974. The evolution of social behavior. *Annual Review of Ecology & Systematics* 5, 325-383.

Austad, S.N. & Rabenold, K.N. 1985. Reproductive enhancement by helpers and an experimental inquiry into its mechanism in the bicolored wren. *Behavioral Ecology & Sociobiology* 17, 19-27.

_____. 1986. Demography and the evolution of cooperative breeding in the bicolored wren, *Campylorhynchus griseus*. *Behaviour* 97, 308-324.

Brown, J.L. 1983. Cooperation - a biologists' dilemma. *Advances in the Study of Behavior* 13, 1-37.

_____. 1987. *Helping and Communal Breeding in Birds: Ecology and Evolution*. Princeton, New Jersey: Princeton University Press.

Brown, J.L. & Brown, E.R. 1980. Reciprocal aid-giving in a communal bird. *Zeitschrift für Tierpsychologie* 53, 313-324.

_____. 1981. Kin selection and individual selection in babblers. In: *Natural Selection and Social Behavior: Recent Research and New Theory*. (ed. by R. D. Alexander and D.W. Tinkle), pp. 244-256. New York: Chiron Press.

Brown, J.L., Brown, E.R., Brown, S.D. & Dow, D.D. 1982. Helpers: Effects of experimental removal on reproductive success. *Science* 215, 421-422.

Brown, J.L., Dow, D.D., Brown, E.R. & Brown, S.D. 1978. Effects of helpers on feeding of nestlings in the grey-crowned babbler (*Pomatostomus temporalis*). *Behavioral Ecology & Sociobiology* 4, 43-59.

Burke, T., Davies, N.B., Bruford, M.W. & Hatchwell, B.J. 1989. Parental care and mating behaviour of polyandrous dunnocks *Prunella modularis* related to paternity by DNA fingerprinting. *Nature* 338, 249-251.

Caraco, T. & Brown, J.L. 1986. A game between communal breeders: When is food-sharing stable? *Journal of Theoretical Biology* 118, 379-393.

Carlisle, T.R. & Zahavi, A. 1986. Helping at the nest, allofeeding and social status in immature Arabian babblers. *Behavioral Ecology & Sociobiology* 18, 339-351.

Clark, A.B. 1978. Sex ratio and local resource competition in a prosimian primate. *Science* 208, 163-165.

Clarke, M.F. 1984. Co-operative breeding by the Australian bell miner *Manorina melanophrys* Latham: A test of kin selection theory. *Behavioral Ecology & Sociobiology* 14, 137-146.

_____. 1989. The pattern of helping in the bell miner (*Manorina melanophrys*). *Ethology* 80, 292-306.

Craig, J.L. 1980. Pair and group breeding behaviour of a communal gallinule, the pukeko, *Porphyrio p. melanotus*. *Animal Behaviour* 28, 593-603.

Curry, R.L. 1988a. Influence of kinship on helping behavior in Galápagos mockingbirds. *Behavioral Ecology & Sociobiology* 22, 141-152.

_____. 1988b. Group structure, within-group conflict and reproductive tactics in cooperatively breeding Galápagos mockingbirds, *Nesomimus parvulus*. *Animal Behaviour* 36, 1708-1728.

Davies, N.B. 1985. Cooperation and conflict among dunnocks, *Prunella modularis*, in a variable mating system. *Animal Behaviour* 33, 628-648.

Dwyer, P.D. 1984. Functionalism and structuralism: Two programs for evolutionary biologists. *American Naturalist* 124, 745-750.

Eden, S.F. 1987. When do helpers help? Food availability and helping in the moorhen, *Gallinula chloropus*. *Behavioral Ecology & Sociobiology* 21, 191-195.

Edwards, S.V. & Naeem, S. 1990. The phylogenetic component of cooperative breeding in passerine birds. Unpublished manuscript.

Emlen, S.T. 1981. Altruism, kinship and reciprocity in the white-fronted bee-eater. In: *Natural Selection and Social Behavior: Recent Research and New Theory*. (ed. by R. D. Alexander & D.W. Tinkle), pp. 217-230. New York: Chiron Press.

_____. 1982. The evolution of helping. I. An ecological constraints model. *American Naturalist* 119, 29-39.

_____. 1984. Cooperative breeding in birds and mammals. In: *Behavioural Ecology: An Evolutionary Approach*, 2nd Edition. (ed. by J.R. Krebs & N.B. Davies), pp. 305-339. Oxford: Blackwell.

Emlen, S.T., Emlen, J. M. & Levin, S.A. 1986. Sex-ratio selection in species with helpers-at-the-nest. *American Naturalist* 127, 1-8.

Emlen, S.T., Ratnieks, F.L.W., Reeve, H.K., Shellman-Reeve, J., Sherman, P.W. & Wrege, P.H. 1990. Selected versus unselected hypotheses for helping behavior. Unpublished manuscript.

Emlen, S.T. & Wrege, P.H. 1986. Forced copulation and intraspecific parasitism: Two costs of social living in the white-fronted bee-eater. *Zeitschrift für Tierpsychologie*, 71, 2-29.

_____. 1988. The role of kinship in helping decisions among white-fronted bee-eaters. *Behavioral Ecology & Sociobiology* 23, 305-315.

_____. 1989. The role of direct and indirect selection in the evolution of helping behavior in white-fronted bee-eaters of Kenya. *Behavioral Ecology & Sociobiology* 25, 303-319.

Faaborg, J. & Bednarz, J.C. 1990. Galápagos and Harris' hawks: Divergent causes of sociality in two raptors. In: *Cooperative Breeding in Birds: Long-term Studies of Ecology and Behavior* (ed. by P.B. Stacey & W.D. Koenig), pp. 357-383. Cambridge: Cambridge University Press.

Fisher, R.A. 1930. *The Genetical Theory of Natural Selection,* 2nd Edition Oxford: Clarendon.

Ford, H.A., Bell, H., Nias, R. & Noske, R. 1988. The relationship between ecology and the incidence of cooperative breeding in Australian birds. *Behavioral Ecology & Sociobiology* 22, 239-249.

Gaston, A.J. 1978. The evolution of group territorial behavior and cooperative breeding. *American Naturalist* 112, 1091-1100.

Gayou, D C. 1986. The social system of the Texas green jay. *Auk* 103, 540-547.

Gibbons, D.W. 1986. Brood parasitism and cooperative nesting in the moorhen, *Gallinula chloropus*. *Behavioral Ecology & Sociobiology* 19, 221-232.

_____. 1987. Juvenile helping in the moorhen *Gallinula chloropus*. *Animal Behaviour* 35, 170-181.

Gittleman, J.L. 1985. Functions of communal care in mammals. In: *Evolution: Essays in Honour of John Maynard Smith* (ed. by P.J. Greenwood, P.H. Harvey & M. Slatkin), pp. 187-205. Cambridge: Cambridge University Press.

Gould, S.J. & Lewontin, R.C. 1979. The spandrels of San Marco and the Panglossian paradigm: A critique of the adaptationist programme. *Proceedings of the Royal Society of London* B205, 581-598.

Gowaty, P.A. & Lennartz, M.R. 1985. Sex ratios of nestlings and fledgling red-cockaded woodpeckers (*Picoides borealis*) favor males. *American Naturalist* 126, 347-353.

Greene, H.W. 1986. Diet and arborality in the emerald monitor, *Varanus prasinus*, with comments on the study of adaptation. *Fieldiana (Zoology) New Series* 31, 1-12.

Heinsohn, R.G., Cockburn, A. & Cunningham, R.B. 1988. Foraging, delayed maturation, and advantages of cooperative breeding in white-winged choughs, *Corcorax melanorhamphos*. *Ethology* 77, 177-186.

Holekamp, K.E. & Sherman, P.W. 1989. Why male ground squirrels disperse. *American Scientist* 77, 232-239.

Hoogland, J.L. & Sherman, P.W. 1976. Advantages and disadvantages of bank swallow (*Riparia riparia*) coloniality. *Ecological Monographs* 46, 33-58.

Horn, A. G. 1990. The levels of confusion. Unpublished manuscript.

Hunter, L.A. 1985. The effects of helpers in cooperatively breeding purple gallinules. *Behavioral Ecology & Sociobiology* 18, 147-153.

____. 1987. Cooperative breeding in purple gallinules: The role of helpers in feeding chicks. *Behavioral Ecology & Sociobiology* 20, 171-177.

Jamieson, I G. 1986. The functional approach to behavior: is it useful? *American Naturalist* 127, 195-208.

____. 1988. Provisioning behaviour in a communal breeder: an epigenetic approach to the study of individual variation in behaviour. *Behaviour* 104, 262-280.

____. 1989a. Behavioral heterochrony and the evolution of birds' helping at the nest: An unselected consequence of communal breeding? *American Naturalist* 133, 394-406.

____. 1989b. Levels of analysis or analyses at the same level. *Animal Behaviour* 37, 696-697.

Jamieson, I.G. & Craig, J.L. 1987. Critique of helping behaviour in birds: A departure from functional explanations. In: *Perspectives in Ethology*, volume 7 (ed. by P. P. G. Bateson & P.H. Klopfer), pp. 79-98. New York: Plenum Press.

Kinnaird, M.F. & Grant, P. R. 1982. Cooperative breeding by the Galápagos mockingbird, *Nesomimus parvulus*. *Behavioral Ecology & Sociobiology* 10, 65-73.

Koenig, W.D. 1990. Opportunity of parentage and egg destruction in polygynandrous acorn woodpeckers: An experimental study. Unpublished manuscript.

Koenig, W.D. & Mumme, R. L. 1987. *Population Ecology of the Cooperatively Breeding Acorn Woodpecker*. Princeton, New Jersey: Princeton University Press.

Koenig, W.D., Mumme, R. L. & Pitelka, F. A. 1983. Female roles in cooperatively breeding acorn woodpeckers. In: *Social Behavior of Female Vertebrates* , (ed. by S. K. Wasser), pp. 235-261. New York: Academic Press.

____. 1984. The breeding system of the acorn woodpecker in central coastal California. *Zeitschrift für Tierpsychologie* 65, 289-308.

Koenig, W.D. & Pitelka, F. A. 1981. Ecological factors and kin selection in the evolution of cooperative breeding in birds. In:

Natural Selection and Social Behavior: Recent Research and New Theory (ed. by R. D. Alexander & D.W. Tinkle), pp. 261-280. New York: Chiron Press.

Lack, D. 1968. *Ecological Adaptations for Breeding in Birds.* Methuen: London.

Lande, R. & Arnold, S. J. 1983. The measurement of selection on correlated characters. *Evolution* 37, 1210-1226.

Lawton, M.F. & Guindon, C.F. 1981. Flock composition, breeding success, and learning in the brown jay. *Condor* 82, 27-33.

Leonard, M.L., Horn, A.G. & Eden, S.F. 1989. Does juvenile helping enhance breeder reproductive success? A removal experiment. *Behavioral Ecology & Sociobiology* 25, 357-361.

Lessells, C.M. & Avery, M.I. 1987. Sex-ratio selection in species with helpers at the nest: some extensions of the repayment model. *American Naturalist* 129, 610-620.

Lewis, D.M. 1982. Cooperative breeding in a population of white-browed sparrow weavers *Plocepasser mahali. Ibis* 124, 511-522.

Ligon, J.D. & Ligon, S.H. 1978a. Communal breeding in green woodhoopoes as a case for reciprocity. *Nature* 276, 496-498.

_____. 1978b. The communal social system of the green woodhoopoe in Kenya. *Living Bird* 17, 159-198.

_____. 1983. Reciprocity in the green woodhoopoe (*Phoeniculus purpureus*). *Animal Behaviour* 31, 480-489.

Ligon, J.D. & Stacey, P.B. 1989. On the significance of helping behavior in birds. *Auk* 106, 700-705.

Mader, W.J. 1979. Breeding behavior of a polyandrous trio of Harris' hawks in southern Arizona. *Auk* 96, 776-788.

McGowan, K.J. & Woolfenden, G.E. 1989. A sentinel system in the Florida scrub jay. *Animal Behaviour* 37, 1000-1007.

McKaye, K.R. 1977. Defense of a predator's young by a herbivorous fish: an unusual strategy. *American Naturalist* 111, 301-315.

Mumme, R. L. & de Queiroz, A. 1985. Individual contributions to cooperative behaviour in the acorn woodpecker: Effects of reproductive status, sex, and group size. *Behaviour* 95, 290-313.

Mumme, R.L., Koenig, W.D. & Ratnieks, F.L.W. 1989. Helping behaviour, reproductive value, and the future component of indirect fitness. *Animal Behaviour* 38, 331-343.

Nias, R.C. 1986. Nest-site characteristics and reproductive success in the superb fairy-wren. *Emu* 86, 139-144.

Owens, D.D. & Owens, M.J. 1984. Helping behaviour in brown hyenas. *Nature* 308, 843-845.

Payne, R.B. 1977. The ecology of brood parasitism in birds. *Annual Review of Ecology & Systematics* 8, 1-28.

Payne, R.B., Payne, L.L. & Rowley, I. 1985. Splendid wren *Malurus splendens* response to cuckoos: An experimental test of social organization in a communal bird. *Behaviour* 94, 108-127.

Price, T., Millington, S. & Grant, P. 1983. Helping at the nest in Darwin's finches as misdirected parental care. *Auk* 100, 192-194.

Rabenold, K.N. 1984. Cooperative enhancement of reproductive success in tropical wren societies. *Ecology* 63, 871-885.

____. 1985. Cooperation in breeding by nonreproductive wrens: Kinship, reciprocity, and demography. *Behavioral Ecology & Sociobiology* 17, 1-17.

Raitt, R.J., Winterstein, S.R. & Hardy, J.W. 1984. Structure and dynamics of communal groups in the Beechey jay. *Wilson Bulletin* 96, 206-227.

Reyer, H.-U. 1980. Flexible helper structure as an ecological adaptation in the pied kingfisher (*Ceryle rudis rudis* L.). *Behavioral Ecology & Sociobiology* 6, 219-227.

____. 1984. Investment and relatedness: A cost/benefit analysis of breeding and helping in the pied kingfisher (*Ceryle rudis*). *Animal Behaviour* 32, 1163-1178.

____. 1986. Breeder-helper-interactions in the pied kingfisher reflect the costs and benefits of cooperative breeding. *Behaviour* 96, 277-303.

Reyer, H.-U. & Westerterp, K. 1985. Parental energy expenditure: a proximate cause of helper recruitment in the pied kingfisher (*Ceryle rudis*). *Behavioral Ecology & Sociobiology* 17, 363-369.

Rowley, I. 1978. Communal activities among white-winged choughs *Corcorax melanorhamphus*. *Ibis* 120, 178-197.

Rowley, I. & Russell, E. 1990. Splendid fairy-wrens: demonstrating the importance of longevity. In: *Cooperative Breeding in Birds: Long-term Studies of Ecology and Behavior*. (ed. by P.B. Stacey & W.D. Koenig), pp. 1-30. Cambridge: Cambridge University Press.

Russell, E.M. 1989. Co-operative breeding - a Gondwanan perspective. *Emu* 89, 61-62.

Russell, E. & Rowley, I. 1988. Helper contributions to reproductive success in the splendid fairy-wren (*Malurus splendens*). *Behavioral Ecology & Sociobiology* 22, 131-140.

Sherman, P.W. 1988. The levels of analysis. *Animal Behaviour* 36, 616-619.

_____. 1989. The clitoris debate and the levels of analysis. *Animal Behaviour* 37, 697-698.

Shy, M.M. 1982. Interspecific feeding among birds: A review. *Journal of Field Ornithology* 53, 370-393.

Skutch, A.F. 1961. Helpers among birds. *Condor* 63, 198-226.

Smith, N.G. 1968. The advantage of being parasitized. *Nature* 219, 690-694.

Stacey, P.B. 1979. Kinship, promiscuity, and communal breeding in the acorn woodpecker. *Behavioral Ecology & Sociobiology* 6, 53-66.

Stacey, P.B. & Koenig, W.D. (eds.) 1990. *Cooperative Breeding in Birds: Long-term Studies of Ecology and Behavior.* Cambridge: Cambridge University Press.

Stacey, P.B. & Ligon, J.D. 1987. Territory quality and dispersal options in the acorn woodpecker, and a challenge to the habitat-saturation model of cooperative breeding. *American Naturalist* 130, 654-676.

Stallcup, J.A. & Woolfenden, G.E. 1978. Family status and contribution to breeding by Florida scrub jays. *Animal Behaviour* 26, 1144-1156.

Taborsky, M. 1984. Broodcare helpers in the cichlid fish *Lamprologus brichardi*: Their costs and benefits. *Animal Behaviour* 32, 1236-1252.

_____. 1985. Breeder-helper conflict in a cichlid fish with broodcare helpers: an experimental analysis. *Behaviour* 95, 45-75.

Taborsky, M. & Limberger, D. 1981. Helpers in fish. *Behavioral Ecology & Sociobiology* 8, 143-145.

Tinbergen, N. 1963. On aims and methods of ethology. *Zeitschrift für Tierpsychologie* 20, 410-433.

Vehrencamp, S.L. 1978. The adaptive significance of communal nesting in groove-billed anis (*Crotophaga sulcirostris*). *Behavioral Ecology & Sociobiology* 4, 1-33.

Verbeek, N.A.M. & Butler, R.W. 1981. Cooperative breeding of the northwestern crow *Corvus caurinus*. *Ibis* 123, 183-189.

Wiley, R.H. & Rabenold, K.N. 1984. The evolution of cooperative breeding by delayed reciprocity and queuing for favorable social positions. *Evolution* 38, 609-621.

Williams, G.C. 1966. *Adaptation and Natural Selection.* Princeton, New Jersey: Princeton University Press.

Winkler, D.W. 1987. A general model for parental care. *American Naturalist* 130, 526-543.

Woolfenden, G.E. & Fitzpatrick, J.W. 1984. *The Florida Scrub Jay: Demography of a Cooperative-breeding Bird.* Princeton, New Jersey: Princeton University Press.

Zack, S. & Ligon, J.D. 1985. Cooperative breeding in *Lanius* shrikes. II. Maintenance of group-living in a nonsaturated habitat. *Auk* 102, 766-773.

12. Use of Body Mass and Sex Ratio to Interpret the Behavioral Ecology of Richardson's Ground Squirrels

Gail R. Michener

North American ground squirrels (Family Sciuridae, Genus *Spermophilus*) are medium-sized (>100 to <1000 grams [g]) rodents found in habitats ranging from desert (e.g. round-tailed ground squirrels, *S. tereticaudus*), to prairie (e.g. Richardson's ground squirrels, *S. richardsonii*), to mountain meadows (e.g. Belding's ground squirrels, *S. beldingi*), to arctic tundra (Arctic ground squirrels, *S. parryii*). All species are diurnal, most species exhibit obligate hibernation, most species are found in open habitats in which the animals are readily visible, and many species are easily trapped. These characteristics make ground squirrels popular subjects for studies of behavioral ecology, particularly sociality, mating systems, reproduction, population dynamics, and life-history strategies. Additionally, ground squirrels are an ideal group for interspecific comparisons of the effects of parameters such as body size and habitat on behavioral ecology, social organization, and life history (see review articles in Murie & Michener 1984; also Armitage 1981; Dobson & Murie 1987; Michener 1983a). Interspecific comparisons require that the data being compared are equivalent, but variation among researchers in methods of collection and analysis of data impose limitations on such comparisons. Unless differences in methodologies are recognized, inclusion of nonequivalent data may result in weak or misleading interpretation of the interrelationships among parameters.

Two parameters often reported in field studies of ground squirrels are body mass and sex ratio. Early studies (e.g. McKeever 1963; Neal 1965; Nellis 1969; Skryja & Clark 1970; Sheppard 1972) frequently obtained such information from sacrificed animals, such that different individuals constituted each sample and each sample provided a static representation of the population at one point in time. Kill sampling is sometimes still used to document life-history parameters (e.g. Zammuto & Millar

1985), but live-capture of the same individuals repetitively throughout their lifetimes is more common. Although simple techniques are sufficient to record an animal's body mass and sex, collection of data from many individuals and on many days requires that the data be collated and pooled to detect underlying patterns. Appropriate analysis of such data can provide a detailed and dynamic view of seasonal and sexual variation in body mass cycles and above-ground activity (e.g. Masman et al. 1990; Michener & Locklear 1990a) that illuminates understanding of the behavior of the species and facilitates interspecific comparisons. The aim of this article is to indicate how I have used body mass and sex ratio to interpret sexual differences in mating strategies, reproductive expenditures, and costs of reproduction of Richardson's ground squirrels. In particular, I will compare two methods of pooling data on body mass to demonstrate that more information can be gleaned when the data are analyzed relative to major events in the annual cycle of the individual squirrel rather than relative to calendar date. Because the body mass of hibernating ground squirrels undergoes dramatic fluctuations through the annual cycle, I will also address the issue of using body mass as an indicator of species size in interspecific comparisons. Additionally, I will compare two methods of analyzing sex ratios to show that operational sex ratios provide a clearer understanding of mating strategies than do population sex ratios. Analyses of body mass relative to biological events and sex ratio relative to availability of estrous females share a common theme because both methods focus interpretation of behavioral ecology and life history on the individual rather than the population.

DATE-BASED VERSUS EVENT-BASED ANALYSIS OF BODY MASS

Repetitive capture of the same animals provides a data set that can be analyzed in two ways: 1) date-based analysis which pools data collected on the same date without regard to the identity of the individuals composing the sample or 2) event-based analysis which pools data collected from individuals of the same reproductive status without regard to the date of collection. For hibernating species of ground squirrels, the annual cycle of adult females is composed of a predictable sequence of events (emergence from hibernation, mating, gestation, lactation, prehibernatory fattening, immergence into hibernation) with considerable synchrony among individuals, suggesting that analysis of body mass by calendar date would adequately reveal information about the effects of reproductive status and season on body mass. Body mass of ground squirrels is frequently reported and data are generally pooled by date of collection (e.g. Boag & Murie 1981a; McLean & Towns 1981; Rickart 1982; Choromanski-Norris et al. 1986; Fagerstone 1988; Holekamp & Nunes 1989; Kenagy et al. 1989;). Although such presentation of data by calendar date does reveal a general trend for mass to increase throughout the annual cycle, important details that can illuminate understanding of behavior and life history are lost in the noise generated by pooling masses from females at different reproductive stages. For example, analysis of body mass of female Richardson's ground squirrels relative to their estrous condition indicated that females often exhibited a brief interruption in posthibernation weight gain during estrus (Michener 1985). Although most females (61% of 115 yearlings and 73% of 66 yearlings) failed to gain mass on the day they mated (Michener & Locklear 1990a), the change in mass was small (0 to -10 g for most individuals) and was not detectable in data pooled by calendar date because different females experienced the interruption in weight gain on different days depending on when they were in estrus. Failure to maintain weight gain during estrus suggested that females increased energy expenditure and decreased food intake when in estrus, a prediction that was subsequently confirmed by observational data (McLean personal communication) and indicated an energetic cost to mating activity of females. Thus, detailed analysis of body mass can direct research by suggesting behavioral patterns that merit study.

Date-based analyses of body mass are weaker than event-based analyses when reproductive status has a significant effect on body mass and when individuals in the sample are not at identical stages of reproduction. To demonstrate the different conclusions reached from date-based versus event-based pooling of data, I used both methods to analyze the same data set for body mass of adult female Richardson's ground squirrels (Figure 1). Known-aged squirrels were captured in unbaited livetraps on a 16-hectare (ca. 300 m x 525 m) portion of a study site located near Picture Butte, Alberta in 1984 (see Michener 1984 for descriptions of the study site and capture techniques). I attempted to capture each female on each of the following 10 events in her annual cycle: 1) *emergence* from hibernation (<24 h after resumption of above-ground activity), 2) *estrus* (day of mating), 3) *first trimester* of pregnancy (7-9 days postestrus), 4) *second trimester* of

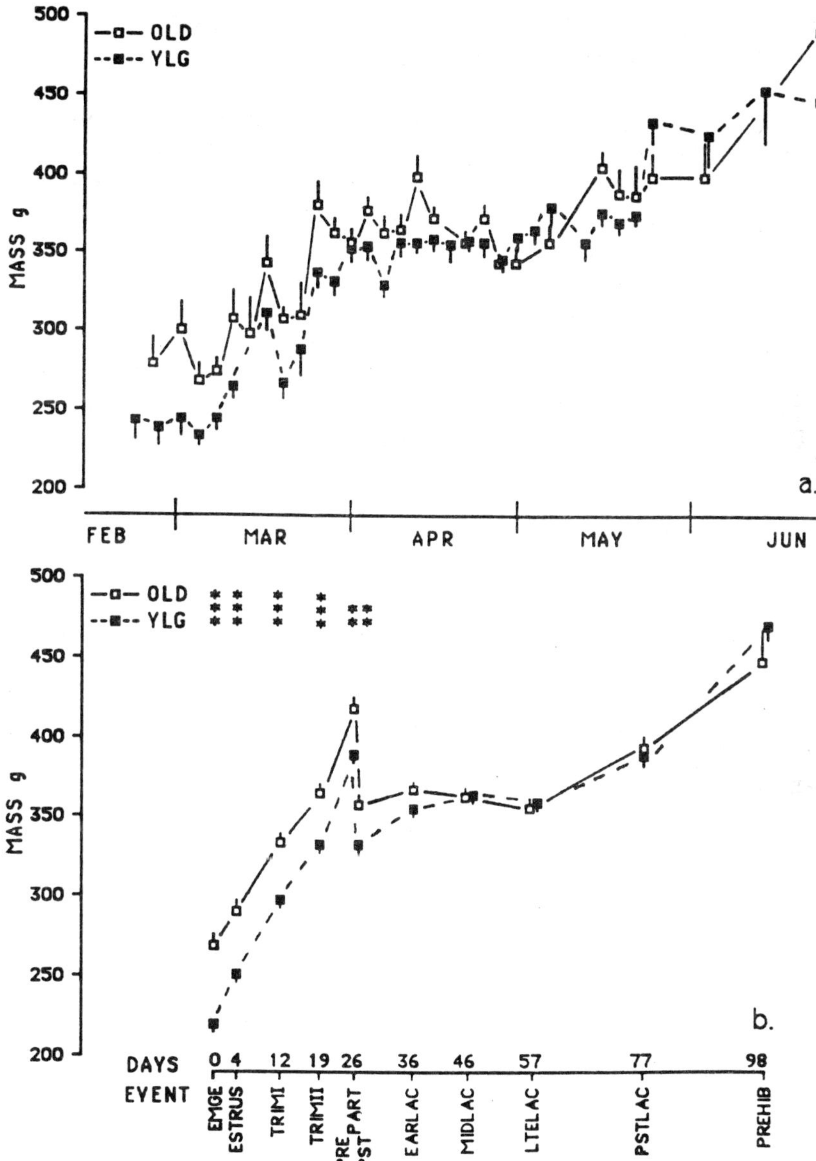

OLD
YLG
MASS g
500
450
400
350
300
250
200
a.
FEB
MAR
APR
MAY
JUN
OLD
YLG
MASS g
b.
DAYS
0 4 12 19 26 36 46 57 77 98
EVENT
EMGE
ESTRUS
TRIMI
TRIMII
PRE
PART
PST
EARLAC
MIDLAC
LTELAC
PSTLAC
PREHIB

Figure 1

Figure 1. Body masses of yearling and older female Richardson's ground squirrels in 1984 presented a) by calendar date and b) by reproductive status of the female. In Figure 1a, data are pooled over 3-day intervals from February through May and over 10-day intervals thereafter. In Figure 1b, data points for 11 events are simultaneously plotted on the mean date for that event and relative to the mean date of emergence from hibernation. Data are presented as mean - 1SE for yearlings and mean + 1SE for older females. In Figure 1a, average sample sizes per data point are 22 for yearlings and 11 for older females. In Figure 1b, average sample sizes per data point for the first 10 events are 68 for yearlings and 31 for older females and sample sizes for prehibernation are 25 and 8 for yearling and older females. Asterisks above data points in Figure 1b indicate that body masses of yearling and older females differ significantly for that event (*t*-tests, *** $P < 0.001$; ** $P < 0.01$).

pregnancy (14-16 days postestrus), 5) *preparturition* (<24 h before parturition), 6) *postparturition* (<24 h after parturition), 7) *early lactation* (8-12 days postpartum), 8) *mid-lactation* (18-22 days postpartum), 9) *late lactation* (28-32 days postpartum = time of litter emergence), and 10) *postreproduction* (47-53 days postpartum). For females resident on the eastern 11 hectares, body mass was also recorded at an 11th event, *prehibernation* (within 10 days of immergence). Further descriptions of how biological events were defined and how they were recognized in the field appear in Michener (1983b, 1985, 1989).

Analysis of body mass by date (Figure 1a) indicated a general trend for body mass of adult females to increase throughout the active season, as is expected for a hibernator that stores fat; the same trend was also clear from the event-based analysis of the same data (Figure 1b). Date-based analysis of mass suggested that yearling females were initially lighter than older females, and did not consistently maintain mass similar to older females until 7 weeks postemergence. Event-based analysis confirmed that yearlings were significantly lighter than older females, but revealed that the weight difference steadily diminished from 49 g at emergence to <12 g by early lactation such that the age groups were of similar mass within 5 weeks of emergence. The date-based analysis (Figure 1a) suggested that 1) females did not increase in mass during late February and early March, 2) females lost mass in the third week of March, and 3) females maintained mass through April. The event-based analysis (Figure 1b) revealed that only the latter pattern was real; the other two patterns were artifacts resulting from biases in the samples collected on those dates. Females increased in mass as soon as they resumed above-ground activity, but this was not apparent in the date-based analysis because inclusion of newly-emerged females (on average, 7 new females appeared each day from 22 February to 9 March) obscured the weight gains made by females that had been active for several days. After a lull in emergence during mid-March, the appearance of 33 new females during 18-22 March resulted in an apparent decline in mass during that period, but inspection of data from individuals revealed that females were steadily increasing in mass through March. The body mass trajectory that resulted from date-based analysis was misleading when interpretation exceeded the limits of what could be extracted validly. Date-based data indicated general trends in body mass at the population level, but did not yield reliable information on seasonal changes in body mass of individuals. In contrast, event-based analysis of body mass permitted a detailed interpretation of reproductive effort by yearling and older Richardson's ground squirrels because it focussed on changes in body mass of individual females relative to their reproductive status (see also Michener 1989). The following sections describe types of information obtained from the event-based data. Means are presented with ± 1 SD.

Growth by Yearling Females

Event-based analysis revealed that yearling female Richardson's ground squirrels immediately and steadily increased in mass postemergence (Figure 1b) and yielded the following conclusions with regard to growth patterns of female squirrels. 1) Yearling females weighed only 80% of the mass of older females on emergence, indicating that females had not grown to adult size as juveniles in their first summer. This conclusion was confirmed in a study of body composition in 1985-1986 (Michener & Locklear 1990b) which showed that, on immergence, the lean dry mass (LDM, a measure of structural size) of juvenile females was only 75% that of adult females (60 ± 3 g, n = 5 versus 80 ± 7 g, n = 15). 2) Females completed growth to adult size as yearlings during gestation and early lactation. This conclusion was also confirmed from body composition (Michener & Locklear unpublished data). Whereas the LDM of yearling females was significantly lower than that of older females on emergence (45 ± 3 g, n = 11 versus 57 ± 3 g, n = 7, $t = 7.72$, $P < 0.001$), LDM did not differ significantly by late gestation (66 ± 4 g for both 12 yearlings and 9 older females). 3) By completing growth in the first five weeks of the active season, yearlings attained adult size before the period of high energy demands in the last week of lactation, when litter mass exceeded maternal mass (Michener 1989). Completion of growth to adult size after mating indicates that mating success for females is not size-dependent; all females mated regardless of age and size, and females did not compete with each other for mating opportunities.

The steady posthibernatory increase in mass made by females was not apparent when data were pooled by date of collection (Figure 1b) because animals weighed on the same date ranged from those that were newly emerged to those that were in late pregnancy. The following example, based on the body masses of three sibling yearling Richardson's ground squirrels, exemplifies the extent to which differences in reproductive status mask similarity of body mass. The sisters, identified for convenience as F1, F2, and F3, differed little in body mass at emergence (205, 210, 210 g) but they emerged over a 4-week period (22 February, 3 March, 20 March 1984) even though they resided within 35 m of each other. On 21 March, when F1 was due to give birth, F2 was entering the last trimester of pregnancy, and F3 was in estrus, their body masses varied by 130 g (340, 280, 210 g) in direct relation to time since emergence from hibernation, hence time since mating. By mid-April, when F1 and F2 were lactating and F3 was due to give birth, F3 was heaviest (320, 300, 370 g). Not until late April, when F1 and F2 were terminating lactation and F3 was lactating, did the similarity of their body masses become apparent (325, 320, 335 g). If used without knowledge of their reproductive status, inspection of the body masses of these sisters by calendar date alone would suggest substantial within-litter variation in mass and, depending on the date selected, would provide contrary information about which sister was largest or healthiest as judged by body mass. For body mass to be used as a meaningful indicator of size and condition, only females at the same stage of their reproductive cycle should be compared.

Production by Yearling and Older Squirrels

Event-based analysis of body mass yielded information on reproductive output (Michener 1989) that could not be extracted from date-based data. Because parturition occurred in an underground nest, litter mass at birth could not be obtained directly by weighing infants. However, the change in mass of females weighed on consecutive days encompassing the time of parturition provided an estimate of litter mass at birth. The weight change recorded over the 24 hour period in which parturition occurred was primarily attributable to loss of mass due to birth of embryos, but also included loss due to expulsion of extra-embryonic membranes and changes in mass of gut contents. Because the sources of error incurred in estimating litter mass from weight loss were the same for yearling and older squirrels, I considered the estimate suitable for comparison of the gestational effort made by yearling and older squirrels. Estimated litter mass at birth (mean ± SD) in 1984 was similar for both age groups (53 ± 15 g for 56 yearlings and 56 ± 16 g for 26 older females), suggesting that the smaller size and mass of yearlings at emergence did not consign them to lower productivity. Similarity of reproductive output for the two age groups was confirmed by the similarity of litters, both in number of young (6.4 ± 2.1 for 55 yearlings, 6.3 ± 2.1 for 26 older mothers) and mass (534 ± 140 g for 32 yearlings, 573 ± 188 g for 23 older mothers), when they emerged from the natal burrow at ca. 30 days of age. The event-based analysis of mass revealed that despite having emerged from hibernation at only 80% of adult mass and having completed growth to adult size coincident with gestation and early lactation, yearling Richardson's ground squirrels achieved the same fecundity as older females. Litter mass at birth could not be estimated from the date-based presentation of body mass (Figure 1a) because individual variation in the date of parturition obscured both the timing and magnitude of the parturitional weight loss.

Gestational and Lactational Reproductive Effort

Event-based analysis of maternal and litter growth permitted comparison of reproductive effort during gestation and lactation. Because maternal mass was known for individual squirrels at the beginning and end of gestation and at the beginning of lactation, the total gain made in 1984 by 46 yearlings and 18 older Richardson's ground squirrels during the 23-day gestation period could be partitioned into gain attributable to embryos and gain attributable to personal growth by the female. Yearlings made a significantly greater gain during gestation than older mothers (139 ± 28 versus 123 ± 26 g; $t = 2.11$, $P < 0.05$), with the difference arising from greater deposition of personal mass by yearlings (86 ± 26 versus 65 ± 24 g, $t = 2.92$, $P < 0.01$) rather than greater investment in embryos (53 ± 15 versus 58 ± 18 g). For both age classes of females, less than half the mass desposited during gestation was consigned to the litter; 61% of the gestational gain of yearlings and 52% of the gestational gain of older squirrels accrued as an increase in the mother's own mass (Figure 1b). This analysis indicated that females met the energy requirements of growing embryos while increasing their own mass and, further, that yearlings achieved the same reproductive performance as older females while making a larger investment in personal growth. That pregnant females did not trade off their own growth against that of the developing litter was further indicated by the weak nonsignificant association between birth mass of the litter and increase in maternal mass during gestation ($r = -0.16$ for 46 yearlings, $r = -0.25$ for 18 older females). Furthermore, females retained fat stores throughout gestation; in 1986, fat content was 33 ± 9 g (n = 18) at emergence from hibernation, 36 ± 9 g (n = 20) in midgestation, and 40 ± 12 g (n = 21) in late gestation. Together, these observations indicate that the ability of Richardson's ground squirrels to invest in the litter is not energy-limited during gestation. The ability of yearlings to grow without sacrificing reproductive output suggests that foraging behavior, activity budgets, and energy balance of females merit attention to determine the roles of diet, time spent feeding, and digestive efficiency on the ability of yearlings to accumulate more mass than older females during pregnancy.

An equivalent analysis of changes in maternal and litter mass during lactation revealed that virtually all the mass deposited by lactating females was devoted to the litter (Michener 1989). During the ca. 30-day period from birth to litter emergence, young were solely dependent on maternal nutrition. Based on changes in maternal mass of individual squirrels from postpartum to late lactation and on changes in litter mass from birth (estimated from maternal weight loss at parturition) to emergence (determined by weighing newly-emerged juveniles), litters of 27 yearling and 12 older mothers in 1984 increased by 498 ± 130 g and 480 ± 126 g, respectively. The mothers' masses changed by only 26 ± 24 g and 0 ± 15 g, respectively, in this time period. Thus, <5% of the net change in mass of the mother-litter unit accrued to the mother during lactation compared with >50% during gestation. That maternal weight gain was less likely for a lactating than a pregnant female was predictable given the relative sizes of litters at the end of gestation and the end of lactation in 1984. At birth, litters weighed only 16 ± 4% (54 ± 15 g, n = 85) of concurrent maternal mass, whereas at emergence from the natal burrow, litters weighed 154 ± 41% (555 ± 161 g, n = 55) of concurrent maternal mass.

Several lines of evidence indicated that females were energy-limited during lactation (Michener 1989). Whereas females retained their fat reserves through gestation, fat was depleted during lactation (Michener & Locklear 1990a), indicating that lactating females did not meet the energy demands of milk production through foraging alone. Field-caught females that reared litters in captivity weaned litters of greater mass (719 ± 138 g for 60 captive litters), but not greater size, than field litters, but did not themselves exhibit an increase in personal mass (7 ± 31 g weight change for 45 females weighed 0 and 30 days postpartum) despite access to ad libitum food and reduced energy costs associated with thermoregulation and foraging (Michener 1989). Presumably females in the field were unable to wean litters with mass equivalent to captive litters because of energy limitations. The need to augment food intake with fat reserves to meet lactational demands may result from physical and physiological limitations on the female's capacity to eat and process food or from behavioral limitations determining the tradeoff between time available to forage and time spent suckling and caring for young.

Collection of Data for Event-Based Analysis

The limitations of date-based analyses were recognized by Hohn & Marshall (1966 Figure 3) and Morton (1975 Figure 5) who presented data for one female thirteen-lined ground squirrel (*S. tridecemlineatus*) and five female Belding's ground squirrels, respectively, to demonstrate abrupt changes in mass associated with parturition and prehibernatory fattening that were not apparent in their date-based analyses. Morton's (1975 Figure 5) simultaneous presentation of data for five females graphically reveals asynchrony of up to 3 weeks among individual Belding's ground squirrels for phenomena such as gestational increase in mass, parturition weight loss, and prehibernatory fattening. Despite these indications of the utility of inspecting data in the knowledge of the animal's reproductive status, most reports of hibernating sciurids present body mass by date (e.g. Boag & Murie 1981a; Choromanski-Norris et al. 1986; Fagerstone 1988; Holekamp & Nunes 1989; Kenagy & Barnes 1988; McLean & Towns 1981; Rickart 1982) rather than by event (Michener 1984; Michener 1989; Michener & Locklear 1990a; Masman et al. 1990).

The paucity of event-based analyses of data presumably reflects the logistical demands of obtaining such data. Traditional methods of studying small mammals involve operating traps at predetermined intervals (e.g. one day every week) such that the researcher obtains a picture of the population on selected dates without following the precise reproductive status of each individual. Event-based analyses require that the time of at least one major reproductive event (estrus, parturition, or litter emergence) be known for each individual in the population, so the researcher must monitor animals at frequent intervals then capture the appropriate females on target dates. Ideally, each major reproductive event would be detected for each animal. However, provided the occurrence of one event is known and the intervals between events (gestation and juvenile age at emergence) are known, back-dating or forward-dating can be used to estimate the time of occurrence of the other events for that individual. Estrus and parturition are brief events that occur on one day, so knowledge of these dates permits precise determination of the female's reproductive status such that data from different females can be aligned exactly. Litters may differ in age at first emergence by several days (Michener 1985; Kenagy et al. 1989), but in the absence of other information, litter emergence dates provide better synchronization of data than calendar dates. Because information on litter emergence can be obtained from daily observation, this event may be the most useful marker of reproductive chronology for species that are difficult to trap at the time of estrus or parturition. Litter emergence is readily detectable for species in open habitats, but for species occupying habitats with reduced visibility, time and place of litter emergence must be anticipated to ensure litters are detected as soon as they commence above-ground activity. An alternative method of synchronizing data, again with less perfection than if estrous or parturition dates are known, is to use known dates of emergence from hibernation to align each female's data by the number of days since she emerged. Females of many species of ground squirrels mate within several days of emergence (Michener 1985; Murie & Harris 1982), so synchronization of data by emergence date approximates synchronization by estrous date. Regardless of which events are used to align data, event-based analyses depend on knowledge of the reproductive status of individuals.

Intervals between emergence and estrus, estrus and parturition, and parturition and litter emergence have been reported for remarkably few North American ground squirrels (Michener 1985). Gestation length varies by up to 20% among species of ground squirrels (Michener 1989 Table 5), yet no attention has been directed to the significance of such variation in terms of developmental state of neonates, physical limits to passage of embryos through the pelvic canal, or tradeoffs between prenatal and postnatal growth.

Here I have shown that the method of analysis of the same data set influences the interpretation of those data. The overall shape of the date-based curve for body mass of female Richardson's ground squirrels was a wave composed of two periods of weight maintenance followed by two periods of weight gain, with gradual transitions between each phase (Figure 1a). In contrast, the event-based curve was composed of a single period of weight maintenance preceded and followed by periods of weight gain, with distinct transitions between each phase (Figure 1b). The former curve is not biologically interpretable but is a statistical artifact that results when females of different reproductive status are treated as a common sample by virtue of being captured on the same date. The latter curve is biologically meaningful and coincides with three reproductive conditions, gestation, lactation, and prehibernation. Alignment of data by event rather than date provides information about growth, litter mass at birth, and partitioning of mass between the mother and her litter. Event-based data can also be used for interannual comparisons of body mass because females of similar reproductive status can be compared even when dates of reproductive events differ by several weeks between years (Michener 1984; Michener & Locklear 1990a). The utility of event-based analyses is not limited to ecological data. Physiological parameters such as metabolic rates and hormone titers, and behavioral parameters such as social interaction rates and activity budgets are best understood in reference to the animal's reproductive status. In addition to facilitating within-sex comparisons, event-based analyses also provide data suitable for intersexual and interspecific comparisons of body mass, life-history patterns, and behavior.

SEXUAL DIFFERENCES IN BODY MASS

Reproduction by female ground squirrels is marked by two brief events, estrus and parturition, each lasting <4 hours, that unambiguously define their reproductive status and which influence parameters such as body mass, time and activity budgets, and social interaction rates. Equivalent events do not occur for male ground squirrels. However, the active season of males is marked by an event lasting several weeks, the mating season, during which estrous females are available. Knowledge about the timing and intensity of the mating season illuminates interpretation of date-based body mass trajectories of males (Figure 2) and seasonal variation in behavior of males.

Although adult male and female Richardson's ground squirrels both exhibited an overall tendency to increase in mass through the active season, their body mass trajectories differed substantially in detail (Figures 1 and 2, see also Michener 1984; Michener & Locklear 1990a). Body mass of adult males was recorded at approximately weekly intervals in 1984. Males showed a posthibernatory increase in mass, as did females, but then they lost mass coincident with the mating season, slowly increased in mass after the mating season, and rapidly increased in mass before immerging. Males attained their lowest mass for the active season several weeks after emergence from hibernation whereas females were at their lowest mass on emergence, and males experienced a greater scope of weight change through the annual cycle (from 392 g during mating to 688 g at immergence, Figure 2) than females (from 219 g for yearlings and 268 g for older females at emergence to 462 g at immergence, Figure 1b).

The mating season in 1984 was unusual in that a snowfall and subzero °C air temperatures in mid-March delayed emergence and mating of some females such that the availability of estrous females exhibited a bimodal pattern instead of the typical unimodal pattern (Figure 3). In the four intervals 6-12 March, 13-18 March, 19-26 March, and 27 March-2 April, the average numbers of females in estrus per day were 10.3, 1.8, 4.9, and 0.7 and the average changes in mass of males reweighed within each interval were -31 g (n = 10), +14 g (10), -13 g (6), and +23 g (5). Thus, males exhibited a bimodal pattern of weight loss that corresponded to the fluctuating availability of potential mates (Figure 2).

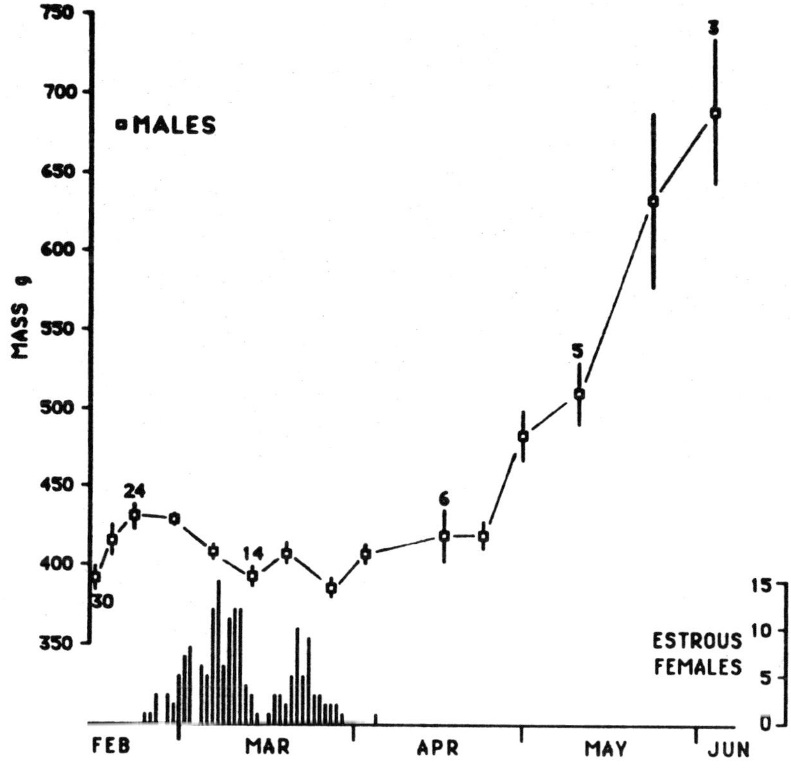

Figure 2. Mean ± SE body masses of adult (≥1 year) male Richardson's ground squirrels in 1984, and numbers of females in estrus on each day of the mating season. Body masses are plotted by calendar date of capture except for the first and last points which give mean emergence and prehibernation masses and are plotted on the mean dates for those events. Due to attrition of males, particularly during the mating season (see Figure 3), sample sizes declined. Sample sizes are indicated beside representative data points at approximately monthly intervals.

Loss of mass coincident with mating activity, which occurred in all years (Michener 1984; Michener & Locklear 1990a), suggested that the energetic costs of mating were not met by foraging and that males subsidized these costs by mobilizing energy stores. These assumptions were confirmed by studies of activity budgets (McLean unpublished) and body composition (Michener & Locklear 1990a,b). The amount of time males spent moving and interacting was higher in the mating season (34%) than before or after (17% and 16%) but the amount of time foraging was reduced (24% versus 31% and 43%). Males emerged from hibernation in 1986 with substantial fat reserves (74 ± 15 g, n = 11), but fat content had declined to 47 ± 13 g (n = 7) a week later when one third of the females had mated, and to 9 ± 8 g (n = 10) another 9 days later when 91% of females had mated. Lean dry mass did not exhibit a corresponding decline (78 ± 6, 83 ± 4, and 85 ± 6 g at emergence, and early and late in the mating season, respectively), so loss of mass in the mating season resulted primarily from depletion of residual fat stores during periods when time spent foraging was insufficient to meet energetic costs. Females did not exhibit a similar dependency on fat during the first 3 weeks of activity (Michener & Locklear 1990a).

By considering mass of male Richardson's ground squirrels relative to the availability of estrous females (Figure 2), the analysis, although date-based in format, effectively became event-based. Masman et al. (in press) presented mass of male Cascade golden-mantled ground squirrels (*S. saturatus*) relative to the mean date of parturition of females in the population and thus likewise imposed an event-based perspective on date-based data. Backdating Masman et al.'s body mass curves by the length of gestation gives body mass of males on the mean date of estrus but, without an indication of the degree of synchrony of mating by females, details of the relationship between availability of mates and changes in mass cannot be assessed. Potential future refinements for interpreting the relationship of body mass to mating effort in male ground squirrels involve monitoring the mating success of individual males, their corresponding changes in mass, and their use of fat stores. Walsberg (1988) described a method of determining fat content on live animals that would permit fat to be assessed repetitively on the same individual throughout the mating season.

Sexual differences in body mass trajectories and use of residual fat stores indicated that mating effort for male

Richardson's ground squirrels was more energetically costly than either mating or maternal effort for females (Michener & Locklear 1990a), thereby raising the question of whether males incurred greater costs than females in terms of survival. Analysis of seasonal mortality schedules showed that the reproductive season (from emergence to the end of the mating season) was much riskier for males than either the prehibernation (from the end of reproduction to immergence) or overwinter seasons; average (n = 4 years) percentages of males disappearing within each period were 64, 28, and 31% for reproduction, prehibernation, and overwinter. The corresponding values for females were 16, 7, and 25% for reproduction (from emergence to above-ground appearance of the litter), prehibernation, and overwinter. When adjusted for duration of the period, males disappeared at a rate of 1.56%/day during reproduction compared with 0.28%/day for females. Thus, adult males were more likely to disappear during reproduction than females, establishing that reproduction incurred not only a greater proximate (energetic) cost but also a greater ultimate (survival) cost on males than females (Michener & Locklear 1990a). Event-based body mass trajectories, when viewed from the perspective of detailed knowledge of the mating season, thus reflected reproductive strategies of males and females.

Masman et al. (in press) used the annual body mass cycle of Cascade golden-mantled ground squirrels to explore differences in reproductive and hibernation strategies of males and females. They also found sexual differences in body mass trajectories. Body mass of female golden-mantled ground squirrels followed a similar annual trajectory to that of female Richardson's ground squirrels, with three phases corresponding to gestation, lactation, and prehibernatory fattening. However, the body mass trajectory of males differed substantially between the species; adult male golden-mantled ground squirrels increased in mass by only 18% throughout the active season compared with 75% for Richardson's ground squirrels. Male golden-mantled ground squirrels did not store sufficient fat to meet overwinter costs of thermogenesis and homeothermy; apparently they depended on food caches to provide additional energy during hibernation. Explanation of such interspecific differences in body mass trajectories awaits comparable information from other species to decipher the roles of factors such as habitat, diet, and social structure on strategies of energy storage by males for winter and for the mating season.

Because body masses of adult male and female Richardson's

ground squirrels followed different trajectories, particularly in the two months after emergence from hibernation, the magnitude of the sexual difference in body mass varied substantially throughout the active season (Michener & Locklear 1990a). Thus, if body mass is used as an indicator of body size, allowance must be made for the reproductive status of individuals being compared. Although newly emerged males weighed significantly more than newly emerged females (392 ± 40 g for 30 males versus 268 ± 34 g for 41 old females and 219 ± 30 g for 103 yearling females in 1984), by late March, adult males were lighter than prepartum old females and of similar mass to prepartum yearlings (386 ± 16 g for 7 males versus 427 ± 51 g for 8 old females and 377 ± 37 for 21 yearlings). Fat depletion by males coincided with weight gains by females (due both to personal growth and developing embryos), thereby reducing the sexual difference in body mass and obscuring the underlying sexual difference in body size. That males were still structurally larger than females was revealed by an analysis of body composition. Males collected late in the mating season of 1986 weighed 355 ± 18 g and contained 85 ± 6 g of LDM (n = 10); although prepartum females weighed more (395 ± 18 g including embryos), their LDM (excluding embryos) was only 66 ± 4 g (n = 21) (see also Michener & Locklear 1990a for data from 1985). Because the timing of energetically-demanding reproductive events differs for male and female ground squirrels, sexual comparisons of body mass by calendar date will be biased according the reproductive status of squirrels in the samples. Such bias can be minimized by comparing masses at equivalent events instead of by date; postpartum females and premating males, for example, are cohorts of equivalent status in that these animals are prepared for the upcoming period of high energy demands.

BODY MASS VERSUS BODY SIZE

Although body mass can be measured accurately, it is not necessarily a precise indicator of an animal's size or condition. Structural size in species with determinant growth is essentially constant once adult size is achieved, but mass of adults continues to vary with season, nutritional status, and reproductive status. Daily variation in magnitude of the gut contents may confound recognition of variation in mass due to factors such as reproductive status. Mass of the gut contents cannot be readily evaluated in the live animal, though repetitive capture of the same individual can reveal the range of daily variation in body mass. Hohn & Marshall (1966) reported that three thirteen-lined ground squirrels increased in mass by 29-31 g during the day and decreased by 19-26 g overnight. Methods of minimizing such daily variation in body mass include capturing animals at the same time of day relative to their foraging pattern (for example, as they resume foraging in the morning for a diurnal mammal such as the Richardson's ground squirrel) and obtaining large samples. These techniques enabled me to distinguish minor trends in body mass, such as the weight loss commonly associated with estrus in female Richardson's ground squirrels, from variation due to mass of gut contents (Michener 1984).

Seasonal variation in mass of the reproductive tract also confounds interpretation of body mass. The mass of the reproductive tract (ovaries, uteri and contents, and vagina) of female Richardson's ground squirrels increased from 0.6 ± 0.2 g (0.27 ± 0.07% of body mass, n = 18) at emergence, to 6.1 ± 1.9 g (1.9 ± 0.5%, n = 20) 12 days postestrus in midpregnancy, to 57.6 ± 12.6 g (15.1 ± 3.2%, n = 21) 21 days postestrus in late pregnancy. Variation in mass attributable to reproductive condition can be minimized by normalizing data according to each female's date of estrus or date of parturition (Figure 1b). In contrast to females, variation in combined mass of the left and right testes of adult male Richardson's ground squirrels contributed inconsequentially to seasonal variation in body mass. Mass of the paired testes ranged from 2.34 ± 0.35 g (0.58 ± 0.07% of body mass, n = 11) at emergence from hibernation to 0.33 ± 0.06g (0.06 ± 0.01%, n = 5) at immergence. Seasonal changes in mass of males was primarily attributable to changes in fat content.

Even when adjustment has been made for the contribution of the contents of the gut and reproductive tract to mass, similarity of body mass does not necessarily indicate the animals are of similar size; a large lean animal and a small fat animal may weigh the same. Researchers rarely attempt to distinguish size from mass on live animals, but techniques such as measurement of a skeletal parameter (e.g. ulna length, Young 1988) and fat content (e.g. Walsberg 1988) could permit this discrimination to be made.

BODY MASS IN INTERSPECIFIC COMPARISONS

Interspecific comparisons of life-history parameters, such as offspring or litter mass at birth and at weaning, traditionally report these parameters relative to adult mass (e.g. Leitch et al. 1959; Leutenegger 1976; Millar 1977; Western 1977; Blueweiss et al. 1978; Stearns 1983; Modi 1984; Kurta & Kuntz 1987). In such comparisons, mass is being used as an indicator of species size. The advantages of using mass rather than a skeletal parameter as an indicator of species size include the ease with which mass is measured on live animals, equivalence with the units used to measure reproductive output (offspring mass), and avoidance of the confounding problems of taxonomic variation in shape (e.g. fossorial mammals are shorter limbed than cursorial mammals). Leitch et al. (1959) indicated that the ideal body masses to use in interspecific comparisons of reproductive effort are those based on individually paired data for a female and her offspring, and they recommended using maternal mass at mating or at parturition as the measure of adult mass. Although such paired data are occasionally reported (e.g. Michener 1989), more commonly reproductive effort is calculated using average data for each parameter, sometimes with the data for offspring mass and maternal mass coming from different sources (e.g. Armitage 1981). Because maternal mass at mating or at parturition is not available in many field studies, some other mass must be used to represent species size. Millar (1977) suggested that the mass of nonbreeding adult females be used as the reference adult mass in comparisons of reproductive effort because the energetic requirements of such females indicate maintenance needs. Although some authors of interspecific comparisons of life-history parameters define adult mass clearly and restrictively (e.g. Kurta & Kunz [1987] used postpartum mass when available and never included mass of males), others provide no definition (e.g. Modi 1984) or indicate use of species averages that potentially include individuals of both sexes and of different reproductive status (e.g. Stearns 1983). For species in which body mass varies little through adulthood, use of a generalized average mass, particularly when restricted to females that are not visibly pregnant, probably produces little error in interpretations of interspecific life-history patterns. Iskjaer et al. (1989) showed that for six species of small, nonhibernating rodents, body mass is a reasonable measure of body size. For hibernating sciurids, which show dramatic changes in body mass throughout the annual cycle, defining a species-typical mass is problematical.

Body masses of yearling female Richardson's ground squirrels increase by >100% during the active season. Some of this increase is attributable to growth. However, even older females who have attained adult skeletal size, increase in mass by ca. 75% during the active season (Michener 1989). Thus, body mass of hibernators is not a reliable indicator of species size. Because prehibernatory fat storage results in body masses not attained by nonhibernators of equivalent skeletal size and because immergence mass is substantially higher than mass at the time reproductive effort is occurring, immergence mass overestimates species size of hibernators and is not a suitable candidate for interspecific comparisons of life-history parameters. For males, emergence mass approximates their condition as they enter the mating season, so emergence mass is a better indicator of size than immergence mass. However, most reproductive parameters are scaled to the mass of adult females. Because females of hibernating species are at their minimum mass for the annual cycle on emergence and increase in personal mass during gestation, emergence mass underestimates the condition of the female as she commences the energetically expensive function of lactation. Millar's (1977) suggestion of using the mass of nonbreeders as the reference mass is not suitable for hibernators because nonbreeding individuals also exhibit an annual cycle of fat deposition and fat mobilization both in the field (Michener 1978; Choromanski-Norris et al. 1986) and under constant conditions in captivity (e.g. Melnyk 1983; Joy 1984; Phillips 1984). Thus, hibernating species present a special problem in selection of a body mass that is representative of species size, hence appropriate for interspecific comparisons of reproductive effort.

In their study of reproductive effort by Cascade golden-mantled ground squirrels, Kenagy & Barnes (1988: 279) used as a reference "the overall mean of values distributed evenly over the active season and excluding pregnant females." Thus their reference mass was a composite of data from newly-emerged, lactating, and prehibernation animals, and so was a statistically derived mass rather than a mass related to a specific stage in the annual cycle. The relative proportions of animals in such composite samples from each stage of the annual cycle will strongly influence the overall mean. Additionally, inclusion of repetitive data from the same individual may bias the data set if certain animals (for example, light weight animals attracted to baited traps) are more prone to capture than others. Consequently, unless researchers use similar criteria to include values in the overall mean, composite masses for species that undergo marked annual variation in mass will not be comparable between studies.

The mass that is most frequently reported in different studies of hibernating sciurids and which seems to be collected in the most consistent manner is emergence mass in spring. Newly emerged ground squirrels can be identified by a combination of frequent observation to detect new animals and physical appearance such as the presence of dry flakes of skin in the pelage (Murie & Harris 1982; Michener 1983b). Some workers trapped squirrels as frequently as daily during the period of emergence (e.g. Murie & Harris 1982; Michener 1984; Fagerstone 1988), so masses were obtained for individual animals within one or two days of their first appearance above ground. Others used first capture masses of individual animals with an associated criterion such as including only animals caught during the first week of the active season (Choromanski-Norris et al. 1986), only the first 25 animals for that cohort (Knopf & Balph 1977), or only animals known from observation to have been active <8 days (Dobson & Kjelgaard 1985a), thereby restricting the data set to animals that were weighed within a few days of emergence.

In an 11-species comparison of gestational effort among hibernating ground squirrels (Michener 1989, Table 5), I expressed neonate mass and litter mass at birth relative to emergence mass of adult females. Relative neonate mass ranged from 1.8 to 4.3% of maternal mass and relative litter mass ranged from 10.8 to 28.3%. Relative birth mass was negatively correlated with maternal emergence mass (for neonate mass r_S = -0.67, $P < 0.05$; for litter mass r_S = -0.76, $P < 0.01$). Because females are at their lowest body mass for the year on emergence and increase in mass by parturition, relative masses obtained from calculations with emergence mass in the denominator are larger than if concurrent maternal mass is used (Table 1). Furthermore, changing the denominator not only alters the calculated value but also changes the relationship between yearling and older squirrels from one in which yearlings appear to make a greater relative reproductive effort than older squirrels to one in which age does not affect reproductive effort (Table 1). When interspecific comparisons are restricted to hibernators and a consistent reference mass, such as emergence mass, is used, the

Table 1. Litter masses of yearling and older Richardson's ground squirrels at birth and at first emergence above ground relative to maternal mass. Litter masses are calculated as per cent of maternal mass using 1) mother's mass at emergence from hibernation and 2) mother's mass at the end of the period of total infant dependence on the mother, i.e. postpartum mass and late lactation mass. Data are pooled for 1982-1985 (Michener 1989) and are presented as mean ± SD. *** indicates $P < 0.001$; NS indicates $P > 0.20$.

	N	1) Litter/Maternal	N	2) Litter/Maternal
Maternal mass at:		Emergence		Postpartum
Litter mass at birth				
Yearling	154	26 ± 6%	165	17 ± 4%
		***		NS
Older	61	22 ± 6%	72	17 ± 4%
Maternal mass at:		Emergence		Late lactation
Litter mass at emergence				
Yearling	117	268 ± 62%	112	156 ± 29%
		***		NS
Older	63	221 ± 62%	70	159 ± 38%

resulting values are directly comparable even though they may not be an exact index of reproductive effort relative to the female's concurrent mass at the time she makes that effort. Difficulties arise in intertaxonomic comparisons involving mammals that do not hibernate because indices based on emergence mass of hibernators tend to overestimate reproductive effort by hibernators.

Because litter mass at weaning is a measure of the total reproductive output of the female and because litters impose their maximum energy demands on the mother just before weaning, presentation of weaned litter mass relative to maternal mass in late lactation provides an index of the maternal effort required to complete reproductive effort. In attempting to perform such an analysis for ground squirrels (Michener 1989), I discovered that few studies report either masses of juveniles known to be recently emerged from the natal burrow or maternal mass at the time of litter emergence. Most published juvenile masses from field studies are based on data pooled over a period of a week or more from captures conducted after juveniles become readily trappable, so they tend to be heavier than mass at first emergence based on selective capture of known newly emerged young. Captive studies do not provide an accurate indication of mass at first emergence from the natal burrow because captive juveniles typically weigh more than same-aged field animals (Michener 1989; Morton et al. 1974; Turner et al. 1976). Several factors contribute to lack of information about maternal mass at late lactation: Some species are not readily trapped (Rickart 1982), some workers are reluctant to trap or handle females during lactation lest capture affect their maternal behavior or provide an opportunity for conspecifics to harm unprotected young (Dobson & Kjelgaard 1985b), and most studies report mass by calendar date not by reproductive status of the female. Because of such limitations, I restricted the interspecific comparison of litter mass at weaning to six species of ground squirrels for which both juvenile mass at emergence from the natal burrow and maternal mass at emergence from hibernation were available. Unlike relative litter mass at birth, relative litter mass at weaning was not significantly negatively correlated with maternal emergence mass ($r_s = -0.31$, $P > 0.25$). Among the six species compared, Columbian ground squirrels wean an unusually small litter mass for maternal mass and Richardson's ground squirrels wean an unusually large litter mass for maternal mass (85% versus 252%), suggesting that these species merit further study of their reproductive effort.

Kenagy & Barnes (1988) presented an interspecific comparison of relative litter mass for three ground-dwelling squirrels, but as their reference mass was a statistically derived value based on captures throughout the year, their results cannot be compared conveniently with values from other studies. They reported that litter mass at first capture after emergence from the burrow was 170% of maternal mass for Cascade golden-mantled ground squirrels. This differs markedly from the value of 113% that I calculated (Michener 1989) because they used 1) litter size at birth not emergence from the burrow (4.1 versus 2.9), 2) mass of juveniles at first capture not the mass of juveniles known to be recently emerged (85 versus 66 g), and 3) composite maternal mass not emergence mass (208 versus 169 g).

Comparisons of life-history patterns among hibernating ground squirrels are hampered by a combination of lack of suitable information, inadequate descriptions of the criteria used to include data, and differences among researchers in the mass used as a reference to adjust for species differences in size. Comparisons of life-history patterns of hibernators and nonhibernators are further hampered by the difficulty of identifying a body mass for hibernators that is biologically equivalent to the mass used to represent size for nonhibernators, viz. the mass of nonbreeding females. The large differences in relative litter mass introduced by changing the denominator (Table 1) exemplify the problem of selecting an appropriate maternal mass for interspecific comparisons. Given these problems, clarity is required in initial reporting of data for hibernators and caution is required in extracting such information for use in interspecific comparisons.

POPULATION AND OPERATIONAL SEX RATIOS

Sex ratio of adults is generally an easily obtained and commonly reported parameter in field studies of mammalian populations. For many species of ground squirrels, adult females are more abundant than adult males (Boag & Murie 1981b Table 10) and, because all females are inseminated, polygyny is a statistical necessity. Although the numerical abundance of females might suggest that males need not engage in intense competition for access to mates, several observations contradict this assumption. For example, male Belding's and Richardson's ground squirrels lose mass, sustain serious injuries, and disappear at a greater rate than females during the mating season (Michener 1983b; 1985; Sherman & Morton 1984; Michener & Locklear 1990a). The apparent contradiction between females being numerically abundant and males needing to compete for these females is resolved when sex ratios are calculated on the basis of the number of females in estrus rather than the number present in the population. Emlen & Oring (1977) used the term "Operational Sex Ratio" (*OSR)* for the number of fertilizable females per sexually active male.

Because Richardson's ground squirrels are diurnally active and readily trappable, daily counts of the numbers of males and estrous females can be made such that the OSR is known for each day of the mating season. I considered a female to be in behavioral estrus if she mated, and I recognized the occurrence of mating from a combination of the following: swollen pink vulva with an associated vaginal lavage primarily of cornified epithelial cells with no sperm, consorting with males in the late afternoon and then copulating above ground or going underground with a male, and presence of a copulatory plug in the vagina or sperm in the vaginal lavage the next morning. Matings were considered fertile if the female gave birth 23 days later. During a 5-year study, 1982-1986, most females (91-96% annually) mated on only one day each year during which time they were in behavioral estrus for <4 hours and mated with 1-4 males. Females that mated on a second day usually did so either on the next day (8/43) or 5-9 days later (27/43). In calculating OSR, I included both first and subsequent estrous periods for females that mated on >1 day.

Table 2. Sex ratio (females per male) among adult Richardson's ground squirrels. The total population comprised all squirrels present on the study site for at least one day in spring. The post-mating population included males still present on the site the day after the last female mated and females still present in the last trimester of pregnancy. Operational sex ratios (OSRs; estrous females per male) were calculated for each day of the mating season and are presented as mean ± SD and (n).

	1982	1983	1984	1985	1986
Area (ha)	18.0	18.0	16.0	3.7	3.7
Total population					
Females	97	111	172	107	148
Males	23	24	33	30	35
Sex ratio	4.2	4.6	5.2	3.6	4.2
Post-mating population					
Females	87	104	145	97	130
Males	10	12	7	7	13
Sex ratio	8.7	8.7	20.7	13.9	10.0
Operational sex ratio					
	0.21 ± 0.23	0.17 ± 0.33	0.23 ± 0.22	0.38 ± 0.55	0.32 ± 0.25
	(34)	(35)	(42)	(27)	(22)

Based on the numbers of adult Richardson's squirrels present on the study site for at least one day in spring, females outnumbered males 4- to 5-fold each year (Table 2). Attrition of males greatly exceeded that of females (Michener & Locklear 1990a), so that the sex ratio among adults at the end of the mating season was 9-21 females per male. Despite the abundance of females, males outnumbered estrous females on almost all days of the mating season (Figure 3). With the exceptions of 1 day in 1982, 2 days in 1983, 1 day in 1984, and 3 days in 1985, OSR was <1 estrous female per male throughout the mating season. Male-biased OSRs indicated that males were competing for access to a limited resource. Full appreciation of the problems males surmounted in gaining access to estrous females requires consideration not only of the numbers of females in estrus on a given day but also the abundance of estrous females relative to all females in the above-ground population (Figure 4). Early in the mating season, when few females have emerged, most of the females that are above ground are either in estrus or 1-2 days pre-estrus. Such females tolerate the presence of males and allow males to approach and sniff them. As the mating season proceeds, not only does the number of females above ground increase, but an increasing proportion of those females have mated and are of no further reproductive value to males that year. Mated females are intolerant of the proximity of males, including their mate(s), and chase males that approach them or that come near them during approaches to other, unmated, females. Thus the social interactions of males involve detecting the appearance of newly emerged females who will become potential mates, locating and remaining near females that emerged 1-3 days ago and are in estrus, avoiding females that have been active for several days and are pregnant, and driving off males that are potential competitors.

Comparison of daily OSR and population sex ratio (*PSR*) provides a perspective on the social milieu in which males must locate estrous females (Figure 4). Before 12 March 1985, males outnumbered females and only one female mated, so both the PSR and OSR were male biased. Thereafter, the PSR was female biased. On 14 March, females outnumbered males (40 females:11 males), estrous females were less abundant than males (7 estrous females:11 males), and 18% of the females present were in estrus. By 17 March, the PSR was even more female-biased (78 females:11 males) and the OSR was also female-biased (23 estrous females:11 males), with 30% of all females in estrus. However, after the peak period of mating, estrous females comprised a small fraction of females. From 21 March onwards, females were >10 times as numerous as males, but OSR was <1 estrous female per male and <4% of females were available for mating. For example, on 26 March, eight males were seeking two estrous females in a population of 99 females. Because the area under study in 1985 was only 3.7 hectares, each male could potentially have inspected the entire site. In practice, each male's ability to seek estrous females was constrained by the aggression of neighboring males and of pregnant females and by the male's tendency to guard estrous females. These constraints were offset to some extent by the propensity of females to undertake excursions outside their normal areas of residence during their several-hour estrus. Daily OSRs provide an average representation of the availability of mates to males, but do not take into account the temporal and spatial distribution of estrous females that probably results in males having differential access to potential mates. Furthermore, multiple mating by females means that the same female may be available as a mate to several males, but her reproductive value to each male is prorated by the extent to which sperm competition determines the proportion of young sired by each mate.

The comparison of population and operational sex ratios for Richardson's ground squirrels indicates the importance of distinguishing between the presence of females and the availability of those females as mates. Knowledge of OSR enriches interpretation of sexually divergent features such as growth, body mass, fat content, and survival (Michener & Locklear 1990a). Whereas females attain about 80% of adult size as juveniles and complete growth as yearlings during pregnancy and early lactation, males virtually complete growth to adult size as juveniles before entering their first hibernation. Furthermore, males are structurally larger than females as adults (Michener 1989; Michener & Locklear 1990a,b). These sexual differences in growth patterns and adult size correspond with the physical prowess required by males, but not females, during mating activity. Males emerge from hibernation with fat reserves that are substantially larger than those of females and which are depleted much earlier in the active season. These sexual differences in fat content reflect differences in timing of reproductive effort and in activity budgets, particularly effort expended to acquire mates.

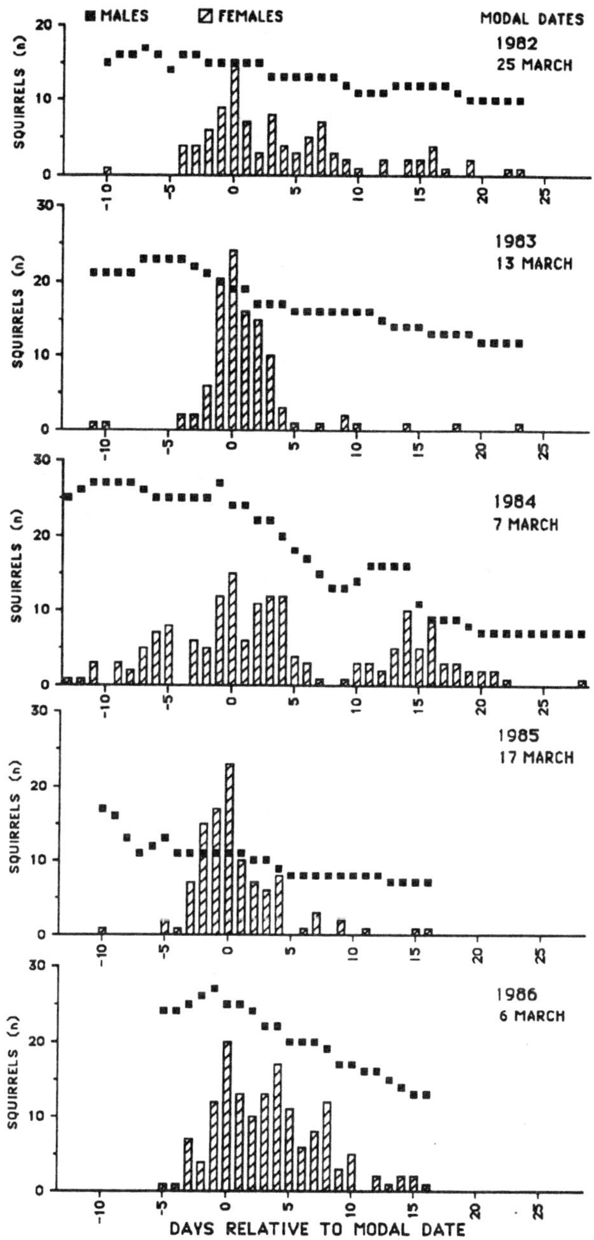
MALES
FEMALES
MODAL DATES
1982
25 MARCH
1983
13 MARCH
1984
7 MARCH
1985
17 MARCH
1986
6 MARCH
SQUIRRELS (n)
DAYS RELATIVE TO MODAL DATE

Figure 3

Figure 3. Numbers of female Richardson's ground squirrels in estrus on each day of the mating seasons in 1982-1986 and numbers of males present each day. The distribution of estrous dates is given relative to the modal estrous date (day 0 = the day on which the maximum number of females were simultaneously in estrus) in each year.

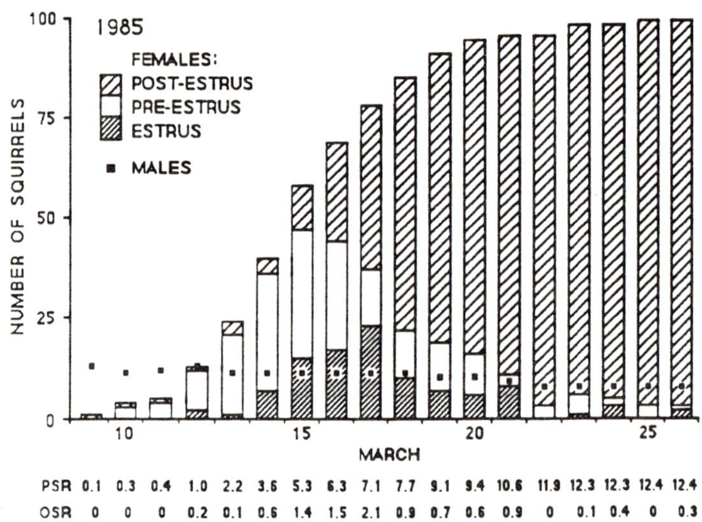

Figure 4. Numbers of male and female Richardson's ground squirrels present during the mating season of 1985. Each female was classed as being pre-estrus, estrus, or post-estrus on each day. PSR = population sex ratio (females per male) based on all squirrels present on the site that day. OSR = operational sex ratio (estrous females per male) based on all males present that day and those females that mated that day. Data are presented for 9-26 March, the period in which 96% (97/101) of females mated (see Figure 3).

CONCLUSION

In my studies of Richardson's ground squirrels, I have relied extensively on using easily obtained data such as body mass and sex ratio in conjunction with knowledge of each female's reproductive status to reveal sexual differences in behavioral ecology and to indicate directions for future research. Richardson's ground squirrels are particularly amenable subjects for such studies because of their trappability. For species that are difficult to livetrap, much information can be gleaned from careful observation of the behavior and appearance of squirrels without handling the animals. Time of estrus can be determined by observation of social interactions between females and males. Time of parturition can be estimated by noting the increased prominence of the nipples (visible when the female assumes a fully upright alert posture) once suckling begins. Time of litter emergence can be determined by observing the burrows that females have provisioned with dry vegetation. Provided reference information is available on the length of gestation and on the age of juveniles at emergence from the natal burrow (Michener 1985), detection of any one major reproductive event for a female enables estimation of the time of occurrence of the others. Regardless of what field methods are used to assess each individual's reproductive status, such knowledge allows greater flexibility and scope in use of data because analyses are no longer restricted to phenomena at the population level. Knowledge of an individual's reproductive status is of special value in behavioral studies because the partitioning of time among activities such as seeking mates, defending territories and nests, foraging, and parental care vary according to reproductive condition. Focussing analysis on individuals also facilitates intra- and interspecific comparisons because analysis of data by event minimizes variation due to different field techniques, eliminates variation attributable to differences in reproductive status, and exposes underlying biological patterns.

Hibernating species of ground squirrels present a particular problem in interspecific comparisons that use body mass as an indicator of size because mass of adults may vary seasonally by >100%. Interspecific comparisons of life-history parameters of hibernators are usually referenced to mass of females on emergence from hibernation (when females are at their minimum mass for the year and substantially lighter than when they make their maximal reproductive effort in late lactation). Thus, in intertaxonomic comparisons, reproductive parameters that are scaled to maternal mass tend to be overestimated for hibernators when compared with nonhibernators. Caution is required when conducting comparisons among hibernators and between hibernators and nonhibernators to insure that interpretation is not biased by the analytical methods used to pool and summarize data.

Methods of data analysis likewise influence interpretation of sex ratio data. When viewed from the perspective of the whole population, male Richardson's ground squirrels are outnumbered by females, but when viewed from the perspective of the reproductive condition of individual females, males outnumber estrous females. Males attune their behavior to locating estrous females, so the operational sex ratio provides a clearer understanding of mating strategies of males than does the population sex ratio.

Because selection acts on individuals, not populations, life-history and behavioral parameters are most appropriately analyzed at the individual level. Although the field effort is more demanding, monitoring individuals to obtain serial data is preferable to taking cross-sectional views of the entire population because the resultant information enables stronger biological inferences to be be made.

ACKNOWLEDGEMENTS

This study was supported by Natural Sciences and Engineering Research Council of Canada operating grants. I thank D. Burnett, S. Dobson, K. Godwin, I. McLean, and G. Pelle for dedicated field assistance in all weather and T. Despins, D. Stix, and K. VandeLigt for assistance with trapping juveniles. J. O. Murie and M. Bekoff provided helpful comments on the manuscript.

LITERATURE CITED

Armitage, K.B. 1981. Sociality as a life-history tactic of ground squirrels. *Oecologia* 48, 36-49.

Blueweiss, L., Fox, H., Kudzma, V., Nakashima, D., Peters, R. & Sams, S. 1978. Relationships between body size and some life history parameters. *Oecologia* 37, 257-272.

Boag, D.A. & Murie, J.O. 1981a. Weight in relation to sex, age, and season in Columbian ground squirrels (Sciuridae:

Rodentia). *Canadian Journal of Zoology* 59, 999-1004.
_____. 1981b. Population ecology of Columbian ground squirrels in southwestern Alberta. *Canadian Journal of Zoology* 59, 2230-2240.

Choromanski-Norris, J., Fritzell, E.K. & Sargeant, A.B. 1986. Seasonal activity cycle and weight changes of the Franklin's ground squirrel. *American Midland Naturalist* 116, 101-107.

Dobson, F.S. & Kjelgaard, J.D. 1985a. The influence of food resources on life history in Columbian ground squirrels. *Canadian Journal of Zoology* 63, 2105-2109.

_____. 1985b. The influence of food resources on population dynamics in Columbian ground squirrels. *Canadian Journal of Zoology* 63, 2095-2104.

Dobson, F.S. & Murie, J.O. 1987. Interpretation of intraspecific life history patterns: Evidence from Columbian ground squirrels. *American Naturalist* 129, 382-397.

Emlen, S.T. & Oring, L.W. 1977. Ecology, sexual selection, and the evolution of mating systems. *Science* 197, 215-223.

Fagerstone, K.A. 1988. The annual cycle of Wyoming ground squirrels in Colorado. *Journal of Mammalogy* 69, 678-687.

Hohn, B.M. & Marshall, W.M. 1966. Annual and seasonal weight changes in a thirteen-lined ground squirrel population, Itasca State Park, Minnesota. *Journal of the Minnesota Academy of Sciences* 3, 102-106.

Holekamp, K.E. & Nunes, S. 1989. Seasonal variation in body weight, fat, and behavior of California ground squirrels (*Spermophilus beecheyi*). *Canadian Journal of Zoology* 67,1425-1433.

Iskjaer, C., Slade, N.A., Childs, J.E., Glass, G.E., & Korch, G.W. 1989. Body mass as a measure of body size in small mammals. *Journal of Mammalogy* 70, 662-667.

Joy, J.E. 1984. Population differences in circannual cycles of thirteen-lined ground squirrels. In: *The biology of ground-dwelling squirrels: Annual cycles, behavioral ecology, and sociality* (ed. by J.O. Murie & G.R. Michener), pp. 125-141. Lincoln, Nebraska: University of Nebraska Press.

Kenagy, G.J. & Barnes, B.M. 1988. Seasonal reproductive patterns in four coexisting rodent species from the Cascade Mountains, Washington. *Journal of Mammalogy* 69, 274-292.

Kenagy, G.J., Sharbaugh, S.M., & Nagy, K.A. 1989. Annual cycle of energy and time expenditure in a golden-mantled ground squirrel population. *Oecologia* 78, 269-282.

Knopf, F.L. & Balph, D.F. 1977. Annual periodicity of Uinta ground squirrels. *Southwestern Naturalist* 22, 213-224.

Kurta, A. & Kunz, T.K. 1987. Size of bats at birth and maternal investment during pregnancy. *Symposium of the Zoological Society of London* 57, 79-107.

Leitch, I., Hytten, F.E. & Billewicz, W.Z. 1959. The maternal and neonatal weights of some mammalia. *Proceedings of the Zoological Society of London* 133, 11-28.

Leutenegger, W. 1976. Allometry of neonatal size in eutherian mammals. *Nature* 263, 229-230.

Masman, D., Kenagy, G.J. & Sharbaugh, S.M. 1990. Sexual strategies reflected in contrasting annual cycles of body mass and energy balance in golden-mantled ground squirrels. *American Naturalist.*

McKeever, S. 1963. Seasonal changes in body weight, reproductive organs, pituitary, adrenal glands, thyroid gland, and spleen of the Belding ground squirrel (*Citellus beldingi*). *American Journal of Anatomy* 113, 153-174.

McLean, I.G. & Towns, A.J. 1981. Differences in weight changes and the annual cycle of male and female Arctic ground squirrels. *Arctic* 34, 249-254.

Melnyk, R.B. 1983. Accelerated circannual cycles in ground squirrels, *Spermophilus richardsonii*, kept in constant conditions. *Canadian Journal of Zoology* 61, 1765-1770.

Michener, G.R. 1978. Effect of age and parity on weight gain and entry into hibernation in Richardson's ground squirrels. *Canadian Journal of Zoology* 56, 2573-2577.

_____. 1983a. Kin identification, matriarchies, and the evolution of sociality in ground-dwelling sciurids. In: *Advances in the Study of Mammalian Behavior* (ed. by J.F. Eisenberg & D.G. Kleiman). *Special Publication of the American Society of Mammalogy* 7, 528-572.

_____. 1983b. Spring emergence schedules and vernal behavior of Richardson's ground squirrels: Why do males emerge from hibernation before females? *Behavioral Ecology and Sociobiology* 14, 29-38.

_____. 1984. Sexual differences in body weight patterns of Richardson's ground squirrels during the breeding season. *Journal of Mammalogy* 65, 59-66.

_____. 1985. Chronology of reproductive events for female Richardson's ground squirrels. *Journal of Mammalogy* 66, 280-288.

_____. 1989. Reproductive effort during gestation and lactation by Richardson's ground squirrels. *Oecologia* 78, 77-86.

Michener, G.R. & Locklear, L.L. 1990a. Differential costs of reproductive effort for male and female Richardson's ground squirrels. *Ecology*.

_____. 1990b. Over-winter weight loss by Richardson's ground squirrels in relation to sexual differences in mating effort. *Journal of Mammalogy*.

Millar, J.S. 1977. Adaptive features of mammalian reproduction. *Evolution* 31, 370-386.

Modi, W.S. 1984. Reproductive tactics among deer mice of the genus *Peromyscus*. *Canadian Journal of Zoology* 62, 2576-2581.

Morton, M.L. 1975. Seasonal cycles of body weights and lipids in Belding ground squirrels. *Bulletin Southern California Academy of Sciences* 74, 128-143.

Morton, M.L., Maxwell, C.S. & Wade, C.E. 1974. Body size, body composition, and behavior of juvenile Belding ground squirrels. *Great Basin Naturalist* 34, 121-134.

Murie, J.O. & Michener, G.R. 1984. *The Biology of Ground-Dwelling Squirrels: Annual Cycles, Behavioral Ecology, and Sociality*. Lincoln, Nebraska: University of Nebraska Press. 459 pp.

Murie, J.O. & Harris, M.A. 1982. Annual variation of spring emergence and breeding in Columbian ground squirrels (*Spermophilus columbianus*). *Journal of Mammalogy* 63, 431-439.

Neal, B.J. 1965. Seasonal changes in body weights, fat depositions, adrenal glands and temperatures of *Citellus tereticaudus* and *Citellus harrisii* (Rodentia). *Southwestern Naturalist* 10, 156-166.

Nellis, C.H. 1969. Productivity of Richardson's ground squirrels near Rochester, Alberta. *Canadian Field-Naturalist* 83, 246-250.

Phillips, J.A. 1984. Environmental influences on reproduction in the golden-mantled ground squirrel. In: *The Biology of Ground-Dwelling Squirrels: Annual Cycles, Behavioral Ecology, and Sociality*. (ed. by J.O. Murie & G.R. Michener), pp. 108-124. Lincoln, Nebraska: University of Nebraska Press.

Rickart, E.A. 1982. Annual cycles of activity and body composition in *Spermophilus townsendii mollis*. *Canadian Journal of Zoology* 60, 3298-3306.

Sheppard, D.H. 1972. Reproduction of Richardson's ground squirrel (*Spermophilus richardsonii*) in southern Saskatchewan. *Canadian Journal of Zoology* 50, 1577-1581.

Sherman, P.W. & Morton, M.L. 1984. Demography of Belding's ground squirrels. *Ecology* 65, 1617-1628.

Skryja, D.D. & Clark, T.W. 1970. Reproduction, seasonal changes in body weight, fat deposition, spleen and adrenal gland weight of the golden-mantled ground squirrel, *Spermophilus lateralis lateralis* (Sciuridae) in the Laramie Mountains, Wyoming. *Southwestern Naturalist* 15, 201-208.

Stearns, S.C. 1983. The influence of size and phylogeny on patterns of covariation among life-history traits in the mammals. *Oikos* 41, 173-187.

Turner, B.N., Iverson, S.L. & Severson, K.L. 1976. Postnatal growth and development of captive Franklin's ground squirrels (*Spermophilus franklinii*). *American Midland Naturalist* 95, 93-102.

Walsberg, G.E. 1988. Evaluation of a nondestructive method for determining fat stores in small birds and mammals. *Physiological Zoology* 61, 153-159.

Western, D. 1979. Size, life history and ecology in mammals. *African Journal of Ecology* 17, 185-204.

Young, P.J. 1988. *Ecological and Energetic Aspects of Hibernation of Columbian Ground Squirrels in Relation to Patterns of Over-winter Survival.* Unpublished Ph.D. dissertation, University of Alberta, Edmonton.

Zammuto, R.M. & Millar, J.S. 1985. Environmental predictability, variability, and *Spermophilus columbianus* life history over an elevational gradient. *Ecology* 66, 1784-1794.

III
Moral Dimensions

Introduction

In recent years it has become increasingly clear that science, like all human practices and institutions, is a proper subject for moral scrutiny. Value commitments, both conscious and unconscious, affect our observations and theories (see Bernard Rollin's chapter in section IV of Volume I).

In the heyday of logical empiricism (circa 1930-1960), science was seen as the most pure of human activities; the model was almost one of "disembodied knowledge acquisition." There was a single thing that was "the scientific method"; observations were distinct from and unaffected by theoretical commitments; theories were "sets of sentences" that made no essential reference to knowers; explanation and prediction were regarded as formal relations between sentences that in principle could be made mechanical; and all of this theorizing, explaining, and predicting was thought to be uncontaminated by values (Braithwaite 1953; Nagel 1961; Hempel 1966).

While many scientists continue to give obeisance to some such picture, it has long since met its demise in the broader intellectual community. Beginning with Quine (1951), influential philosophers argued forcefully against almost every tenet of logical empiricism (Hanson 1958; Feyerabend 1962; Kuhn 1962). Next came historians, abetted by the testimony of important scientific figures. Books such as Watson (1968) painted a very different picture of scientific discovery than was suggested by the logical empiricist model. Scientists were seen to be sometimes selfish, irrational, motivated by power and prestige rather than the Pursuit of Truth - in other words, all too human. In the wake of these critiques have come moral philosophers, sociologists, and most recently feminists (see Lori Gruen's chapter in section I of Volume I). Although many scientists may not be aware of it, there is a swirl of activity addressing almost every feature of the scientific life. As a result, it is a safe prediction that twenty-first century science will be importantly different from science as it is currently perceived and practiced.

One area in which the rubber meets the road (science meets the public) concerns the treatment of animals in scientific contexts. Susan Finsen opens this section with a discussion of "moderate" views regarding the ethics of animal experimentation. She points out that the views of philosophers like Bernard Rollin (1981), Tom Regan (1983), S. F. Sapontzis (1987), and Peter Singer (1990) have not been widely shared by the scientific establishments, if for no other reason than they believe that they have too much to lose. In response to what are sometimes regarded as extreme positions, moderate views that renounce the "anything goes" practices of the Bad Old Days but stop short of accepting an "animal rights" position have emerged. Finsen shows that commonly accepted "humane treatment and necessary use" policies are not as distant from those of the Bad Old Days as might be thought; and that cost/benefit approaches face serious problems of application. Finsen leaves open the possibility of a genuine moderate view (one perhaps like our policies with respect to human experimentation) that takes individual animals seriously and forbids some research when their interests are seriously threatened (see Bekoff & Jamieson 1990).

One area in which our relations with nonhuman animals are morally problematic is that of domestication (see Derr 1990). Thomas Daniels & Marc Bekoff review its history and identify its major features. They disapprove of some of what has been done in the past and their outlook for the future is even bleaker. In domestication, humans are the "doers" or initiators and nonhumans are the recipients of their actions. Daniels & Bekoff argue that humans (to some extent unconsciously) are turning the earth into one large "game reserve" in which virtually no animals are free of human-imposed selection pressures. In such a world it may be that no animal can be truly wild and free to live its own life. In this respect, an important part of nature may be coming to an end (McKibben 1989).

Lawson Crowe is concerned with ethical issues in genetics. Here, humans are the "doers" but humans also are the recipients. Although Crowe's focus is more on humans than nonhumans, there are a number of important points of connection. The "eugenics" movement, whose history Crowe traces, often identified the poor, the nonwhite, the nonprotestant Christian as "subhumans" or "beasts" (see Midgley 1973). Moreover, animals are often used as "instruments" - laboratory equipment - in genetics research. In addition, as we learn more about genetics powerful forces will be intent on producing "new" animals for economic exploitation, a point that is also made by Daniels & Bekoff.

Finally, Michael W. Fox sketches a vision for the future. On his view anthropomorphizing (also see John Andrew Fisher's chapter in section I of Volume I), which arrogantly assimilates animals to us, and objectification, which treats them as mere matter to be molded in accordance with our desires, are two sides of the same coin. For Fox morality and good science go hand-in-hand. In order to understand other animals we must empathize with them; and if we empathize with them, we will not treat them in the ways that we do (see also Stephen Clark's chapter in section IV of Volume I). In a world in which humans and nonhumans live together in an empathetic community, laws and regulations will be unnecessary.

Fox's utopian vision is attractive and he may be right about the ideals that we should strive for. But until we reach that golden age we will have to grapple with difficult moral issues and all of the messy tradeoffs that are identified by Finsen, Daniels & Bekoff, and Crowe. And by now, one thing should be clear: Scientists can no longer plead that their activities are value-free manifestations of decontextualized knowledge. They must come out of the peanut gallery and onto the stage. They must engage in the dialogue.

LITERATURE CITED

Bekoff, M. & Jamieson, D. 1990. Reflective ethology, applied philosophy, and the moral status of animals. *Perspectives in Ethology* 9, in press.

Braithwaite, R.B. 1953. *Scientific Explanation.* Cambridge: Cambridge University Press.

Derr, M. 1990. The politics of dogs. *The Atlantic Monthly* March, 49-72.

Feyerabend, P. 1962. Explanation, reduction and empiricism. In: *Scientific Explanation: Space and Time* (ed. by H. Feigl & G. Maxwell), pp. 231-272. Minneapolis, Minnesota: University of Minnesota Press.

Hanson, N.R. 1958. *Patterns of Discovery.* Cambridge: Cambridge University Press.

Hempel, C. 1966. *Philosophy of Natural Science.* Englewood Cliffs, New Jersey: Prentice-Hall.

Kuhn, T.S. 1962. *The Structure of Scientific Revolutions.* Chicago, Illinois: University of Chicago Press.

McKibbin, B. 1989. *The End of Nature.* New York: Randon House.

Midgley, M. 1973. The concept of beastliness. *Philosophy* 48, 111-135.
Nagel, E. 1961. *The Structure of Science.* London: Routledge and Kegan Paul.
Quine, W.V.O. 1951. Two dogmas of empiricism. *Philosophical Review* 60, 20-43.
Regan, T. 1983. *The Case for Animal Rights.* Berkeley, California: University of California Press.
Rollin, B. 1981. *Animal Rights and Human Morality.* Buffalo, New York: Prometheus Books.
Sapontzis, S.F. 1987. *Morals, Reasons, and Animals.* Philadelphia, Pennsylvania: Temple University Press.
Singer, P. 1990. *Animal Liberation,* 2nd Edition. New York: New York Review of Books.
Watson, J.D. 1968. *The Double Helix.* New York: Antheneum.

13. Domestication, Exploitation, and Rights

Thomas J. Daniels and Marc Bekoff

This is an exciting time to be a biologist, especially on two fronts: (1) ecology (the study of the delicate balance between species and the environment) and (2) the ethics of our interactions with the rest of the biosphere. Ways in which humans interact and interfere with "natural systems," and reasons for their actions, are being questioned, facile explanations are being reanalyzed and revised, and consequences of the careless exploitation of biotic and abiotic environments are being assessed, all on an unprecedented level (Myers 1979; Regan 1984; Stone 1987; Barnett 1988; Rolston 1988; Callicott 1989; Nash 1989; Westra 1989). Although some believe that what is right and wrong is perhaps less clear than in the past, human responsibilities to develop strict guidelines that are consistent with stringent moral and ethical principles have never been greater.

Against the backdrop of heightened human interest in how we interact with the rest of the world, we will attempt to (1) characterize the process of nonhuman animal (hereafter, and nonpejoratively, referred to as "animal") domestication, concentrating on those groups in which humans have played an integral role shaping, molding, and supporting, (2) discuss the methods that have been and are currently being used to produce domesticated species, and (3) consider the ethical implications of human interference, in the form of artificial selection, in the breeding biology of other species. We will employ Charles Darwin's typology of different forms of selection.

Understanding the process of domestication permits humans to view how they have, and do, interact with species with which they have the closest contact. As will become clear, the process of domestication is much broader and more pervasive than we generally think. And, just what constitutes a domestic animal,and precisely what moral obligations humans have to domestic versus wild animals (Callicott 1980, 1989; Johnson 1981; Rolston 1988), are issues that demand careful scrutiny.

The concept of artificial selection, which is the guiding force of domestication, is an important one that played a key role in the development of Darwin's ideas about natural selection and the evolution of behavior (Waters 1986; see also Richards 1987). Furthermore, interpreting data that are collected on the behavior of domestic animals, and subsequently using this information to explain (or to argue by inference about) the behavior of wild relatives, requires that we understand how domestication has influenced at least the behavior patterns with which a given study is concerned.

Although domestication is not restricted solely to human-animal interactions (see below), we will focus on what we call *human-facilitated domestication.* Domestication speaks to us of our place on earth, of how we interact with our cohabitants and ecosystems. It is a millennia-old social exercise that should, as any other, have rules and guiding principles. Whereas a pride of lions, for instance, has no obligation to consider ethical issues in its relationships with a herd of gazelles, we believe that there are right and wrong ways of interacting with animals *by* animals (e.g. human moral agents) that know right and wrong. There are right and wrong ways of dealing with populations of animals, in this case, domestic populations. The question of obligations owed to *groups* of organisms is an important one receiving much attention today (for discussion see Myers 1979; Aiken 1984; Regan 1984, 1985a; Jamieson 1985, 1990; Norton 1986, 1987; Vermeij 1986; Rachels 1987; also see Koshland 1988) with respect to threatened and endangered species, but not with the vastly more abundant domestic animals. Obligations owed to domestic populations will be addressed herein.

WHAT IS DOMESTICATION?

> A domestic animal is a slave with which one amuses oneself, which one makes use of, which one abuses, corrupts, removes from its natural element and perverts. (Buffon, as translated in de Luce & Wilder 1983: 8)

Domestication is a process characterized by a symbiotic relationship that *evolves over time* in which one group of organisms takes advantage of another group (parasitism) or in which both groups receive some benefit (mutualism). Domestication is not synonymous with *socialization*, the nonevolutionary developmental process by which individual animals form social bonds with other individuals, although amiable contact with humans was probably important in early stages of domestication. However, humans are not the only domesticators. Domestication systems have evolved in taxa quite distant from *Homo sapiens*. For example, some highly social species of ants tend or farm other insects. The "domesticated" species are suckers of plant juices and subsequently, providers of sweet, nutritious droplets of "honeydew" (partially digested excrement) to the ants. Ants protect (drive away predators), move (put out to pasture), build shelters for, and in general care for their domesticants (Way 1963; Wilson 1971, 1975; Reed 1980; see Reed 1977 for summary of the literature on ant-"slave" interactions).

Humans have domesticated many zoological groups ranging from insects to mammals; over 3,000 mammal breeds and strains are partially or wholly domesticated (Hafez 1968). Domestication is a broad subject encompassing such disparate fields of inquiry as archaeology, organismic and molecular biology, sociology, anthropology, and philosophy. Investigators in each discipline attempt to fit together small pieces of information regarding the time, location, purpose, and species involved. The picture remains incomplete for most species; what is known has been reviewed extensively by Downs (1960), Zeuner (1963), Leeds & Vayda (1965); Price & King (1968); Hale (1969), Ucko & Dimbleby (1969); Hyams (1972); Boice (1973); Cole & Garrett (1980); Clutton-Brock (1981); and Price (1984).

Ratner & Boice (1975) characterized domestication as involving the removal of organisms from some natural selection pressures over generations resulting in changes due to both environmental and genetic influences. Hale (1969) defined domestication as a condition under which the breeding, care, and feeding of animals are more or less controlled by humans, resulting in morphological, physiological, and/or behavioral changes in the animals. Bokonyi (1969: 219) noted that domestication is both a complicated process and "nothing more than man's special interference in the life of certain animal species." Hyams (1972) observed that domestication is a process composed of a large number of small modifications in the behavior of man toward certain animals and, likewise, in theirs to him. As such, domestication is not a specific single event, but an ongoing evolutionary process (Ucko & Dimbleby 1969; Hyams 1972) which, despite the connotation of slow speed (Darwin

1859/1968; Haldane 1949), can take place relatively quickly, with genetic changes that are evident in just a few generations (Anderson 1952; Donaldson & Menasveta 1961; Baker & Manwell 1981).

Recently, Price (1984: 3) defined domestication as "that process by which a population of animals becomes adapted to man and to the captive environment by some combination of genetic changes occurring over generations and environmentally induced developmental events reoccurring during each generation" (see also Bokonyi 1969). This definition formally included the concept of ontogeny, which was always implicitly understood (Daniels & Bekoff 1989), but it still relied on a somewhat restrictive view of people's role as one of actively maintaining animals in a captive environment.

Daniels & Bekoff (1989) highlighted the relationship between artificial and natural selection by suggesting that domestication be viewed as an evolutionary process in which the relative importance of artificial (see below) selection becomes much greater than that of natural selection. Building on that interpretation, we now suggest that domestication be defined as *an evolutionary process in which one population of organisms imposes a selection regimen onto another resulting in a decrease in the relative importance of natural selection to the domesticant population, and phenotypic (usually) and genotypic changes that mark increased adaptation to life with the domesticator population.*

ORIGINS OF DOMESTICATION

It should be noted at the outset that a potential methodological bias may color our interpretation of when domestication actually began. Ways of discerning domestic from nondomestic animals, particularly early in the domestication process when morphological change is minimal, may not be reliable (Ducos 1969). Domestication is usually not recognized until animals are bred actively under artificial conditions, when real isolation from the wild has occurred (Grigson 1969). This shortcoming in our ability to identify animals resulting from early domestication may partly explain the apparent long lag between the time humans began domestication and the appearance of domestic animals that could be clearly identified as such on the basis of archaeological remains. At present, explanations for the delay are speculative.

Despite possible methodological limitations, it seems fairly certain that animal and plant domestication occurred independently in several places (Helbaek 1959; Hawkes 1969). People's earliest success with domesticating animals apparently involved the wolf, most probably *Canis lupus pallipes* (Fox & Bekoff 1975), which produced the domestic dog (*C. familiaris*) some 10,000 to 14,000 years BP (before present) (Turnbull & Reed 1974; Davis & Valla 1978). Because humans were hunter-gatherers at this time, and generally nomadic, it is likely that early contact with wolves was based on competition for food. The precise way in which this contact developed into the domesticator-domesticant relationship is unknown, but may have involved packs of wolves following human bands to scavenge food remains, followed by stealing and raising of wolf pups by humans. It is often suggested that the original motive for domesticating dogs was to serve as an aid in hunting (Downs 1960; Zeuner 1963; Scott & Fuller 1965; Fox 1978) but it is just as likely that motives varied in different locations (Manwell & Baker 1984), and that perhaps pet-keeping arose long before humans made use of these animals for other purposes (Serpell 1986). Bonds with animals are emotionally fulfilling (Serpell 1986) and the finding of a 12,000 year old burial site containing the remains of a human with its hand on the shoulder of a 5 month-old (apparent) dog (Davis & Valla 1978) offers a "timeless and eloquent gesture of attachment." (Serpell 1986: 58)

Because the benefits of animal domestication to humans are apparent, it is easy to assume that the process began as one in which the ultimate goals were clearly established in the minds of domesticators. In fact, early domestication probably occurred without humans being aware of what was happening (Hyams 1972). The first domesticators could not have foreseen any uses for those animals other than the ones they traditionally had: providers of bones, skins, and meat (Reed 1980). Thus, the secondary uses of animals, those typically thought of as goals of domestication such as milk and wool production, sources of locomotion, and as instruments of prestige, sport, and war, would only be realized after a long period of contact, an intensification of a sedentary lifestyle, and an accumulation of random mutations in the potential domesticant population (Reed 1980). During these periods, which would vary in length from site to site and among species, humans developed and modified methods of animal acquisition (e.g. herd, corral, chase, trap), care and maintenance, and most important, animal husbandry (Bokonyi 1969; also see Cranstone [1969] for a discussion of animal management techniques).

Early domestications primarily involved members of the order Artiodactyla, such as pigs (*Sus scrofa*), sheep (*Ovis aries*), goats (*Capra hircus*), and cattle (*Bos taurus*). However, these efforts represented a significant leap in people's ability to domesticate other species and, as such, pose an interesting theoretical question: What took so long to begin the process? Although the record of manufactured tools goes back more than one million years, evidence of domesticated plants and animals only starts near the end of the last Ice Age (12,000 years BP; Reed 1969). Humans had not changed in physical appearance for nearly 30,000 years, and the animals domesticated between 14,000 and 8,000 years BP had not changed for an even longer period (Reed 1980). Thus, animals were "ready" for domestication prior to this period, but humans were not. The necessary changes were cultural in origin and correlated with major climatic changes facing man and animal. For instance, the end of the Ice Age meant climatic warming and a slow but continuous shift of vegetational zones to higher latitudes (Reed 1969). Animal populations likewise followed the shift to maintain themselves in hospitable environments. Biological and environmental changes may have stimulated cultural innovations. In the Near East, where the oldest evidence of domestication has been gathered, people began moving out of caves to build villages, and reap and mill cereals (Reed 1969) within a thousand years of such changes. Even so, the cultural revolution was fairly gradual; early attempts at agriculture were often halting and varied greatly in extent and degree of sophistication (Ucko & Dimbleby 1969).

THE EFFECTS OF ANIMAL DOMESTICATION ON HUMANS

With an increase in a sedentary lifestyle as a result of the need to tend and harvest crops, humans came to depend on a relatively stable food source that could be dried and stored. In fact, the development of food storage technology may have been the key to domestication of the ruminants and pigs that parallels the rise of village life. Hyams (1972) suggests that the accumulation of food in one localized site probably attracted potential domesticant species and increased their contact with humans. These species likely provided a surplus of food "on the hoof" (Flannery 1969).

An interesting assessment of the domestication process has been offered by Flannery (1969), who notes that early irrigation practices in southwest Asia around 9,500 to 7,000 BP modified people's reliance on the weather for successful crop growth, but also aggravated environmental destruction to the point that a return to a wild resource economy became virtually impossible. Furthermore, although prehistoric populations tended to remain stable at a density below the point of resource depletion, changes in demography associated with village life increased local population numbers to carrying capacity (Flannery 1969). Thus, changes leading to extensive food production may have been responses to disturbances in human density equilibrium. As animal domestication proceeded with the removal of a small number of representative species from the wild and artificial increases in the number of individuals kept under people's control, natural processes that had kept animal numbers in check were no longer important (Flannery 1969) and the road back to the predomestication and preagriculture lifestyle was no longer feasible.

In addition to increasing the potential for environmental destruction, domestication changed the means of production in society, in that it made possible a division of labor that was previously uncharacteristic of hunter-gatherers, and it laid the foundations for social stratification while permitting the human population to expand at a geometric rate (Flannery 1969). Along with the increase in human population size came an increased ability to exploit previously uninhabited or marginally inhabited regions (Aschmann 1965). Reindeer (*Rangifer tarandus*) served as horse/cow substitutes in the tundra, camels (*Camelus dromedarius, C. bactrianus*) allowed exploitation of desert regions by humans (Sweet 1965), and yaks (*Bos grunniens*) played a similar role in mountainous regions of Asia (Hyams 1972). Some indication of the importance of a domestic animal species to a human population can be gained from the number of specialized terms relating to those animals. Lapp reindeer herders, for example, have 50 names for colors and the distribution of patches of color, nearly as many for the varieties of antler form, and dozens for age and sex combinations (Cranstone 1969). Domestic animals are still a very important source of transport/motive power in much of the world (Cole 1980), though most attention in recent years has focused on the roles of animals in commercial food production and in research (see "Domestication Today and Tomorrow" below).

THE EFFECTS OF DOMESTICATION ON THE ANIMALS INVOLVED

Humans' impact on the animals that were domesticated has been significant in at least three ways. First, domestic populations typically are numerically much larger than nondomestic populations. This is likely to be a function of how useful the animals are to people, the quantity of resources available to feed and house them, and the behavior of the animals. It is no coincidence that most domestic species are highly social animals that not only interact positively with conspecifics, but also generalize bond formation to the domesticator (for example, see Scott & Fuller 1965 and Fox & Bekoff 1975 for discussions of socialization in dogs).

Second, domestication in many cases has significantly altered the physical appearance of the animals involved. While some overall generalizations can be made about the types of changes that have been induced in domestic animals (Table 1; Zeuner 1963; Kruska 1988), the directions of change are not uniform across taxa. Price (1984) noted that body size has increased in some species (e.g. horse [*Equus caballus*] and rabbit [*Oryctolagus cuniculus*]) but has decreased in others (e.g. sheep and cattle).

Variability among species is also evident in comparisons of domestic and nondomestic animals within the same genus. Domestication may lead ultimately to speciation, as in dogs and chickens (*Gallus gallus*), or to an animal that is taxonomically indistinct from its nondomestic counterpart, as in the case of the reindeer (domestic) and caribou (nondomestic), both of which are classified as *Rangifer tarandus*. Therefore, it may be incorrect to conclude that domestication is any more advanced in cases in which domesticants vary from nondomestic forms than in cases where the two populations are physically similar (Daniels & Bekoff 1989). Species classifications historically have been questioned given that a large number of recognized species readily interbreed (Gray 1971) and may produce fertile offspring (e.g. Mengel 1971; Gipson 1972; Mahan et al. 1978). Also, depending on the domesticators' goals and the particular gene pool involved, retention of the physical similarities between domestic and nondomestic forms might be favored selectively.

Although Berry (1969) noted that it is not possible to recognize any traits which inevitably accompany domestication, a number of genetic disorders often accompanies breed formation within a domestic species, accumulating over the course of its evolutionary history as more and more profound changes in the

Table 1. Phenotypic changes that accompany domestication.

Character	Domesticant Phenotype
Body Size	Usually smaller than nondomesticants
Color	More variable than nondomesticants
Skull	Facial region shortened relative to cranial Teeth smaller Horns reduced in size
Post-cranial Skeleton	Smaller muscle ridges on bones Poorly defined facets of joints Number of caudal vertebrae reduced
Hair	Change in underlying shorter hairs
Soft Anatomy	Greater local fat accumulation Decreased brain size Increased or decrease in musculature
General	New characteristics rarely produced Pathological conditions favored Growth rates affected

Modified from Zeuner 1963.

original ancestral form are made through artificial selection. Van Vleck (1980) listed 14 genetic disorders of domestic horses and noted that 20% of all foals are born with some form of abnormality; approximately one-half of these are severely handicapped or born dead. Greater still, each of about 300 breeds of dog (Fiorone 1973) suffers from some genetic disorder; beagles, German shepherds, and boxers lead the way with 24, 23, and 18 disorders, respectively, potentially afflicting them (Patterson 1986).

These disorders generally are unintentional byproducts of artificial selection for desirable traits (although animals with specific disorders are also routinely bred for medical research purposes). For example, as many as 90% of collies suffer from "collie eye anomaly" or chorioretinal dysplasia, an autosomal recessive disorder that varies in severity from nearly normal vision to complete blindness. The high incidence of the condition indicates that it has been maintained by artificial breeding (Van Vleck 1980). The power of selection is quite apparent when we consider the range of body size (weight) within *C. familiaris*; the largest breed (St. Bernard) may be 470 times larger than the smallest (Chihuahua) (Van Vleck 1980).

Some researchers have claimed, because of the number of genetic abnormalities that afflict domestic species, that they are "degenerate." The so-called "degeneracy hypothesis" (Price 1984) states that domestic animals are inferior to their nondomestic relatives as a consequence of artificial selection for traits that have little survival value in the wild (Vincent 1960; Ortega y Gasset 1972; Eibl-Eibesfeldt 1975; Kruska 1988). However, it is important to note that degeneracy is relative to the habitat in which the animals are found. Because most domestic animals do not live in the wild (but see Daniels & Bekoff 1989) and are quite well-suited for life in captivity, the degeneracy argument has been questioned (Boice 1973; Miller 1977; Baker & Manwell 1981; Price 1984; Kruska 1988).

The third major impact that humans have had on domestic populations is perhaps the most widely recognized and is seen in the ease with which the animals involved can be tamed (Daniels & Bekoff 1989). Manageability, or ease in manipulating the behavior of domesticants, has always been a primary goal of domesticators. Vertebrate domestication apparently has enhanced the generalization of bond formation to species other than the domesticant's own (under appropriate conditions) and has led to a lengthening of the socialization period during which social relationships are established. In essence, domestication involves a loss of social and other inhibitions such as those involved in approaching novel stimuli (Lorenz 1965) by raising thresholds for avoidance or for submissive responses (Price 1984). This is because it is difficult to form close bonds with animals that avoid contact or are exceptionally submissive.

Likewise, domestic animals' evolutionary trend toward greater neoteny, or the retention of juvenile characteristics into adulthood (Fox 1968), is the result of selection for shifts in rates of development (Gould 1977). In turn, this may result in greater behavioral plasticity (Boice 1973), although further comparisons between domestic and nondomestic animals under various environmental and experimental conditions would be needed to verify this (Ratner & Boice 1975; Price 1984). Tameness, or the lack of avoidance responses by the animal when humans approach (Hediger 1964, 1968), is a learned trait (Price 1984) and may be viewed as an ontogenetic phenomenon facilitated by artificial selection. Thus, an important result of domestication is the *modifiability* of behavior, though not necessarily its *modification* (Daniels & Bekoff 1989). For example, the form of the main behavior patterns performed by domestic dogs is relatively resistant to change, whereas frequencies or rates of performance of certain acts have been modified (Scott & Fuller 1965; Bekoff 1972; Scott & Causey 1973).

TYPES OF SELECTION: WHERE TO DRAW THE LINE

Darwin (1859/1968, 1868) was among the first to examine seriously the subject of domestication by comparing human-mediated selection and the subsequent production of numerous animal breeds to natural selection and speciation. He (1868) recognized three forms of selection (also see Bateson Volume I), each varying in the degree of human control involved in the process. Aside from natural selection, which is free of human interference, *methodical selection* and *unconscious selection* together compose what is generally referred to as *artificial selection,* the driving force of domestication.

According to Darwin animal domestication likely began with unconscious selection; although humans took an active part in deciding which individual animals bred and passed genes to the next generation, selection was not goal-oriented. This gave rise over time to deliberate or methodical selection in which some ultimate phenotype was envisioned and strived for.

Although Darwin (1868) identified three forms of selection, his treatment of the subject suggests a view that artificial and natural selection occupy different ends of a continuum; where on the scale a particular selection event occurs depends on the level of humans' interference and intentions (Rheinberger & McLaughlin 1984). But the concept of selection as a continuum is often obscured; even suggestions that all selection imposed on captive populations which cannot be ascribed to artificial selection must be natural selection (Price & King 1968; Wright 1977) do not clarify whether selection processes are continuous or discrete.

This situation is largely one of necessity. For selection to be defined as natural or artificial, a line must be drawn somewhere along the continuous scale. However, it may be exceedingly difficult to identify examples of unconscious selection that fall somewhere between the two extremes. Once a decision is made as to "type of selection," the processes are no longer seen as continuous, but rather as discrete phenomena. Thus, in an attempt to categorize selection events as human-made or not, we are forced to move one step away from Darwin's views. Consequently, in identifying selection processes as primarily human-mediated or natural, it is imperative that the contribution of the secondary, though probably critical, process be recognized as part of the selection regime.

An argument can be made that all selection, whether instituted by humans or not, should be called natural (Rheinberger & McLaughlin 1984). In the case of any other animal species, one population adapting to the action of another over time would be a normal part of the natural selection process. Although anthropocentrism has played an important part in shaping our view of biological processes and people's role in nature (e.g. Dobzhansky 1967), the fact remains that our actions are in some ways distinctly different from those exerted by any other taxon, if not always in the *type* of influence (preventing certain organisms from reproducing is essentially the same in both artificial and natural selection), then certainly in the *magnitude* of those actions in the relative short term. The term *artificial selection* is therefore retained in the following discussion to indicate a *human-mediated selection process.*

Domestication Today and Tomorrow: An Additional Selection Process?

Several works have suggested that the practice of domestication today is much reduced compared to the past. Jewell (1969), Hyams (1972), Crawford (1974), and Price (1984), among others, not only observed that relatively few animal species had been domesticated in the past, but concluded that the opportunity to domesticate others, especially for use as food resources, remained available and should not be missed (Wilson 1989). Hyams (1972: 187) noted that several previously domesticated species had been lost and that man was no longer "in the mood" to practice domestication, presumably on a large scale. Indeed, Hyams (1972) concluded that the "Age of Domestications" occurred long before recorded history, approximately 5,000 to 7,500 years BP. These authors were limiting discussions to domestication via methodical selection, a process that is expanding at an ever-increasing rate today but which constitutes only one segment of the domestication process. Regardless, more animal species have been domesticated within the last 2,000 years than in any previous period of the same duration, including the "Age of Domestications" (Figure 1). Thus, claims that humans are not "in the mood" for domestication are not accurate.

Nor is artificial selection what it used to be; the range of processes involved has expanded at both ends of the spectrum. With respect to methodical selection (Point A, Figure 2), sperm banks, artificial insemination, and embryo transplants now preclude some of the potential roadblocks, such as constraints due to mate selection and the difficulties involved in physically mating the larger, potentially dangerous species that previously interfered with breeding desirable phenotypes. Because of long-term sperm storage capabilities, the new husbandry methods also permit increased selection pressure on the fewer animals needed to provide gametes (Pollak 1980).

Recent advances in biotechnology leading to direct manipulation of the genome and the creation of transgenic animals (those into which "desirable" genes have been inserted) have radically enhanced the potential to develop new phenotypes. However, the practical goals of curing human genetic diseases by replacing defective genes, engineering animals for more efficient food production or, in essence, turning animals into living pharmaceutical factories that produce large quantities of hard-to-obtain compounds like insulin, may not be achieved easily. Where the inserted gene will implant itself in the organism's genome often is not predictable (Capecchi 1989); it is as likely that a useful gene will be replaced as that the transferred gene will enhance the animal's value to breeders (Pursel et al. 1989).

The implications of human biotechnological advances are important beyond the effect seen on individual animals (Sagoff 1988). In the past, methodical selection generally resulted in organisms that were modified but remained under the care and protection of humans. This may not be the case today and in the future. For instance, recent attempts at enhancing growth rates of some fish species promise hardier, more aggressive fish for sportsmen but they also offer the potential for the new forms to alter radically ecosystems into which they are introduced

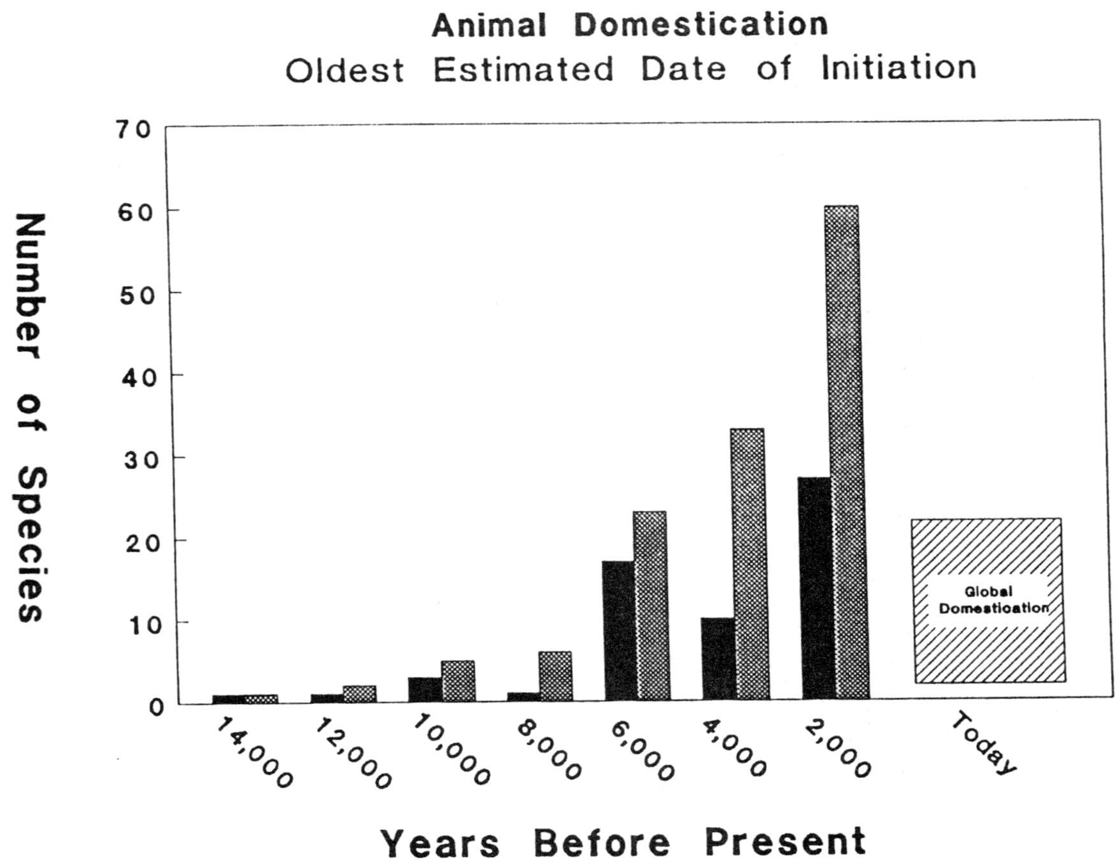
Animal Domestication
Oldest Estimated Date of Initiation
Number of Species
70
60
50
40
30
20
10
0
14,000
12,000
10,000
8,000
6,000
4,000
2,000
Today
Global
Domestication
Years Before Present

Figure 1

Figure 1. An approximation of the number of species domesticated by humans over time. Solid bars indicate the number of new species recruited in that time period; hatched bars indicate the cumulative number of species domesticated by the end of that period. Although different populations of a single species would have been domesticated at different times in various locations, only the oldest estimated date is plotted per species. Global domestication is indicative of modern times (see text) and the number of species formally domesticated is difficult to estimate.

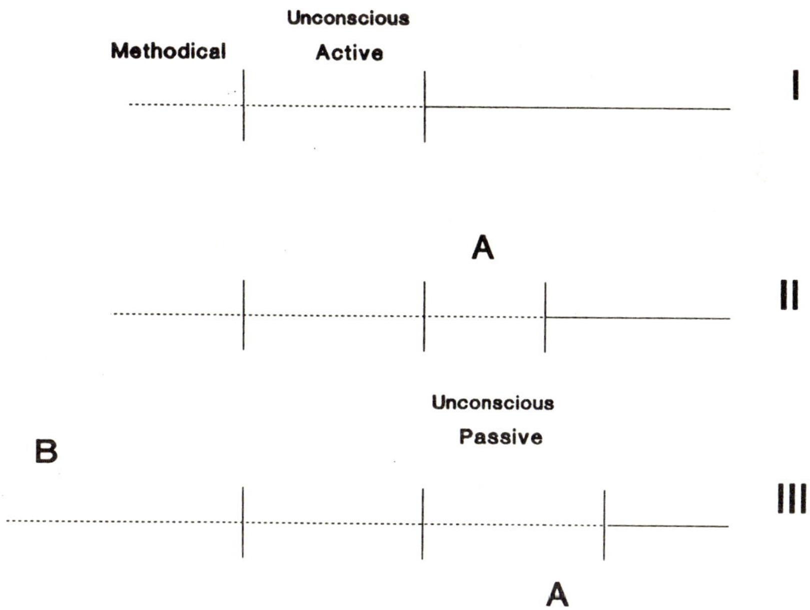

Figure 2. The selection continuum and its modification over time. Solid line = Natural selection, dotted line = Artificial selection. *Period I* (PreIndustrial Revolution): only Natural selection and two forms of Artificial selection practiced. *Period II* (PostIndustrial Revolution): the increased impact of human activity on the environment led to an extension of Artificial selection termed Unconscious Passive selection (A). *Period III* (Modern): an increase in the rate of Methodical selection (B) due to current technology has led to a further extension of Artificial selection. Additionally, Unconscious Passive selection has increased (A').

(Vermeij 1986). In addition, the rate at which new phenotypes can be produced has increased dramatically.

Our ability to manipulate life via methodical selection carries an extraordinary responsibility to care for the organisms produced, to consider the impact of species introductions on native fauna before they are conducted, and to respect evolutionary lineages. But our moral responsibility is just as great, if less understood, at the other end of the artificial selection curve (Figure 2, Point B). We suggest that an additional form of artificial selection is now in place and, in fact, has been at least since 1760 when the Industrial Revolution began in England. This selection process should be considered an *unconscious passive* one compared to Darwin's complementary processes of methodical (*conscious active*) and unconscious (*unconscious active*) selection.

Prior to the Industrial Revolution the worlds of nature and humans were both perceptually and effectively different. Human impact on the planet was easy to identify and separate from the reciprocal process of natural selection. With the rise in technology that characterized the late eighteenth century, a paradoxical relationship between humans and nature developed that continues today. Humankind is now the earth's pilot (Dobzhansky 1967) and it is becoming increasingly more difficult to separate artificial and natural selection. *Unconscious passive* artificial selection is occurring on a grand scale as people radically alter the environment, the classical domain of natural selection. For example, if lakes become too acidified to support life as a result of acid rain, it is reasonable to suggest that people have practiced community (or even ecosystem) selection. If fair-skinned humans exhibit a significant increase in the number of malignant melanomas as a result of atmospheric ozone depletion, we might infer selection against that phenotype. Any number of examples can be imagined; the point is that artificial selection so defined is not working solely on already-domesticated animals.

Specific examples involving the use of pesticides in the environment are illustrative of unconscious passive selection. Although most pesticides are considered transient (they are unstable and break down over time), they may drastically alter species composition in the immediate application area (McEwen & Stephenson 1979). Some pesticides, like DDT, persist in the environment; even local applications result in much wider distribution (Wurster 1968), although levels often are of no biological significance *per se*. But DDT magnifies biologically in some organisms more than others, having greater impact on a species as it moves along the food chain. Widespread use of DDT in the environment during the 1950s and 1960s resulted in lethal and sublethal (e.g. eggshell thinning, egg breakage, poor reproductive success) effects in a wide range of bird species (McEwen & Stephenson 1979). Eggshell thinning was reported in at least some populations of 39 species of birds, and in 27 species the problem was sufficient to impair reproduction (McEwen & Stephenson 1979). In addition, pesticide use has resulted in both teratogenic (see Clegg & Khera [1973] for review) and mutagenic (see Epstein & Legator [1971] and Durham & Williams [1973] for reviews) effects on animal populations that conceivably could have important evolutionary consequences.

It is clear that *unconscious passive selection*, in which no defined phenotype is actively selected but in which human action plays a critical role in defining an animal's fitness, is an extension of artificial selection to virtually all species on Earth except, perhaps, the most isolated. We do not mean to imply that such selection will always be the only mechanism influencing a population's future genetic constitution, but that unconscious selection for populations and species that can withstand the human assault on their habitats will be important.

The question of human intent was important in previous definitions of domestication. In what might be termed the global domestication of the world's fauna due to unconscious passive selection, humans often have no direct say in choosing what is a desirable phenotype, though there is *de facto* selection of phenotypes best adapted to life with humans.

Further examples of global domestication may be found by examining the plight of endangered species. Habitat destruction usually refers to modification of an area to provide farmland or housing for rapidly growing human populations at the expense of plant and animal life inhabiting those areas. Such modification is the single-most important factor contributing to the rise in the number of endangered species in recent years (Vermeij 1986). Deliberate or accidental introductions of domestic or exotic species into new habitats is another human act that may have a serious impact on resident fauna. Warner (1968) observed that the nearly 85% decline in the number of endemic forest bird species in Hawaii was the result of the spread of bird pox and avian malaria by mosquitos introduced to the island in 1826 (although Atkinson [1977] suggested that introduction of the roof rat (*Rattus rattus*) was the primary cause). Feral cats are endangering the Bahama Amazon parrot (*Amazona leucocephala*)

(R. Gnam personal communication), while feral dogs on the Galápagos Islands have the same effect on marine iguanas (*Amblyrhynchus cristatus*) (Kruuk & Snell 1981). The loss of vegetation cover and eating of plant leaves by feral goats inhabiting some islands has been responsible for the extinction of many endemic plants and indirectly has contributed to the extinction of some animals (Coblentz 1978).

Despite the frequency of such introductions, local populations have rarely gone extinct. One reason may be that introduced species tend to remain in contact with human populations, thus decreasing their interaction with endemic fauna (Vermeij 1986). Even marine organisms introduced into new areas by humans are often found in sheltered bays and estuaries (Vermeij 1978). Where faunal mixing has been more complete (Zaret & Paine 1973; Moyle 1976), extinction of endemic wildlife species has been rare, but shifts in the size and composition of resident populations have been documented (Ben-Yami & Glaser 1974; Vermeij 1986). However, human-induced extinctions due to faunal mixing have occurred and will continue to occur.

Actions in addition to direct introduction of exotic species into new areas also have had an impact on endemic fauna. Christie (1974) concluded that overfishing of native species in the Great Lakes helped to establish introduced populations of smelt (*Osmerus mordax*), lamprey eel (*Petromyzon marinus*), and others that further decreased native populations. Thus, although a key aspect of natural selection is the extinction of species as new ones develop, extinction today is largely a result of human action, specifically, unconscious passive selection, resulting from habitat fragmentation and destruction (Elliot 1986; Soulé 1986; Lande 1988) as well as hunting (Vermeij 1986). Good old-fashioned natural selection as the primary force influencing the faunal composition of the planet no longer accurately depicts what is now happening.

The consequences of extinction extend beyond the obvious ones to the victimized lineage. Disappearance of a species represents changes in the selective environment of remaining local species that may ultimately have wide-reaching harmful effects (Southwick 1983; Jolly 1985). Thus, human influence is magnified to the extent that the defunct group interacted with these other species.

Although artificial selection, and therefore domestication, may be more widespread than commonly thought, it is important to note that what determines the importance of people's effect in supplanting natural selection is how our actions influence those aspects of the animal's life history that actually impinge on its survival and reproduction. For instance, if human activity alters the acidity of a lake, but the limiting factor for fish in that lake is the number of breeding sites, and this is not affected by the lake's acidity, a factor not influenced by people remains the dominant selective force.

THE ETHICS OF DOMESTICATION

It will be clear by now that we feel discussion of only "official" domestic animals is inadequate; virtually every species on the planet is undergoing some degree of domestication. The alternative is extinction and, taking a pragmatic approach to the issue of domestication, we might conclude that survival of a species in a domestic form is preferable to the alternative. We do not believe that human moral obligations to domestic animals are necessarily different from those we owe to wild animals (Callicott 1980; but see Johnson 1981), even if domestic animals are viewed as "artifacts" (Callicott 1980) and not as natural kinds (Rolston 1988).

But we need to ask if the act of domestication itself is morally acceptable with respect to humans' role in the process, not simply with respect to the outcome for the animals involved. Domestication may be acceptable in a system where *neither* population is capable of making a moral decision about right and wrong (e.g. ants and aphids) but unacceptable in systems where *at least one* population can (see Regan 1985a). Furthermore, the issue of quality of life is an important one that needs to be raised. What we regard as being in the interest of a *species* (e.g. survival, increased reproduction) may not be in the best interest of *individual* members of that species (Jamieson 1985). Survival of a species may be a mark of success, but if individual animals cannot maintain themselves without the help of humans, or if animals are in pain due to dysgenic breeding practices of keepers, it is at best a qualified success. What obligations do humans have to their animal charges?

Two major schools of thought address questions of humans' moral obligations to other animals. According to Regan (1985a), nonhuman animals have moral rights (universal, equal, inalienable, not created by acts of humans) that limit what humans are morally permitted to do to them. Such rights include the right not to be harmed, the right to life, and the right of noninterference. In his *"rights"* approach, Regan argues that nonhuman animals have value apart from any human interests and that they are entitled to respect because they are subjects of a life. Regan's ideas are far-ranging and beyond the scope of this paper; his theory of rights is discussed in detail by Jamieson (1990; for further discussions see Dresser 1985; Singer 1985a,b; Regan 1985b, 1987; and Finsen 1988, this volume).

The other approach is a *utilitarian* one, most notably associated with Singer (1990), in which acts are right if they bring about the greatest possible balance of intrinsic good over intrinsic evil for all concerned. Singer (1990) argues that sentient nonhuman animals are worthy of equal consideration with humans and have an interest in not suffering.

Both approaches are nonanthropocentric and both deny rights and interests, respectively, to species or groups of animals. They also deal essentially with sentient animals. However, each approach has its drawbacks. Singer's (1990) argument for vegetarianism, for example, is based on the idea that current animal housing and slaughtering practices inevitably lead to suffering by the animal, rather than on the notion that killing animals is always wrong. Likewise, Singer's version of utilitarianism is difficult to apply to species protection needs, for example, because extermination of a species or its replacement by introduced species need not cause any suffering at all (Gunn 1984).

The rights view is more restrictive in terms of what people can do to animals and rejects traditional roles of animals that may not be abusive but clearly subordinate their rights. For example, human agriculture based on motive power provided by horses, oxen, and other nonhuman animals must be halted (Aiken 1984; Regan 1984), given that these animals have rights equal to humans. However, while it is always wrong to disregard an animal's rights, these rights may be overridden (Gunn 1984). This is a significant challenge to the rights position; ethical decisions often require ranking by priority the interests or rights of different individuals. Once a decision of this sort is called for, the consequences of the act need consideration. In a hypothetical case where one of two animals, dog or human, must be excluded from a lifeboat, for instance, Regan (1985a,b) would base such a decision on the magnitude of the harm of death, which is a function of the number and variety of opportunities for satisfaction that the death would foreclose. But, because it is humans that make the decision about the sum of good and bad consequences, this determination may be inherently biased. Furthermore, the assumption must be made that humans are qualified to adequately judge a nonhuman animal's life and its opportunities, and the ability of people to do this can certainly be questioned.

Domestication and Exploitation

With respect to domestication, a process that deals with populations of animals, individual-oriented (atomistic) theories such as those above may be inadequate. Whether or not the act of domestication itself is inherently wrong (deontological theory) does not free us from consideration of the consequences of the act. At the very least, humans have an obligation to consider where the produced organisms will live, where food will come from, what effect artificial selection will have on the species gene pool, whether or not the animals will be abandoned, and if the animals can survive without human help. Many environmentalists have turned to a different type of nonanthropocentric theory, an ecocentric theory, that emphasizes whole groups, relationships among them, and ecosystems as sources of value (Aiken 1984; Regan 1984; Callicott 1989). Although we are not ready to abandon individual-based theories, we believe additional criteria for consideration may be found in a holistic approach.

However, human responsibilities may vary with the form of artificial selection. The breeding and creation of animals for scientific research through *methodical selection,* even more so than the factory farming of animals for food, is the primary animal use issue taking us into the next century. Not only are population gene pools being manipulated, but the research procedures themselves often are detrimental to the animals. It is critical that we seriously question the morality of creating animals to advance science or to destroy them for food. In the case of methodical selection, humans are aware of the consequences of their acts. Our obligations to these animals include consideration and fulfillment of their lifetime needs, not simply for as long as they are useful in a specific context (Linden 1986; Bekoff & Jamieson 1990a,b). Before heeding calls for increased domestication of different species, particularly as food resources (e.g. Jewell 1969; Price 1984), ethical and ecological energetic arguments against carnivory (Singer 1990) should be considered. With respect to the selection of purebred companion animals, certain practices should be stopped. The deliberate breeding of dogs and other companion animals to accentuate dysgenic characteristics such as bulging eyes, shortened faces, and disproportionately short legs, simply for the sake of fanciers, should be halted when animals suffer as a consequence (Working Party Council for Science and Society 1988).

Methodical selection is also the cornerstone of endangered species preservation programs and the consequences of this form of domestication merit consideration. Some efforts to save these groups from extinction, for example, focus on collecting and freezing sperm samples that represent the genetic future of the species. This facilitates artificial selection of certain phenotypes that will become the future representatives of the species. What impact our decisions of relative fitness of individuals will have on future populations of that species is not known. One often hears that the ultimate goal of preservationists is reintroduction of endangered species into the wild; zoo breeding programs are simply a rest stop along the way. But in fact, reintroduction of many species is highly unlikely and successful reintroduction even less so, given the current rate of environmental destruction (see Jamieson 1985). These animals face a captive life with humans. We may ask then if domestication is good in this case. Assuming we are aware of these animals' needs and care for them, the answer must be yes; survival is success. Clearly though, their lives will not be natural and if we assume that they have moral rights (e.g. Regan 1984), we must weigh the subordination of their right to a natural life against their extinction, given conditions as they exist today.

It is somewhat ironic that the holistic approach to animal welfare has been adopted by environmentalists seeking to account for the added value endangered species have over populations of more plentiful animals (Aiken 1984). Endangered populations more and more are being taken out of the context of that environment; their roles in the ecosystem are reduced daily. Here, atomistic theories of animal welfare (utilitarian, rightist) are more appropriate perhaps than they are for populations of feral domestic animals who are playing an ever greater role in the dynamics of different ecosystems.

Unconscious active selection historically has been the start of domestication episodes with virtually every species and accounts for most of the overpopulation today of companion animals. It is what the average human does who either wishes to breed the family dog with the neighbor's dog or who doesn't actively prevent this from happening, and generally does not involve a commercial interest in the animals. The result often is that individual animals chosen for inclusion in a human family receive more than adequate care while the vast majority receive little or none. Although animal population control efforts aimed at reducing birth rates by spaying and neutering pets may be considered interference in the animal's ability to live its own life (e.g. Regan 1984), long term impacts of such large populations on the environment and native fauna must be considered.

Feral populations are the result of either escapes or intentional releases of domestic animals such that the relative importance of natural selection increases (Daniels 1987; Daniels & Bekoff 1989). Insofar as humans can deduce the outcome of such releases in terms of both potential suffering of the individuals involved and long term impacts on the environment, our obligation lies in maintaining a level of active artificial selection and provision of resources to these animals instead of abandonment, which is often under the pretense of setting the animals "free" to live their own lives.

Our greatest obligation, though, lies in the direction of the greatest unknown. *Unconscious passive selection* is having profound effects on nature's biotic and abiotic components. Ignorance of these effects does not abrogate our responsibility to the animals influenced by artificial selection processes. Domestication of this sort is morally least acceptable because of our ignorance of its consequences and, perhaps, our inability to alter the outcome. Logically, serious worldwide efforts must be made to reverse the trend of wholesale alteration or obliteration of nature and thereby, reduce unconscious passive selection. It is wrong to hold a species hostage to the destructive influence of another and herein lies the great fault of global domestication. Although certain phenotypes best adapted to life with humans are selected for, the overall result clearly has been, and will continue to be, loss of animals and loss of species. This truth must be learned before changes in our treatment of these animals can be expected. Appeals to ethics are not enough.

ANTHROPOCENTRISM REVISITED: HUMAN SOCIOBIOLOGY AND THE WAY OF THE WORLD

Aiken's (1984: 268) comment that "nature gives us our moral norms and tells us how we ought to act" might more accurately be stated if we change the phrase "how we ought to act" to "why we act as we do." Our "moral norms" generally are based on biological truths. Incest avoidance, for example, is the rule in nature and is probably the result of selection for individuals that outbreed and thus whose offspring do not suffer inbreeding depression (but see Shields 1983).

From this perspective, we suggest that an anthropocentric view of life on earth, rather than being a selfish strategy that only undermines our efforts to achieve higher moral ideals, is, if not the only way to resolve our dilemma, certainly an important one. Humans may not necessarily be more selfish than other sentient species, but the human potential for altruism based on moral principle alone is not always enough to overcome human selfishness. This fundamental truth dictates our present options.

Animals were domesticated because they served a purpose, and domestication continues today at such a high rate because it remains beneficial to *us* at least in the short-run. However, short-term benefits may turn out to be long-term losses. Mistreatment of animals in the broad sense, as a result of improper care or housing or abandonment, also continues because it is perceived as beneficial to do so in the short-term, whether the benefit be financial or otherwise. Where the results of such treatment are immediately apparent (methodical and unconscious active selection), and this will generally involve individual animals, there is some hope that persuasive ethical reasoning will curb these practices. But in cases where the effects of our behavior are not immediately apparent, as in unconscious passive selection, a more effective approach to meeting our obligations to other animals may involve alerting humans to the biological consequences of their actions to themselves. If we continue to foul the planet, to raze forests, and to kill animals indiscriminately, we inevitably bring ourselves closer to extinction. In the long-term, human interests and those of other animals, must be aligned as closely as possible if we are to survive together on this planet.

ACKNOWLEDGEMENTS

We thank Dale Jamieson for his helpful comments on an earlier draft of this paper.

LITERATURE CITED

Aiken, W. 1984. Ethical issues in agriculture. In: *Earthbound: New Introductory Essays in Environmental Ethics* (ed. by T.

Regan), pp. 247-288. Philadelphia, Pennsylvania: Temple University Press.
Anderson, E. 1952. *Plants, Man, and Life.* Berkeley, California: University of California Press.
Aschmann, H. 1965. Comments on "Man, culture, and animals." In: *Man, Culture, and Animals* (ed. by A. Leeds & A.P. Vayda), pp. 259-270. Washington, D. C.: American Association for the Advancement of Science.
Atkinson, I.A.E. 1977. A reassessment of factors, particularly *Rattus rattus* L., that influence the decline of endemic forest birds in the Hawaiian Islands. *Pacific Science* 31, 109-133.
Baker, C.M.A. & Manwell, C. 1981. "Fiercely feral": on the survival of domesticates without care from man. *Zeitschrift für Tierzuchtung und Zuchtungsbiologie* 98, 241-257.
Barnett, S.A. 1988. *Biology and Freedom: An Essay on the Implications of Human Ethology.* New York: Cambridge University Press.
Bekoff, M. 1972. The development of social interaction, play, and metacommunication in mammals: An ethological perspective. *Quarterly Review of Biology* 47, 412-434.
Bekoff, M. & Jamieson, D. 1990a. Reflective ethology, applied philosophy, and the moral status of animals. *Perspectives in Ethology* 9.
_____. 1990b. Cognitive ethology and applied philosophy: The significance of an evolutionary biology of mind. *Trends in Ecology & Evolution* 5, 156-159.
Ben-Yami, M. & Glaser, T. 1974. The invasion of *Saurida undosquamis* (Richardson) into the Levant Basin: An example of biologic effects of interoceanic canals. *Fisheries Bulletin* 72, 359-373.
Berry, R.J. 1969. The genetical implications of domestication in animals. In: *The Domestication and Exploitation of Plants and Animals* (ed. by P.J. Ucko & G.W. Dimbleby), pp. 207-217. Chicago, Illinois: Aldine Publishing.
Boice, R. 1973. Domestication. *Psychological Bulletin* 80, 215-230.
Bokonyi, S. 1969. Archaeological problems and methods of recognizing animal domestication. In: *The Domestication and Exploitation of Plants and Animals* (ed. by P.J. Ucko & G.W. Dimbleby), pp. 219-229. Chicago, Illinois: Aldine Publishing.
Callicott, J.B. 1980. Animal Liberation: A Triangular Affair. *Environmental Ethics* 2, 311-338.

_____. 1989. *In Defense of the Land Ethic: Essays in Environmental Philosophy.* Albany, New York: SUNY Press.

Capecchi, M.R. 1989. Altering the genome by homologous recombination. *Science* 244, 1288-1292.

Christie, W.J. 1974. Changes in the fish species composition of the Great Lakes. *Journal of the Fisheries Research Board of Canada* 31, 827-854. Cited in Vermeij 1986.

Clegg, D.J., & Khera, K.S. 1973. The teratogenicity of pesticides, their metabolites and contaminants. In: *Pesticides and the Environment: a Continuing Controversy* (ed. by W.B. Deichmann), pp. 267-276. New York: Intercontinental Medical Book Corp.

Clutton-Brock, J. 1981. *Domesticated Animals from Early Times.* Austin, Texas: University of Texas Press.

Coblentz, B.E. 1978. The effects of feral goats (*Capra hircus*) on island ecosystems. *Biological Conservation* 13, 279-286.

Cole, H.H. 1980. Work as an animal product. In: *Animal Agriculture: the Biology, Husbandry, and Use of Domestic Animals,* 2nd Edition. (ed. by H.H. Cole & W.N. Garrett), pp. 156-169. San Francisco, California: W.H. Freeman & Co.

Cole, H.H. & Garrett, W.N. (eds.). 1980. *Animal Agriculture: the Biology, Husbandry, and Use of Domestic Animals*, 2nd Edition. San Francisco, California: W. H. Freeman & Co.

Cranstone, B.A.L. 1969. Animal husbandry: The evidence from ethnography. In: *The Domestication and Exploitation of Plants and Animals* (ed. by P.J. Ucko & G.W. Dimbleby), pp. 247-263. Chicago, Illinois: Aldine Publishing Co.

Crawford, M.A. 1974. The case for new domestic animals. *Oryx* 12, 351-360.

Daniels, T.J. 1987. *The Social Ecology and Behavior of Free-Ranging Dogs.* Ph.D. thesis, University of Colorado, Boulder.

Daniels, T.J. & Bekoff, M. 1989. Feralization: the making of a wild domestic animal. *Behavioral Processes* 19, 79-94.

Darwin, C. 1859/1958. *The Origin of Species.* New York: Mentor Books.

_____. 1868. *The Variation of Animals and Plants Under Domestication,* Vols I and II. London: John Murray.

Davis, S.J. & Valla, F.R. 1978. Evidence for domestication of the dog 12,000 years ago in the Natufian of Israel. *Nature* 276, 608-610.

de Luce, J. & Wilder, H.T. 1983. Introduction. In: *Language in Primates: Perspectives and Implications* (ed. by J. de Luce & H.T. Wilder), pp. 1-17. New York: Springer-Verlag.

Dobzhansky, T. 1967. *The Biology of Ultimate Concern.* New York: American Library, Inc.

Donaldson, L.R. & Menasveta, D. 1961. Selective breeding of the Chinook salmon. *Transactions of the American Fisheries Society* 90, 160-164.

Downs, J.F. 1960. Domestication: An examination of the changing social relationships between man and animals. *Kroeber Anthropological Society Papers,* No. 22. University of California, Berkeley.

Dresser, R. 1985. Respecting and Protecting Nonhuman Animals: Regan's *The Case for Animal Rights. American Bar Foundation Research Journal* 1984, 831-850.

Ducos, P. 1969. Methodology and results of the study of the earliest domesticated animals in the Near East (Palestine). In: *The Domestication and Exploitation of Plants and Animals* (ed. by P.J. Ucko & G.W. Dimbleby), pp. 265-275. Chicago, Illinois: Aldine Publishing Co.

Durham, W.F. & Williams, C.H. 1973. Mutagenic, teratogenic, and carcinogenic properties of pesticides. In: *Pesticides and the Environment: a Continuing Controversy* (ed. by W.B. Deichmann), pp. 307-334. New York: Intercontinental Medical Book Corp.

Eibl-Eibesfeldt, I. 1975. *Ethology: the Biology of Behavior,* 2nd Edition. New York: Holt, Rinehart, and Winston.

Elliot, D.K. (ed.). 1986. *Dynamics of Extinction.* New York: John Wiley and Sons.

Epstein, S.S. & Legator, M.S. (eds.). 1971. *Mutagenicity of Pesticides: Concepts and Evaluations.* Cambridge: Massachusetts: MIT Press.

Finsen, S. 1988. Sinking the Research Lifeboat. *The Journal of Medicine and Philosophy* 13, 197-212.

Fiorone, F. 1973. *The Encyclopedia of Dogs* (ed. by P. Lecaldano). London: Hart-Davis. (Cited in Manwell & Baker 1984.)

Flannery, K.V. 1969. Origins and ecological effects of early domestication in Iran and the Near East. In: *The Domestication and Exploitation of Plants and Animals* (ed. by P.J. Ucko & G.W. Dimbleby), pp. 73-100. Chicago, Illinois: Aldine Publishing Co.

Fox, M.W. 1968. The influence of domestication upon behavior of animals. In: *Abnormal Behavior in Animals* (ed. by M.W. Fox), pp. 64-76. Philadelphia, Pennsylvania: W.B. Saunders.

_____. 1978. *The Dog: Its Domestication and Behavior.* New York: Garland Press.
Fox, M.W. & Bekoff, M. 1975. The behaviour of dogs. In: *The Behaviour of Domestic Animals,* 3rd Edition. (ed. by E.S.E. Hafez), pp. 370-409. London: Bailliere Tindall.
Gipson, P.S. 1972. *The Taxonomy, Reproductive Biology, Food Habits, and Range of Wild Canis (Canidae) in Arkansas.* Ph.D. thesis, University of Arkansas, Fayetteville.
Gould, S.J. 1977. *Ontogeny and Phylogeny.* Cambridge, Massachusetts: Harvard University Press.
Gray, A.P. 1971. *Mammalian Hybrids: a Check-list with Bibliography.* Slough, England: Commonwealth Agricultural Bureaux.
Grigson, C. 1969. The uses and limitations of differences in absolute size in the distinction between the bones of aurochs (*Bos primigenius*) and domestic cattle (*Bos taurus*). In: *The Domestication and Exploitation of Plants and Animals* (ed. by P.J. Ucko & G.W. Dimbleby), pp. 277-294. Chicago, Illinois: Aldine Publishing Co.
Gunn, A.S. 1984. Preserving rare species. In: *Earthbound: New Introductory Essays in Environmental Ethics* (ed. by T. Regan), pp. 289-335. Philadelphia, Pennsylvania: Temple University Press.
Hafez, E.S.E. (ed.). 1968. *Adaptation of Domestic Animals.* Philadelphia, Pennsylvania: Lea and Febiger.
Haldane, J.B.S. 1949. Suggestions as to the quantitative measurement of rates of evolution. *Evolution* 3, 51-56.
Hale, E.B. 1969. Domestication and the evolution of behavior. In: *The Behaviour of Domestic Animals,* 2nd Edition. (ed. by E.S.E. Hafez), pp. 22-42. London: Bailliere Tindall.
Hawkes, J.G. 1969. The ecological background of plant domestication. In: *The Domestication and Exploitation of Plants and Animals* (ed. by P.J. Ucko & G.W. Dimbleby), pp. 17-29. Chicago, Illinois: Aldine Publishing Co.
Hediger, H. 1964. *Wild Animals in Captivity.* New York: Dover.
_____. 1968. *The Psychology of Animals in Zoos and circuses.* New York: Dover.
Helbaek, H. 1959. How farming began in the old world. *Archaeology* 22, 183-189.
Hyams, E. 1972. *Animals in the Service of Man: 10,000 Years of Domestication.* London: J.M. Dent and Sons.
Jamieson, D. 1985. Against zoos. In: *In Defense of Animals* (ed. by P. Singer), pp. 108-117. New York: Basil Blackwell.

_____. 1990. Rights, justice, and duties to aid: A critique of Regan's theory of rights. *Ethics* 100, 349-362..

Jewell, P.A. 1969. Wild mammals and their potential for new domestication. In: *The Domestication and Exploitation of Plants and Animals* (ed. by P.J. Ucko & G.W. Dimbleby), pp. 101-109. Chicago, Illinois: Aldine Publishing Co.

Johnson, E. 1981. Animal liberation versus the land ethic. *Environmental Ethics* 3, 265-273.

Jolly, A. 1985. The evolution of primate behavior. *American Scientist* 73, 230-239.

Koshland, D.E. 1988. For whom the bell tolls. *Science* 241, 1405.

Kruska, D. 1988. Mammalian domestication and its effect on brain structure and behavior. In: *Intelligence and Evolutionary Biology* (ed. by H. J. Jerison & I. Jerison), pp. 211-250. Berlin: Springer-Verlag.

Kruuk, H. & Snell, H. 1981. Prey selection by feral dogs from a population of maring iguanas (*Amblyrhynchus cristatus*). *Journal of Applied Ecology* 18, 197-204.

Lande, R. 1988. Genetics and demography in biological conservation. *Science* 241, 1455-1460.

Leeds, A. & Vayda, A.P. (eds.). 1965. *Man, Culture, and Animals.* Washington, D.C.: American Association for the Advancement of Science.

Linden, E. 1986. *Silent Partners.* New York: Random House.

Lorenz, K. 1965. *Evolution and Modification of Behavior.* Chicago, Illinois: University of Chicago Press.

Mahan, R., Gipson, P.S. & Case, R.M. 1978. Characteristics and distribution of coyote x dog hybrids collected in Nebraska. *American Midland Naturalist* 100, 408-415.

Manwell, C. & Baker, C.M.A. 1984. Domestication of the dog: hunter, food, bed-warmer, or emotional object? *Zeitschrift für Tierzuchtung und Zuchtungsbiologie* 101, 241-256.

McEwen, F.L. & Stephenson, G.R. 1979. *The Use and Significance of Pesticides in the Environment.* New York: John Wiley and Sons.

Mengel, R.M. 1971. A study of dog-coyote hybrids and implications concerning hybridization in *Canis. Journal of Mammalogy* 52, 316-336.

Miller, D.B. 1977. Social displays of mallard ducks (*Anas platyrhynchos*): effects of domestication. *Journal of Comparative Physiology and Psychology* 91 221-232.

Moyle, P.B. 1976. Fish introductions in California: history and impact on native fishes. *Biological Conservation* 9 101-118.
Myers, N. 1979. *The Sinking Ark.* New York: Random House.
Nash, R. 1989. *The Rights of Nature: a History of Environmental Ethics.* Madison, Wisconsin: University of Wisconsin Press.
Norton, B.G. (ed.) 1986. *The Preservation of Species: the Value of Biological Diversity.* Princeton, New Jersey: Princeton University Press.
_____. 1987. *Why Preserve Natural Variety?* Princeton, New Jersey: Princeton University Press.
Ortega y Gasset, J. 1972. *Meditations on Hunting.* New York: Scribners.
Patterson, D.F. 1986. A catalogue of congenital and hereditary disorders of dogs (by breed). In: *Current Veterinary Therapy IX* (ed. by R. W. Kirk), pp. 1281-1285. Philadelphia, Pennsylvania: Saunders.
Pollak, E.J. 1980. Selection, mating systems, and the significance of breeds. In: *Animal Agriculture: the Biology, Husbandry, and Use of Domestic Animals* 2nd Edition. (ed. by H. H. Cole & W. N. Garrett), pp. 193-207. San Francisco, California: W. H. Freeman and Company.
Price, E.O. 1984. Behavioral aspects of animal domestication. *Quarterly Review of Biology* 59, 1-32.
Price, E.O. & J.A. King. 1968. Domestication and adaptation. In: *Adaptation of Domestic Animals* (ed. by E.S.E. Hafez), pp. 34-45. Philadelphia, Pennsylvania: Lea and Febiger.
Pursel, V.G., Pinkert, C.A., Miller, K.F., Bolt, D.J., Campbell, R.G., Palmiter, R.D., Brinster, R.L. & Hammer, R.E. 1989. Genetic engineering of livestock. *Science* 294, 1281-1288.
Rachels, J. 1987. Darwin, species, and morality. *The Monist* 70, 98-113.
Ratner, S.C. & Boice, R. 1975. Effects of domestication on behaviour. In: *The Behaviour of Domestic Animals,* 3rd Edition. (ed. by E.S.E. Hafez), pp. 3-19. London: Bailliere Tindall.
Reed, C.A. 1969. The pattern of animal domestication in the prehistoric Near East. In: *The Domestication and Exploitation of Plants and Animals* (ed. by P.J. Ucko & G.W. Dimbleby), pp. 361-380. Chicago, Illinois: Aldine Publishing Co.
_____. (ed.). 1977. *Origins of Agriculture.* The Hague: Mouton.
_____. 1980. The beginnings of animal domestication. In: *Animal Agriculture: the Biology, Husbandry, and Use of Domestic*

Animals, 2nd Edition. (ed. by H. H. Cole & W. N. Garrett), pp. 3-20. San Francisco, California: W. H. Freeman and Co.
Regan, T. 1984. Introduction. In: *Earthbound: New Introductory Essays in Environmental Ethics* (ed. by T. Regan), pp. 3-37. Philadelphia, Pennsylvania: Temple University Press.
_____. 1985a. *The Case for Animal Rights*. Berkeley: University of California Press.
_____. 1985b. The dog in the lifeboat: an exchange. *New York Review of Books* April 25, 56-57.
_____. (ed.) 1987. Animal rights. *The Monist* 70, 1-133.
Rheinberger, H. & McLaughlin, P. 1984. Darwin's experimental natural history. *Journal of the History of Biology* 17 345-368.
Richards, R.J. 1987. *Darwin and the Emergence of Evolutionary Theories of Mind and Behavior.* Chicago, Illinois: University of Chicago Press.
Rolston, H., III. 1988. *Environmental Ethics: Duties to and Values in the Natural World.* Philadelphia, Pennsylvania: Temple University Press.
Sagoff, M. 1988. Ethics, ecology, and the environment: integrating science and law. *Tennessee Law Review* 56, 77-229.
Scott, J.P. & Fuller, J.L. 1965. *Genetics and the Social Behavior of the Dog*. Chicago, Illinois: University of Chicago Press.
Scott, M.D. & Causey, K. 1973. Ecology of feral dogs in Alabama. *Journal of Wildlife Management* 37, 253-265.
Serpell, J. 1986. *In the Company of Animals*. Oxford: Basil Blackwell.
Shields, W.M. 1983. *Philopatry, Inbreeding, and the Evolution of Sex.* Albany, New York: SUNY Press.
Singer, P. 1985a. Ten years of animal liberation. *New York Review of Books* January 17, 46-52.
_____. 1985b. The dog in the lifeboat: an exchange. *New York Review of Books* April 25, 57.
_____. 1990. *Animal Liberation: A New Ethics for Our Treatment of Animals,* 2nd Edition. New York: New York Review of Books.
Soulé, M.E. 1986. *Conservation Biology, the Science of Scarcity and Diversity.* Sunderland, Massachusetts: Sinauer.
Southwick, C.H. (ed.). 1983. *Global Ecology*. Sunderland, Massachusetts: Sinauer.
Stone, C.D. 1987. *Earth and Other Ethics: The Case for Moral Pluralism.* New York: Harper and Row.

Sweet, L.E. 1965. Camel pastoralism in North Arabia and the minimal camping unit. In: *Man, Culture, and Animals* (ed. by A. Leeds & A.P. Vayda), pp. 129-152. Washington, D.C.: American Association for the Advancement of Science.

Turnbull, P.F. & Reed, C.A. 1974. The fauna from the terminal Pleistocene of Palegawra cave, a Zarzian occupation site in northeastern Iraq. *Fieldiana Anthropology* 63, 81-146.

Ucko, P.J. & Dimbleby, G.W. (eds.) 1969. *The Domestication and Exploitation of Plants and Animals.* Chicago, Illinois: Aldine Publishing Co.

Van Vleck, L.D. 1980. Inheritance in horses and dogs. In: *Animal Agriculture: the Biology, Husbandry, and Use of Domestic Animals,* 2nd Edition. (ed. by H. H.Cole & W. N. Garrett), pp. 331-351. San Francisco, California: W. H. Freeman.

Vermeij, G.J. 1978. *Biology and Adaptation: Patterns of Marine Life.* Cambridge, Massachusetts: Harvard University Press.

_____. 1986. The Biology of Human-caused Extinction. In: *The Preservation of Species* (ed. by B. G. Norton), pp. 28-49. Princeton, New Jersey: Princeton University Press.

Vincent, R.E. 1960. Some influences of domestication upon three stocks of brook trout (*Salvelinus fontinalis Mitchill*). *Transactions of the American Fisheries Society* 89, 35-52.

Warner, R.E. 1968. The role of introduced diseases in the extinction of the endemic Hawaiian avifauna. *Condor* 70, 101-120.

Waters, C.K. 1986. Taking analogical inference seriously: Darwin's argument from artificial selection. *Philosophy of Science Association* 1: 502-531.

Way, M.J. 1963. Mutualism between ants and honeydew producing Homoptera. *Annual Review of Entomology* 8, 307-344.

Westra, L. 1989. Ecology and animals: Is there a joint ethic of respect? *Environmental Ethics* 11, 215-230.

Wilson, E.O. 1971. *The Insect Societies.* Cambridge, Massachusetts: Harvard University Press.

_____. 1975. *Sociobiology: the New Synthesis.* Cambridge, Massachusetts: Harvard University Press.

_____. 1989. Threats to biodiversity. *Scientific American* 261, 108-116.

Working Party Council for Science and Society. 1988. *Companion Animals in Society.* Oxford: Oxford University Press.

Wright, S. 1977. *Evolution and the Genetics of Populations*, Vol. 3: Experimental Results and Evolutionary Deductions. Chicago, Illinois: University of Chicago Press.
Wurster, C.F., Jr. 1968. DDT reduces photosynthesis by marine phytoplankton. *Science* 159, 1474-1475.
Zaret, T.M. & Paine, R.T. 1973. Species introduction in a tropical lake. *Science* 182, 449-455.
Zeuner, F.E. 1963. *A History of Domesticated Animals.* London: Hutchinson and Company.

14. Ethical Issues in Genetics: Cautionary Considerations

Lawson Crowe

ADVANCES IN GENETIC KNOWLEDGE

Humans and other animals may benefit from the advance of genetic knowledge, but they are also vulnerable to its misuse and abuse. The history of hereditarian thought is replete with examples of individuals and whole segments of the human and animal communities harmed by incomplete knowledge and faulty understanding of heredity. Because we are inclined to intervene in the workings of nature, humans are perhaps too eager and too willing to find applications for genetic knowledge that may shape in profound and unanticipated ways social policies and practices and even the natural environment.

There is reason to believe that interest in genetic sensitivity to specific substances has encouraged trivial product and drug testing and sometimes useless experiments on nonhuman animals. In addition, geneticists and biologists have the capacity to manipulate the genome of various animals even though the effect of introducing such altered animals into the natural environment is unknown (Daniels & Bekoff this volume).

There is also evidence that businesses and industries are likely to use genetic knowledge to discriminate against individuals in the work place. Although it is yet far from being realized, some fear that an intimate knowledge of a person's genetic profile may place the possessor in a position to manipulate that person's destiny. And in this connection, perhaps the greatest fear is that governments, businesses, and other institutions will attempt to use genetic knowledge of individuals and groups to confirm their political and/or ethnic prejudices and thereby justify discriminatory policies.

At the public policy level, the problems for humans and other animals are essentially the same. Although the capacity for abstract thought is a glory of the human mind, this very capacity for abstraction sometimes renders us ethically insensitive to the concrete reality of other human beings or other sentient creatures. It is the capacity for abstraction that underlies racism, sexism, and speciesism. It is the abstract concept such as "racial purity" or "the superiority of man over beast" or "man over woman" that allows us to ignore individual variability, individual merit, and the actual sufferings of individuals of all species.

Is it not remarkable that whole societies (Nazi Germany, the American white South, etc.) have sometimes been dominated by such abstractions when, in fact, ordinary, commonplace experience has always been of individuals only? One person at a time. For the most part, one animal at a time. Each of these beings possess individual temperament, habits, intelligence, and idiosyncrasies. For human beings who have the capacity to distinguish right conduct from wrong conduct, i.e., who are moral agents, it is the presence of an actual individual that demands moral attention and moral consideration. It is this presence that renders the abstraction of race, sex, or species morally insignificant and makes it merely an interesting fact about the individual. The burden of showing the moral relevance of these abstractions rests on those who assert it.

For the foregoing reasons, I propose first to examine briefly some of the possible ethical problems associated with recent advances in genetic knowledge. Afterwards, I will turn to a longer view of hereditarian thought. In the light of history, I propose to introduce a caution against repeating past mistakes in the future.

NEW OPPORTUNITIES AND NEW ETHICAL PROBLEMS

Since the late 1950s, the revolution in genetic knowledge has created a new set of opportunities and, not surprisingly, new ethical problems. On the one hand, the new genetics has made prenatal diagnosis available to prospective parents whose offspring are at risk for inherited diseases or other deleterious conditions. Chromosomal anomalies indicate that the fetus is defective. The parent or parents must then choose either to abort the fetus or to accept the consequences of carrying it to term. On the other hand, and over the longer range, the new genetics promises to reveal in profound detail a person's hereditary legacy which will then make it possible to assess that person's future health prospects. In addition, the new genetics has awakened fresh scientific interest in discovering the possible biological determinants of individual differences in intelligence, and the possible predisposition to undesirable behaviors such as alcoholism, other drug addictions, and criminality (Holtzman 1989; Milunsky 1989).

The ability to locate on the human chromosome genes responsible for specific diseases opens a new era in clinical medicine. The exact or approximate locations of genes associated with muscular dystrophy, Huntington's disease, some psychiatric disorders, and a variety of cancers have been found. Although the incidence of most genetic diseases is relatively low, gene mappers have already located more than 400 genetic "markers," indicators of genetic disease, on all 46 human chromosomes (Weiss 1989). It is expected that acquisition of this knowledge will accelerate with the mapping of the human genome over the next decade or so. Although it is not entirely clear what the "biological determinants" of behavior would be like, it is expected that somehow the etiology of certain behaviors (alcoholism, drug addiction, criminality) will be located in some combination of genes.

The consequence of this technology will be a vastly improved array of diagnostic tests able to reveal who is at risk for both rare and common diseases, who may be predisposed to develop undesirable behaviors, and who may be susceptible to certain environmental hazards. In addition to identifying persons actually having a disease, it is probable that these diagnostic tests will have predictive value. Diagnostic tests invariably precede therapy. It follows, therefore, that some conditions will be diagnosed for which no therapy exists, a circumstance having its own ethical problems. Whether therapy is available or not, however, the misuse of such information may have a profound impact on a person's access to employment, health care (for example, a person could be declared uninsurable), and social life in general. Unless carefully regulated, mandatory screening for genetic disease, for future risk of genetic disease, or for predisposition to undesirable behaviors may create morally unacceptable discriminatory practices (Holtzman 1989). Sickle cell anemia screening programs of a decade ago caused particular hardship for individual carriers of the sickle cell because employers and insurers did not understand that the carriers themselves were not diseased (President's Commission 1983).

If misused, new genetic knowledge may threaten individual autonomy, that is, our fundamental right to self-determination and the concurrent right to privacy. The ethical issue is who will have access to a person's genetic information and how will it be used? The possibilities for a new kind of discrimination and for development of social policies having adverse consequences for those genetically at risk are considerable. Because of their reproductive function, women would appear to be especially vulnerable. The history of hereditarian thought suggests that such adverse consequences are possible even when our knowledge remains incomplete. Dedication to the ideal of eliminating deleterious alleles from the gene pool or to other ideological or political commitments tends to obscure the rights and interests of those who do not share such goals.

In fact, the societal record for safe and appropriate use of technological and scientific advances is not impressive either in America or in Europe. Leroy Hood, a leader in gene-sequencing technology at California Institute of Technology, was recently quoted by Weiss (1989: 42) as commenting on this problem. Hood said: "What science does is give society opportunities. What we have to do is look at these opportunities and then set up the constraints and the rules that will allow society to benefit in appropriate ways."

No doubt this opinion is correct, but, to mention only a few examples, societal efforts to regulate the generation of nuclear power, the disposal of radioactive and other toxic wastes, and the control of other environmental hazards do not offer much encouragement. In addition, the long history of doubtful or positively harmful medical interventions (unnecessary surgery, toxic drugs, etc.), often enthusiastically embraced by both doctors *and* patients, casts further doubt on our capability to restrain the inappropriate application of new knowledge.

Both scientific and popular hereditarian beliefs have had a profound influence on social behavior and on social policies pursued by governments. Immigration policies, the prohibition of alcohol, educational policies, welfare programs, and prison and penal policies have, in part, been based on commonly held assumptions about the nature of inheritance and its controlling effects on individual ability and destiny. Although recent advances in genetics have discredited many of these assumptions, they have also produced evidence that genes contribute significantly to individual differences in abilities and behavior. It is this evidence that has encouraged contemporary scientists to search for the possible "biological determinants" of undesirable behaviors. The discovery of such determinants could, if indeed they exist, have profound ethical implications for personal autonomy, for personal responsibility for health and welfare, and for therapeutic interventions in persons at risk for alcoholism, other drug addictions, and criminality. In what follows it will be seen that contemporary ethical issues are perhaps more serious and far-reaching than, but are not substantially different from, those generated in the earlier history of hereditarian thought.

EARLY HEREDITARIAN THOUGHT

The idea that children inherit the characteristics of their parents has always been an important part of human self-understanding. It is an ancient observation not only about humans, but also about every living plant and animal species. As many commentators have noticed, the Bible reflects awareness of this commonplace wisdom. For example, the *Book of Genesis* (Chapter 1) speaks of God creating plants and animals "after their own kind," and in the *New Testament*, the Gospel of Matthew represents Jesus in the Sermon on the Mount as asking rhetorically, "Do men gather grapes of thorns or figs of thistles?" (*Matthew*, Chapter 7: 16) In *The Republic*, Plato has Socrates describe in detail the breeding of the Guardians of the state. They are to be compared to guard dogs bred for implacable fierceness toward their enemies and gentleness towards those they protect (*The Republic*, Book II, Para. 375). Many other such references could be cited. In any event, the inheritance of the essential features of the forebears, the descent from clan or tribe, the perpetuation of the line, along with the hereditary transmission of titles and property, from earliest times have been considered of highest significance for those who were beneficiaries of these concerns.

The vast social changes attendant upon the revolutions of the eighteenth and early nineteenth centuries aroused renewed interest in heredity and the privileges and disabilities it conferred. The rising economic and social demands of the working class in Britain and western Europe along with the advent of socialist doctrines, the influx of immigrants into America, the emergence of industrial labor unions in America, and the series of American and European wars all contributed to the reordering of western society and its values. Drunkenness, prostitution, crime, and general immorality among the poor were seen as social evils requiring explanation. For this and similar reasons, hereditarian doctrines were invoked from the mid-nineteenth century onward to explain what many of the well-to-do regarded as the decay and degeneration of late nineteenth and early twentieth century western European and American society.

Francis Galton (1822-1911), perhaps the last truly great English dilettante, had interests ranging from African travel, to development of a device for the measurement of female beauty, to a study of the efficacy of prayer, to the development of statistical methods, and to the breeding of animals for specific characteristics. Galton was profoundly influenced by his cousin, Charles Darwin. With Darwin's publication of *On The Origin of Species* in 1859, Galton was led to pursue new inquiries into the nature of heredity, a word which he apparently invented (F. Darwin 1914/1968: 10). In subsequent writings in *MacMillans Magazine* in 1865 and in *Hereditary Genius* in 1869, he argued that genius itself rose from innate inherited differences between individuals. Genius did not occur as a consequence of birth to privilege, money, or power (Suzuki & Knudtson 1989). Still later, in *Human Faculty* published in 1884, Galton used the term "Eugenics" to describe the science of the improvement of the human condition (F. Darwin 1914/1968). He argued for "judicious matings" that would produce more suitable progeny that would prevail over the less suitable (Suzuki & Knutson 1989). Galton suggested that eugenics would become "an orthodox religious tenet" of the future. Reflecting his Darwinist perspective, he wrote: "...Eugenics cooperates with the workings of nature by securing that humanity shall be represented by the fittest races. What Nature does blindly, slowly, and ruthlessly, man may do providentially, quickly, and kindly." (as quoted in F. Darwin 1914/1968) As will be seen below, Galton's words appear benign in comparison to the words of some who claimed to be his followers.

In the meantime, various European and American physicians attempted to establish the biological basis for degenerate behavior. They focused their attention on those they classified as "mental defectives," "idiots," and "imbeciles." In particular, they identified the "moral imbecile." These were persons who could not help doing evil because they suffered from an inherited moral incapacity (Scheerenberger 1983). The subjects of these studies were primarily inmates of prisons, insane asylums, work houses for the poor, and homes for the "feebleminded." For example, Cesare Lombroso and his followers attempted to show through surveys of prisoners that physical characteristics (facial features, cranial size, etc.) could be used to predict criminality (Chorover 1979).

Many nineteenth and early twentieth century physicians and temperance writers found that "racial degeneracy" was caused by the use of alcohol. Parents who drank alcohol were said to transmit hereditary diseases and alcoholism to their offspring (Warner & Rosett 1975). Intoxication of the parents at the time of conception was thought to produce defective children. Among these writers, the causal connection between alcohol use and inherited diseases, feeblemindedness, imbecility, etc. was not to be doubted. In 1848, Samuel Howe reported to the Massachusetts legislature that nearly half of the mental defectives he studied in Massachusetts public institutions had intemperate parents. Later studies purported to confirm that the offspring of alcohol-using parents were depraved, perverted, and suffering from "mental degeneration" and "an irresistible craving" for alcohol. As the chief cause of "innate imbecility," alcohol was a major source of crime, prostitution, and other social evils. It was a toxic substance damaging to the nervous and reproductive systems, and this damage was thought to be transmitted directly to offspring (Crowe 1985).

American medical temperance writers expressed similar opinions. With large numbers of immigrants arriving at the turn of the century, T.A. MacNicholl, a New York physician, revealed his ethnic prejudices in sentiments so often expressed by eugenicists: "A vast immigration of inferior peoples, attracted by our great material prosperity and the hope of political liberty, bring with them their vices as well as their virtues, augmenting our drinking classes, furnishing additional soil from which to propagate criminals, thereby increasing our burdens..." (MacNicholl 1907: 397)

One of the most robust expressions of these views may be found in a report on admissions to the insane asylum for the City of Paris and the Department of the Seine covering the period 1867 to 1912. The authors said that more than 20 percent of the admissions were alcohol related and that there was an estimated 33 percent infant mortality rate among children of alcoholics. They wrote: "...among the survivors are counted many idiots, epileptics, and a large number of degenerates, destitute of moral sense, instinctively perverted, impulsive, abnormal, miserable victims of their parent's alcoholism." (Magnan & Fillassier 1912: 379)

Francis Galton was skeptical of these claims and especially of the so-called statistical methods used to support them. In 1910, his follower, Karl Pearson, first director of the Galton Eugenics Laboratory of the University of London, published a statistical study raising serious doubts about the view that alcohol use was causally related to heritable disease in offspring (Elderton & Pearson 1910). Although violent objections by eugenicists followed the appearance of this work, a second study by Pearson and Elderton demolished and ultimately discredited the arguments and statistics of the advocates of the view that alcohol was the cause of "racial degeneracy" (Pearson & Elderton 1910).

THE LATER EUGENICS MOVEMENT

In the first decades of the twentieth century the interests of the Eugenics movement were focused on one primary theme, namely, "the rapid degeneration of the race" (see, e.g. Webb 1909/1968: 73). Societal trends of the time were clearly puzzling to those who accepted Darwin's theory of the survival of the fittest through natural selection and who furthermore understood the theory through the Social Darwinism of Herbert Spencer. Simply stated, it appeared that the "unfit" were reproducing themselves more rapidly than the "fit." From Galton forward this trend was viewed with alarm. Sidney Webb, for example, called for the deliberate manipulation of the environment. He blamed the English Poor Law for providing indiscriminate relief to the destitute, thereby encouraging the birth of the unfit. He wrote: "...we must take steps to prevent the continued procreation of feebleminded, degenerate stocks at the public expense..." He argued that the whole class of degenerates should be placed under control of a civil authority that had the power to "segregate" these people from the rest of society. This must be done, he said, "so that survivors may be of the type which we regard as the highest." (Webb 1909/1968: 73)

The distinction between the "superior" stock of the upper classes and the "better" working classes, on the one hand, and the "inferior" stock of the lowest classes on the other was explicit and regarded as self-evident by those who made it. The problem was how to account for the fecundity of this "inferior stock." Some identified technology as the origin of the problem. Havelock Ellis (1917/1968), for example, cited birth control as a "dysgenic" practice of the superior classes. Ignorant, inferior stocks most in need of the knowledge of contraception were said to propagate freely. He argued that the "capable" must always bear the burdens of the "incapable": "...the vigorous, hardworking, and prudent people assume evergrowing financial and other burdens which limit their powers to do justice to their own children, while rendering it more possible for the lazy, the improvident, and the diseased to live in ease they have not earned, to procreate their own kind, and to escape the natural results of their own laziness, improvidence, and disease" (Ellis 1917/1968: 78). (The reader may wish to regard the foregoing as an early twentieth century version of President Reagan's "Welfare Queen Argument.") Ellis said this process had been going on for a long time, especially in the most civilized and progressive countries. He said that World War I was making matters worse since the total number of "capable" people was being reduced.

In 1921, W. R. Inge, a prominent English churchman, wrote that Western societies face "the survival of the unfittest." He quoted Karl Pearson to the effect that the "better stock" is taxed and penalized by governmental policies encouraging Nature's failures and misfits to increase and multiply. Inge made the contrast between Nature and Nurture explicit and left no doubt as to where he stood (p. 96): "It is the man who makes his environment not the environment which makes the man." Inge identified the eugenics movement with Christianity. The reason to eliminate slum conditions was simply that they were the breeding grounds of "undesirable citizens." Also, he speculated that if slums were removed their inhabitants might not "breed so fast." Inge interpreted Jesus' saying in the Sermon on the Mount (*Matthew*, Chapter 7: 16-20) to the effect that a corrupt tree does not bear good fruit as having eugenic import. He said that "we are breeding from our worst stocks and...our best are being squeezed out of existence." In what must be one of the most fatuous remarks of all time, he compared the "lowest strata" of degenerates, "the worst specimens," with the "higher ranks" in which a thoroughly degenerate stock tends to die out "unless there is great beauty or wealth or a title to act as a makeweight." (Inge 1921/1968: 96)

Because British eugenicists believed in social progress, it became necessary to explain the erosion of the class system. This required a reinterpretation of evolutionary theory. In a lecture entitled "The Future of Our Race: Heredity and Social Progress," Leonard Darwin ascribed the "moral and intellectual contagion" affecting the British nation to "the presence in its ranks of inferior types." (L. Darwin 1924/1968: 105) He also said that social policy should aim at lowering the birth rate of the "grossly unfit," that mental defectives and epileptics should be confined to institutions and segregated by sex, and that habitual criminals, wastrels, and drunkards should be confined for lengthy periods of time in order to reduce their opportunities for procreation. On the other hand, everything possible should be done to increase the birth rate of the better classes and of the great mass of efficient, hard working, law-abiding, and healthy citizens. Otherwise, he said that Britain faced slow and steady deterioration in the qualities of the nation as a whole. He wrote: "...a short time ago it seemed as if all we had to do was to let evolution run its course and human progress would inevitably be the result. Now we see that evolution is a process in which organisms may deteriorate in regard to all the qualities that we wish to promote and indeed in which species or races may disappear altogether." (L. Darwin 1924/1968: 101)

Leonard Darwin concluded, therefore, that human evolution had to be "guided." Such manipulations as might be necessary would require sacrifices on the part of both the unfit and the fit. As a matter of duty, the former would have to forgo reproduction and the latter would have to have children. Although he never spelled out in detail what would have to be done to the unfit to gain their compliance in this program, Darwin seems to have had some fairly Draconian solution in mind. He asked rhetorically: "Ought not some drastic action to be taken to prevent large families from coming into the world if they would be likely to be composed of persons of low morals, poor intelligence, or marked inefficiency? The nation must answer this question in the affirmative if our race is to advance in social well-being as quickly as may be." (L. Darwin 1924/1968: 107)

Similar statements abound in the eugenics literature. More often than not they are accompanied by pious admonitions to treat these unfortunate "inferior types" with as much kindness and gentleness as good eugenic practice will permit.

I have considered eugenicist concerns at length because it is important to recognize the ethical implications of these views. Hereditarian thought in Britain, on the European continent, and in America was expressed as an ideology and a political program. Aside from a few speculations about the nature of inheritance confined mostly to medical temperance writers, and references to a few "statistical" studies of institutional inmates and school children, the leaders of the eugenics movement offered no scientific support for their views (Crowe 1985). The issue of exactly what was inherited and the mechanism by which it affected behavior was simply not addressed. The assumption seems to have been that offspring inherit physical and mental qualities in much the same way as they inherit property.

With respect to the question of whether Nature or Nurture is the primary factor in the development of an individual, most eugenicists were unequivocally on the side of Nature. From the perspective of the latter part of the twentieth century, it is easy to see the self-interest and moral complacency that hereditarian theory supported. The well-to-do, educated eugenicists were at some pains to explain to their peers as well as to those of lesser status that the existing societal arrangements were really the product of Nature, that there was little that Nurture could do about them beyond some modest amelioration made possible through social progress. Dean Inge, for example, as well as Karl Pearson and many others, believed that the nation or race advanced or decayed according to "stern natural laws" or, as they were sometimes described, "biological realities." As a group, both in America and abroad, it seems fair to say that eugenicists had a rather good opinion of themselves, consistently counting themselves among those they regarded as superior. They evidently never grasped the implications of the Darwinian theory of the survival of the fittest through natural selection. They did not understand that the key component of survival was reproductive fitness. Therefore, the fecundity of the "inferior" sort or "types" and the apparent infertility of the "superior" types remained a mystery for many. The rediscovery of Mendel's work in 1901 had no apparent influence on the eugenicist viewpoint except that Mendelian genetics quickly supplanted Galton's and Pearson's so-called biometrical approach to the study of heredity (F. Darwin 1914/1968).

In early twentieth century America, the attention of hereditarian thought focused on the "defective delinquent," formerly known as the "moral imbecile." The mental testing movement, encouraged by the need to classify soldiers in World War I, purported to reveal substantial differences in intelligence between ethnic groups, immigrant groups, and native white Americans. Inmates in various prisons and reformatories were evaluated. Most prostitutes were found to be feebleminded, most criminals mentally retarded. In 1916, Lewis Terman, Professor of Psychology at Stanford University, wrote that not all criminals were feebleminded but that all feebleminded persons were potentially criminals. The alleged innate criminal tendencies of the mentally retarded became a primary justification for state sterilization programs. Without doubt, these programs met the requirement for "drastic action" proposed by Leonard Darwin and others. Between 1907 and 1958 more than 30,000 mentally retarded persons (10,990 males and 20,048 females) were sterilized in the 30 states having the requisite laws (Scheerenberger 1983). In this atmosphere, American eugenicists supported the passage of the Immigration Act of 1924, a highly restrictive law designed to keep out poor, uneducated immigrants from southern and eastern Europe. In large part, the law was justified on the grounds that these immigrants were genetically inferior, prone to criminality and drunkenness.

Both the sterilization programs and the Immigration Act of 1924 were designed to reduce crime and other social evils by preventing the feebleminded from replicating themselves and by preventing those of "genetically inferior stock" from entering the country at all. Both the programs and the Act were based on bad science and questionable values. Both reflected common and unchallenged prejudices against the poor, ethnic minorities, immigrants, and others whose values and tastes did not match those of the privileged. As advanced by eugenicists, hereditarian doctrine, even as modified by new genetic knowledge, was dedicated to the abstract ideal of "racial improvement" and the prevention of "racial decay." With respect to the welfare of the so-called inferior types (also an abstraction), it was callous and mean-spirited.

Many events and circumstances led to the decline of the eugenics movement. The lack of a scientific basis for eugenic theories, and the apparent impracticality of the eugenics program caused many otherwise sympathetic biologists, other scientists, and social reformers to drift away. Preoccupation with hereditary transmission of qualities leading to "racial decay" obscured the impact of social conditions promoting poverty, ignorance, and crime. Eugenicists insisted that forces innate within the individual were more significant than any environmental conditions. It seems fair to say that they seriously misinterpreted the social climate of their time. The growing demands of the labor movement and the campaign for women's suffrage were not, as many thought, signs of "racial degeneration." Rather, they were indications of the further democratization of western Europe, Britain, and America, which with setbacks here and there, has continued to the present day. The Civil Rights Movement of the 1960s and 1970s in the United States, the latest surge of feminism, and the Gay Liberation movement may be the most recent examples of this on-going process.

In any event, hereditarian doctrine made for an easy social conscience and reinforced the moral complacency of the well-to-do. It fostered paternalism and moral insensitivity towards the needs of real people in favor of a commitment to an abstract ideal of "social progress." One can almost hear Dean Inge quoting scripture to the effect that God sends His rain to fall upon the just and the unjust. If the poor suffer, it must be their hereditary destiny.

In this atmosphere, it is not surprising that the traditional anti-Semitism of western Europe combined with eugenic doctrine to produce a program for the systematic extermination of the "genetically unfit." Shortly after the Nazi Party under Adolph Hitler rose to power in 1933, the German government attempted to achieve "racial purity." In 1934 the Nazis sterilized 56,000 people on the grounds that they were "genetically unfit" (Suzuki 1989). In the name of eugenics, Hitler and his followers murdered millions of Jews, Gypsies, and other ethnic minorities. The horror and revulsion felt throughout the world when the extent of Nazi atrocities was finally known was the final blow to the eugenics movement. It was not that racial and ethnic prejudices were finally overcome, but rather that for many people in Europe and America, their overt expression was no longer morally or socially acceptable and could no longer serve as the basis for social policy.

AFTERWORD

In the closing years of the twentieth century, racist ideology within the United States is mostly moribund. Civil rights are protected. Medical practice and scientific research are controlled internally by a body of ethical thought and externally by positive law. In recent years, movements to protect the welfare of other animal species have also emerged.

Nevertheless, as a society we should not be complacent. Sterilization laws are still in effect in many states. And it was only 57 years ago that the United States Public Health Service initiated the Tuskeegee study of 400 black men infected with syphilis. Without their knowledge, treatment was withheld from these men so that Public Health physicians in Macon County, Alabama, could follow the "natural course" of the disease. Critics of this experiment believe that it was largely conceived and carried out in the light of racist assumptions about black male sexuality (Brandt 1978; Rothman 1982).

Other examples serve to make the point as well. In the mid-1960s at the Jewish Chronic Disease Hospital in Queens, New

York, researchers injected live cancer cells beneath the skin of elderly, chronically ill, nonconsenting patients. They wanted to observe the growth and development of cancer cells in these elderly patients who were mostly senile and near the end of their lives (Katz 1971).

From 1956 until 1970, physicians at the Willowbrook State Hospital in Staten Island, New York, infected between 700 to 800 newly admitted mentally retarded children with serum hepatitis in the hope of developing a vaccine against the disease. These children were chosen as subjects because of their physical and mental status and their assignment to a public health institution (Beecher 1976).

In the late 1970s, some scientists and medical researchers in Boston, Massachusetts, undertook a screening program to detect extra Y chromosomes in newborn males. Earlier studies of a prison population in Scotland suggested that an extra Y chromosome in males might be a correlate of violent behavior. The researchers intended to follow the development of these children to see if they were more aggressive and more prone to violence than males without this chromosomal anomaly. The program was terminated under intense methodological and political criticism. The study was based in part on the expectation that a disproportionate number of male prison inmates confined for violent acts would be found to have an extra Y chromosome, but since no one had established the incidence of the extra Y in the general population, prison population numbers could not be interpreted. (Six percent of the male inmates chosen for study in one institution had an extra Y chromosome. Was that a high or a low incidence?) In the meantime, further studies have shown that XYY males are fairly common in the general population and that the incidence of XYY males in prisons is more probably related to low intelligence than to aggressive tendencies (Milunsky 1989). Given our history of racial conflict and the fact that the original studies of this anomaly were conducted in prisons and institutions where a large part of the population was composed of blacks and other ethnic minorities, it is no surprise that some would construe such research as a political program designed to confirm racist convictions (The XYY Controversy 1980).

Other examples of ethically and scientifically questionable research can be given, but these should be sufficient, along with the history of hereditarian thought, to persuade the reader that scientists, being human and sharing the prevailing societal values, are capable of making ethical mistakes and that science as an activity is subject to moral analysis and moral judgment. It must also be acknowledged that popular understanding may promote the future misuse of genetic knowledge. Pseudo-scientific, nonrigorous versions of genetic principles have been used in the past by various groups to appeal to the human tendency to regard one's own racial, social, or religious group as superior to all others. Such views have also played on the anxieties of the dominant majority and have led to policies and actions that were morally reprehensible and scientifically indefensible. As with all human activities, the study of human genetics and the genetic bases of behavior should be conducted in an atmosphere morally sensitive to the autonomy and privacy of the individual and to the potential sufferings of other species. Its findings should not be used to discriminate against any person on the basis of race or gender, political or religious belief.

LITERATURE CITED

Beecher, H.K. 1976. Ethics and clinical research. *New England Journal of Medicine* 274, 1354-1368.

Brandt, A.M. 1978. Racism and research: The case of the Tuskeegee syphilis study. *Hastings Center Report* 8(6), 21-29.

Chorover, S.L. 1979. *From Genesis to Genocide: The Meaning of Human Nature and the Power of Behavior Control.* Cambridge, Massachusetts: The MIT Press.

Crowe, L. 1985. Alcohol and heredity: Theories about the effects of alcohol use on offspring. *Social Biology* 32, 146-161.

Darwin, F. 1914/1968. Francis Galton, 1822-1911. *The Eugenics Review* 60, 3-11.

Darwin, L. 1924/1968. The future of our race: Heredity and social progress. *The Eugenics Review* 60, 99-108.

Elderton, E.M. & Pearson, K. 1910. *A First Study of the Influence of Parental Alcoholism on the Physique and Ability of the Offspring*, 2nd Edition. London: Dulau and Company.

Ellis, H. 1917/1968. Birth control and eugenics. *The Eugenics Review* 60, 76-81.

Holtzman, N.A. 1989. *Proceed with Caution.* Baltimore, Maryland: The Johns Hopkins University Press.

Inge, W.R. 1921/1968. Eugenics and religion. *The Eugenics Review* 60, 92-98.

Katz, J. 1971. *Research Involving Human Subjects.* New York: Russell Sage Foundation.

MacNicholl, T.A. 1907. Alcohol and the disabilities of school children. *Journal of the American Medical Association* February 2, 296-398.

Magnan, M. & Fillassier, A. 1912. Alcoholism and degeneracy. In: *Problems in Eugenics: First International Eugenics Congress held at the University of London, July 25-30, 1912*, pp. 367-379. London: The Eugenics Education Society.

Milunsky, A. 1989. *Choices, Not Chances.* Boston, Massachusetts: Little, Brown and Company.

Pearson, K. & Elderton, E.M. 1910. *A Second Study of the Influence of Parental Alcoholism on the Physique and Ability of the Offspring: Being a Reply to Certain Medical Critics of the First Memoir and an Examination of the Rebutting Evidence Cited by Them.* London: Dulau and Company.

Plato, *The Republic* In: *The Collected Dialogues of Plato* (ed. by E. Hamilton & H. Cairs, 1961). Princeton, New Jersey: Princeton University Press.

President's Commission for the Study of Ethical Problems in Medicine and Biomedical and Behavioral Research. 1983. *Screening and Counseling for Genetic Conditions: A Report on the Ethical, Social and Legal Implications of Genetic Screening, Counseling and Education Programs.* Washington, D.C.

Rothman, D.J. 1982. Were Tuskeegee and Willowbrook "studies in nature?" *Hastings Center Report* 12, 5-7.

Scheerenberger, R.C. 1983. *A History of Mental Retardation.* Baltimore, Maryland: Brookes Publishing Company.

Suzuki, D. & Knudtson, P. 1989. *Genetics.* Cambridge, Massachusetts: Harvard University Press.

The XYY Controversy: Researching Violence and Genetics. 1980. *Hastings Center Report* 10, Special Supplement.

Warner, R. H. & Rosett, H.L. 1975. The effects of drinking on offspring: An historical survey of the American and British literature. *Journal of Studies on Alcohol* 36, 1395-1420.

Webb, S. 1909/1968. Eugenics and the Poor Law: The minority report. *The Eugenics Review* 60, 71-75.

Weiss, R. 1989. Predisposition and prejudice. *Science News* 135, 40-42.

15. On Moderation

Susan Finsen

In recent years the philosophical case for according serious moral status to nonhuman animals has been made quite forcefully from a variety of philosophical perspectives (Regan 1983; Rollin 1981; Singer 1975; Sapontzis 1987). These philosophers have demonstrated that whether we take a deontological, utilitarian, contractarian or broadly commonsense moral framework, it is possible to construct strong arguments showing that our current treatment of animals is inconsistent and ultimately immoral. These arguments provide strong support for the claim that our practices in rearing animals for food and clothing, in laboratory experimentation, hunting, trapping, and entertainment are largely indefensible. Strong cases have been made in particular for the abolition of all animal research (Regan 1983) and for drastic reductions (Rollin 1981; Singer 1975; Sapontzis 1987).

These arguments are by now well-known, and it is not my purpose to review them in what follows. Regardless of how compelling these arguments appear to be, they are often rejected because they are perceived as having conclusions which are too extreme. The fact that these arguments would require radical changes which most people are loath to see come about is surely one factor in this rejection. Until recently, the response of the research community to pressures from animal rights and animal welfare organizations has been mainly defensive. Researchers have sought to justify their use of animals in terms of benefits to humans, and have claimed that the standards of care for laboratory animals are already sufficiently high (King et al. 1988; Miller 1985). This response is still prevalent, and the research community is becoming much more aggressive and sophisticated in battling to maintain the *status quo*. Researchers themselves, along with college administrators, doctors and patients have formed organizations, hired media consultants, made videos and spoken before state and federal legislatures to block animal welfare legislation (Blumenstyk 1989).

But there is also a countervailing trend. Perhaps because of the prolonged and intense pressure from animal rights organizations, perhaps because of public exposure of clear abuses of laboratory animals, or perhaps even because of the force of strong arguments on the part of various philosophers, the research community's posture is changing. It is now widely accepted that when animals are used, especially in invasive or deadly research, this is not a morally neutral activity, but one requiring some ethical justification. The existence of federally mandated animal care committees forces home the point that, every time a research project using animals is undertaken, the ethics of that research must be examined. The existence of these committees has perhaps also moved scientists to begin to take an active role in the discussion of the ethics of animal research. This is a welcome development, since rational discussion is no doubt a prerequisite for moral progress. The views of scientists emerging from this discussion ought to be taken seriously, and it is my goal in this essay to do so. In what follows I will examine the moderate positions of those researchers and their sympathizers who have accepted the idea that animals ought to be given some moral consideration, but who insist that this weight is not enough to cause us to eliminate animal research or even drastically curtail it. Are such positions clear enough to be understood and applied? Are they defended by recognizable moral principles and arguments? Do they offer a view of the value of animal life which accounts for the different standards of treatment accorded nonhumans, as opposed to humans? Are they genuinely moderate positions? These are the questions I hope to answer.

Before turning to the moderate positions, however, I would like to make clear what I take to be reasonable requirements which any ethically defensible position should meet. I am assuming that the fact that a position is a compromise between positions considered "extreme" does not in itself show that a position is defensible. While such compromises may be politically most expedient, this shows nothing about their correctness unless one assumes that moral principles are nothing more than preferences, none being more valid than another. Rather, I assume that an ethically defensible middle ground position must offer and defend principles which are clear enough to show what treatment of research animals is acceptable and what is not. Such a position need not determine answers in all cases, but it must at least indicate what factors are relevant in coming to a decision, and how to weigh them. And any such position ought to offer an account of the value of animal life and its relation to human life, to the extent that it implies different standards of treatment for humans and animals.

While there are many moderate approaches, they can be divided into two broad types. The first type, the "humane treatment" approach, is the approach which is defended by many scientists and appears to underlie current law and research practice, and is thus very important to consider. The second kind of approach is broadly utilitarian or "cost-benefit" in nature, and is also important since it offers suggestions for future reforms with implications both for research and for the welfare of animals. Rather than attempting to discuss the many moderate proposals I will discuss the assumptions and problems common to these two types.

ETHICAL PRINCIPLES: HUMANE TREATMENT AND NECESSARY USE

The humane treatment approach is well-captured by the Stanford University Medical Center Committee on Ethics:

> There is a common-sense view that avoids the extremes of ascribing rights to animals or stripping them of all moral worth and yet allows them to be used in research under certain conditions. Such a position is based on the principle of humane treatment, which places an indirect obligation on humans to prevent the suffering of animals without imposing a direct duty to respect an animal's rights, all things being equal. Applied to laboratory animals, the principle of humane treatment requires, at a minimum, the prohibition of unnecessary pain and suffering. It is a powerful force governing our treatment of animals in research, despite assertions that it has lost its meaning. That some persons view the principle of humane treatment as outdated is the fault of individuals and organizations that use the term too glibly, not of the principle itself or of those who strive to abide by it. (Thomas et al. 1988: 1631)

Whether the principle of humane treatment is a meaningful and powerful force which can be both rationally defended and consistently applied depends in part upon how the notion of "unnecessary pain and suffering" is interpreted. It depends, additionally, of course, upon whether an indirect duty notion of our obligations to animals is defensible, and whether pain and suffering should be the main focus of ethical concern.

According to the Stanford committee on ethics quoted above, the notion of humane treatment is a powerful force governing the treatment of animals in research. The members of this committee provide the following description of how this notion is put into practice:

> At Stanford, before beginning any research using animals, each investigator must submit a rationale for using animals that describes why the species is appropriate, the number of animals needed, and all planned procedures involving animals. Prior to this review, the researcher is directed to ask the following four questions: (1) Have I sought or tried to develop methods that obviate the use of animals? (2) Am I proposing to use an appropriate animal species? (3) Does my protocol call for using the fewest possible animals? (4) Is my protocol in accordance with the principle of humane treatment? (Thomas et al. 1988: 1632)

The proposal, together with a complete description of the protocol, is then submitted to the Administrative Panel on Laboratory Animal Care.

No description is given of any additional rules used by the Administrative Panel, so it is reasonable to assume that they attempt to determine whether the researcher has indeed adequately answered the four questions, as well as following the provisions of the Federal Animal Welfare Act and the National Institutes of Health (NIH) Guidelines.

One notable feature of these questions is that the first three of them would be important to consider even if no issue of animal welfare were involved. The committee earlier notes that, "Virtually no economic consideration favors the use of animals over alternative methods. Research using animals is extremely expensive, perhaps more so than any other type of research." (Thomas et al. 1988: 1630) Thus, questions (1) and (3) are clearly as much prudential and economic as they are ethical. Indeed, in light of the economic expediency of using fewer animals, the "three Rs" of research reform -reduction, refinement and replacement - are well motivated independent of concern for animal welfare. Question (2) regards appropriate research design, which is also as much concerned with good science as with humane concern for animals. These principles are illustrative of ways in which scientific and humane interests can coincide. The fewer animals used, and the better the design of research, the better for both animals and science. Some proponents of moderation go so far as to claim that good science and humane treatment of animals go hand in hand (Perrie Adams 1982). It is the "three Rs" that most clearly support this claim, together with the undeniable fact that the stress of mishandling can interfere with results by inducing unintended physiological and psychological changes. But clearly this claim is not generally supportable since the good husbandry necessary to assure valid scientific results is compatible with severe amounts of social deprivation, pain, boredom and fatigue, especially where the research design is intended to produce one of these states. Questions 1 through 3 are far from sufficient to ensure humane treatment of laboratory animals, and a great deal hinges on the final question. For, it is a reasonable expectation regarding moderate positions that they should not only do what is both in the interests of science and animal welfare, but that they should in some cases be willing to consider the interests of animals even when some sacrifice is involved.

Question 4 is the only one that appears to deal solely with the interests of animals, and it is also most in need of clarification. Being in accordance with the principle of humane treatment means "...the prohibition of unnecessary pain and suffering." Of course, everything hinges on the interpretation of "necessary" here. Far from being a vague or meaningless notion in these contexts, as some have charged, I will argue that this term has a fairly straightforward, though seldom articulated, meaning: a procedure is necessary if (a) there exists a desired benefit (e.g. knowledge) and (b) there is no way to achieve this benefit without the use of the procedure. In other words, a procedure is unnecessary if we could have achieved the *same* result (benefit) in some other way.

This notion of necessity can be found in the writings of Albert Schweitzer, whose views on the reverence for life some researchers quote with approval. While Schweitzer says much in his writings which is not so easy to reconcile with current scientific treatment of animals, the following accords well:

> Whenever I injure life of any kind I must be quite clear as to whether this is necessary or not. I ought never to

> pass the limits of the unavoidable, even in apparently insignificant cases. The countryman who has mowed down a thousand blossoms in his meadow as fodder for his cows should take care that on the way home he does not, in wanton pastime, switch off the head of a single flower growing on the edge of the road, for in so doing he injures life without being forced to do so by necessity. (Schweitzer 1929/1989: 36)

Schweitzer's notion of necessity requires that there be a purpose to the destruction of life, but not that the purpose is itself justifiable. Thus, Schweitzer avoided the unnecessary deaths of insects by working with his windows closed at night, but at the same time he allowed himself the pleasures of meat, since eating meat has a purpose. This use of the term "necessary" I will term "Schweitzer's necessity." This is meant to contrast with another more usual use of this term, in which the benefits themselves are claimed to be essential. That is, the normal use contrasts necessary benefits with unnecessary benefits, not with nonexistent benefits. It is Schweitzer's notion which I claim is currently operative in decisions about whether particular uses of animals in research are justified. From from a legal point of view, this is clearly the case. Consider, for example, the wording of the Canadian Criminal Code, s. 402 (1) (a):

> "Without necessity" does not mean that man, when a thing is susceptible of causing pain to an animal, must abstain unless it be necessary, but means that man in the pursuit of his purposes as a superior being, in the pursuit of his well-being, is obliged not to inflict on animals pain, suffering or injury which is not inevitable taking into account the purpose sought and the circumstances of the particular case. In effect, even if it not be necessary for man to eat meat and if he could abstain from doing so, as many in fact do, it is the privilege of man to eat it. (Linden & Barnes 1988: 10)

The Canadian Criminal code is currently undergoing revisions, and the notion of necessity just articulated is carried over into the new code: "...the pain and injury caused must be justifiable in terms of the object pursued. Where a significant scientific or medical benefit is sought, considerable pain may be justified; where the research is pointless or trivial, very little is justified and the exemption may be lost. The animal experimentation must also be a "reasonably necessary means," that is to say, it must be reasonably unavoidable because no alternative research technique is possible.

Further examples are offered; hoisting and slitting the throats of hogs out of mere sadism would constitute unnecessary cruelty, but no offense is committed where a common slaughtering method is used to kill animals for food. Even tying a rodeo horse with a flank strap or bucking strap does not involve unnecessary cruelty, according to this interpretation. In summary, pain is only "unnecessary" when the benefit sought could have been achieved without it. No offense is committed provided there is a socially recognized purpose and proper means are used, no reasonable alternative being available. Compliance with accepted business standards usually will satisfy these tests (Linden & Barnes 1988: 103).

The Federal Animal Welfare Act (AWA) implicitly adopts the same view of necessity. That is, the Act does not place prohibitions on research, but rather promulgates regulations concerning the purchase, care, handling and disposal of laboratory animals. Recent revisions of the AWA have mandated that Institutional Animal Care and Use Committees (IACUCs) must be formed at "research facilities" (defined in the AWA as institutions that either accept federal money for their animal research or purchase or transport animals across state lines; Sec. 2132e). NIH makes use of such committees a prerequisite for NIH funding. It might be supposed that these committees weigh the necessity of research against costs to animals using the usual notion of necessity. But in fact this is not the mandate of these committees, who are charged with ensuring employment of "...the least traumatic techniques feasible *that would allow the research question to be adequately evaluated.*" (Goy 1982) Once again, "necessity" means, "unavoidable, relative to some desired benefit." It is true that certain procedures, such as multiple recovery surgery and prolonged restraint, are flagged by the AWA and NIH Guides as requiring special attention; in these cases in particular it must be demonstrated beyond a doubt that the research protocol can in no way be achieved without use of the procedure.

While legally, IACUCs are only charged with determining whether the research objectives are achieved with the least possible pain and suffering, there is always the possibility that the usual notion of necessity will enter unofficially as well. The investigator proposing the research, for example, might ask himself whether the expected gain in knowledge is worth the sacrifice of the comfort and lives of the animals involved, and the members of the committee may exercise moral, if not legal influence, in asking questions about the design and expected outcome of the project. Nevertheless, it seems questionable that these issues can be adequately addressed in these ways. Scientists cannot be expected to be objective about the value of their own research, and the members of IACUCs are not likely to press the issue either. Under the revised (1985) NIH policies, IACUCs must have a minimum of five members, including a veterinarian, a practicing scientist, a nonscientist, and someone unaffiliated with the institution and unrelated to anyone affiliated with it. NIH policy states that a single individual may fulfill more than one position, so that a single individual could serve as both the nonscientist and the unaffiliated member. In virtually all cases the majority of people on these committees are scientists, who may be reluctant to challenge the importance of a colleague's work. Laypersons on these committees may not be qualified to judge the importance of research, and in any case will generally not be perceived to be so qualified by the scientists who make up the majority of the committee. Thus, it seems unavoidable that Schweitzer's notion of necessity is the one which will be used in assessing what can and cannot be done to laboratory animals, as a matter of legal mandate and practical reality.

A number of consequences of this use of Schweitzer's necessity should be noted. First, this use of the term is not part of a "cost-benefit" approach since it does not require a weighing of benefits to humans against costs to animals, but assumes that human beings are justified in securing scientific benefits regardless of costs to animals. Thus, implicitly, human good (or scientific knowledge) is taken to be an absolute. As a result the view can hardly be called a "moderate" position and certainly not a "middle ground" with respect to the welfare of animals, in spite of its appearance of moderation. Those adopting such a position must hold that human interests and human value are incomparably more important than animal interests and value, and they owe an account of this qualitative value difference. Adopting Schweitzer's definition of "necessity" simply begs the question of which benefits are essential and which can be foregone. Ultimately, we still need to know, for example, why the desire for fresh air does not justify the sacrifice of the lives of moths, but the desire for meat does justify the sacrifice of cattle. We also need to know whether differences between species make a difference in the application of this principle. Is the issue just biological life, as Schweitzer seems to suggest, or are there morally relevant qualitative distinctions to be made between cows and moths, or ants and primates, as we seem to make between humans and all other animal species? And in the case of animal research, we require an account of why the knowledge gained from some proposed animal experiment justifies the sacrifice of the lives and well being of the animals involved. It is highly counterintuitive to suppose that any increment in knowledge, however slight, can justify the sacrifice of animal lives and well-being. Thus, the Principle of Humane Treatment, given that it rests upon an implausibly weak notion of "unnecessary pain and suffering," is itself an implausibly weak principle. The principle appears plausible because most people who hear it assume the usual use of the notion of "necessity," according to which only the most essential benefits, such as lifesaving medical research, would be allowed.

Aside from the Principle of Humane Treatment with its attendant notion of necessary use, current law and practice embody some further assumptions which ought to be made explicit. First, there is the assumption that the primary focus of ethical concern regarding laboratory animals ought to be on pain and suffering. This is very clear from an examination of the AWA and NIH *Guide*. For example, muscle relaxants or paralytic drugs must not be used alone for surgical restraint (NIH 1985: 37). Physical restraint should be "the minimum required to accomplish research objectives. Prolonged restraint for any reason must be approved by the committee." (NIH 1985: 9) Multiple surgeries are discouraged, except under special circumstances, when they are related components of a research project; under such circumstances they might be permitted with the approval of the IACUC. Further, "[i]f a painful procedure must be conducted without the use of an anesthetic, analgesic or tranquilizer - because such use would defeat the purpose of an experiment - the procedure must be approved by the committee and supervised directly by the responsible investigator." (NIH 1985: 37)

Some institutions have imposed further restrictions. For example, infliction of trauma by beating or crushing is prohibited by the University of Southern California (Dresser 1988: 127). Infliction of severe burn or trauma on unanesthetized animals, production of psychotic-like behavior, infliction of inescapable severe or terminal stress, and use of strychnine or microwave ovens designed for kitchens to kill experimental animals have been forbidden at a number of institutions (Dresser 1988: 127).

The vast majority of laws and guidelines pertaining to laboratory animals deal not with the conduct of research, however, but with the lives of animals while they are not undergoing experimental procedures. These laws are quite extensive, covering every phase from purchase and transport to methods of killing. Obviously, attention to pain and suffering must be a central concern, but there are other ways in which animals can be harmed which ought to be taken into account. These concern an animal's social needs, needs to exercise, and generally to live lives that are normal for their species. Although these needs are not as obvious visually, their lack may lead to as much or more suffering than direct infliction of pain (Bekoff 1976: 31).

The l985 amendments to the AWA include some provision for the psychological and social needs of select species; specifically, exercise for dogs and a physical environment "adequate to promote the psychological well-being of primates." Obviously, psychological and social deprivation can amount to harms comparable to direct infliction of physical pain in psychologically complex and social animals, but this fact is only minimally acknowledged, even for primates and dogs. Researchers at NIH, for example, complain that now that the dogs must be in an open area to satisfy their need to exercise, they are more stressed by laboratory procedures, since they are no longer socialized by being handled during routine maintenance of their cages (discussion session NIH/MRC conference 1988). Obviously, the dogs need both. The sort of environment "adequate to promote the psychological well-being of primates" is undoubtedly lacking in most research settings, but the group caging in open spaces provided at some facilities and breeding colonies goes some way toward providing it. But such group housing can and is dispensed with when it is "necessary," given the research protocol. For example, primates at NIH infected with Simian AIDS are kept in small isolation chambers. It is well to keep in mind that provisions for housing and care can in general be contravened when "necessary" to accomplish the research objective. For example, provisions that animals receive fresh food and water can be contravened in psychologicalgical studies where food and water will serve as rewards.

The minimal provision for psychological well-being in the case of primates and dogs only serves to call attention to the conspicuous lack of such provisions for the vast majority of animals used in research, such as rats and mice. These species, and many others, may still be housed in small, utterly barren cages designed chiefly for ease of cleaning and convenience in observing and handling the animals.

The reasons for the primary focus to date on pain and suffering are undoubtedly emotional, political and economic. Visible physical pain evokes concern more than boredom and social deprivation, and has thus called forth more demand for legislation. In addition, the amelioration of pain through prevention of certain procedures or demands for anesthesia or analgesics is much less expensive than the building of group enclosures.

A second implicit assumption of importance is the current attitude toward the deaths of laboratory animals. It seems obvious that, in general, death is a harm to an animal, unless that animal is leading a miserable life which is not likely to improve. And yet, the research community treats death as either a neutral event or a positive good for laboratory animals. Animals are routinely killed at the termination of experiments, whether or not they have been subjected to invasive or otherwise debilitating procedures. At my campus, a lab session is routinely offered, called, "How to Kill a Rat," in which students have the opportunity to kill a healthy rat. Researchers at my institution consider it inhumane to keep laboratory animals around for too long when they are not being used. Thus, they must not think the life of a laboratory animal is worth living. Alternatively, perhaps they do not think that the lives of such animals are worth anything in general. Either position reveals how far from a middle ground involving serious consideration of the lives of animals the research community remains.

Current law and practice based upon the principle of humane treatment is not as plausible as it initially sounds. Nor is it consistent with the *status quo* to maintain that animal life is given serious weight; the principle of humane treatment dictates otherwise. In fact, if we take the proponents of this principle at their word, there is only an indirect obligation upon researchers to prevent suffering. "Indirect obligation" means that we only owe obligations to animals derivatively, because of their connections to other individuals (presumably humans) to whom we owe direct obligations. This leaves animals without any inherent value at all. As I will argue below it is unlikely that a justification of such a qualitatively disparate rating of the value of animal life in comparison with human benefits can be offered.

In spite of the serious difficulties with the *status quo*, recent revisions in law are in some respects encouraging. In particular, the creation of IACUCs is an important advance. By creating pressure to eliminate outright bad research on animals, such committees may reduce the total amount of animal suffering and death, and thus effect a pragmatic improvement by reducing the number of animals used. On the other hand, by eliminating poorly designed or trivial research these committees also make room for better designed research on animals, so there is no guarantee that the absolute number of animals used will go down through this process. Further, the committees provide a mechanism for the enforcement of the AWA and NIH Guidelines, and in the absence of adequate funding for inspection of research facilities, these committees constitute the most important mechanism for the enforcement of current regulations.

These committees also provide a forum for the discussion of the ethics of animal research, and this may well be the most valuable feature of such committees. Such discussions can sensitize individuals to issues of animal welfare, can encourage scientists to reflect upon the value of their research, and allow dissemination of information on better methods of housing and handling. The clear simple acknowledgement of the ethical dimension of animal research, which these committees affirm, is worth having.

PROPOSED REFORMS: COST/BENEFIT APPROACHES

We have already seen that there is a stronger and more plausible reading of the idea of necessary animal research than the one which is currently operative. According to this notion, the value of scientific knowledge is not absolute, but must be weighed against the costs in animal suffering and death in each case. For the reasons discussed above it seems unlikely that such weighing is a routine feature of current decisions regarding research. Some researchers may nevertheless take a cost/benefit approach in thinking about their use of animals, and some even propose that such weighing should be carried out in all cases. While initially plausible, this approach also has its difficulties, once one attempts to spell out just how the weighing should be done. This can best be seen in relation to a concrete proposal. Driscoll & Bateson (1988) have proposed a decision cube for deciding when research can be justified (see Figure 1). They describe the cube as follows:

> It is not strictly a cost-benefit model since it does not depend on a common currency or on balancing incommensurable properties. It is a set of rules which can be helpful in determining whether or not a particular piece of research should be done. For the purposes of making decisions, three variables are considered: first, the scientific quality of the research; second, the likelihood of animal suffering; and third, the probability of human benefit from the research. These three variables are considered together as illustrated in Figure 1. Animal suffering would be tolerated only when both research quality and certainty of benefit were very high. Moreover, certain levels of animal suffering would be unacceptable regardless of the quality of the research or its probable benefit. Also, the decision rules used in this model would permit research of high quality involving little or no animal suffering even if the work had no obvious potential benefit to humans. This feature takes note of the concern of scientists who want to understand phenomena which have no immediate and obvious benefit for humans. (Driscoll & Bateson 1988: 1572)

One of the most immediately evident features of Driscoll & Bateson's decision cube is the fact that two of its three dimensions deal with human concerns, with "animal suffering" as the one dimension of concern for laboratory animals. This is a rather clear illustration of the wide gap which currently exists between the rapidly expanding knowledge of the complexity of animal cognition and awareness (Griffin 1984; and many articles in this volume) and the assimilation by the research community of the ethical implications of this knowledge (Bekoff & Jamieson 1990). As noted earlier, many dimensions of animal consciousness which cannot be appropriately described as pain or "suffering," including boredom, social deprivation, loss of autonomy and death are important harms which there is no justifiable reason for omitting from calculations involving "costs" to animals.

A particularly ironic example of the failure to see the ethical implications of the cognitive sophistication of animals is provided by Davis (1988). In this study Davis attempted to study the moral development of rats. The fact that experimental psychologists would take seriously the idea of applying moral categories to rats is, of course, quite remarkable in itself. (It is

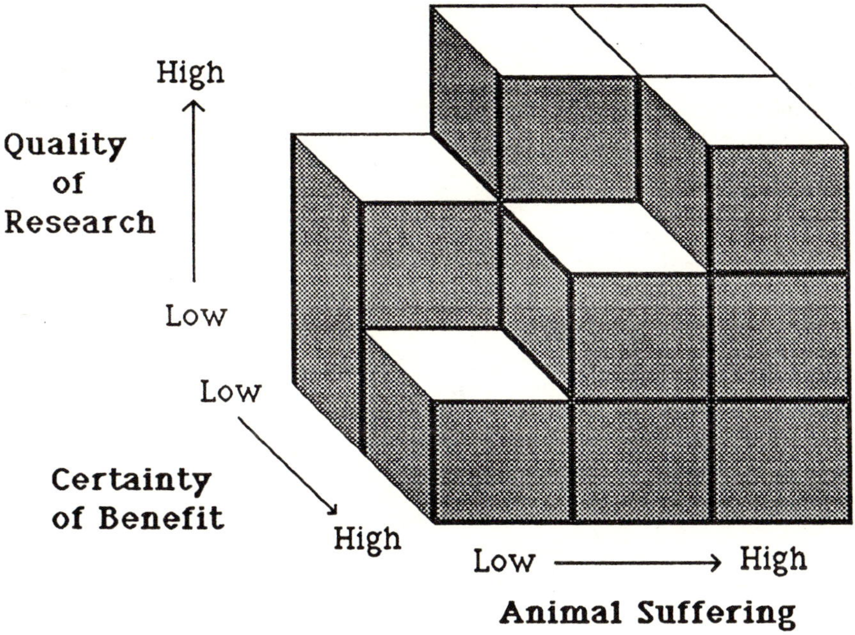

Figure 1. Driscoll & Bateson's (1988) "decision cube" (redrawn).

not entirely clear that the article is intended seriously, in fact, but it is safe to assume that the research reported was in fact carried out.) What is ironic is that Davis felt it appropriate to study such moral development by training rats only to take food under certain conditions, and tested their moral development by observing whether they would try to "sneak" extra food when the trainer was out of the room. It never occurred to Davis that the animals in question might simply regard the researcher as an oppressive warden, to whom no ethical restraint is owed. In other words, the ethical dimensions of Davis' own conduct were completely overlooked. The rats might well have behaved quite differently toward their own conspecifics in an uncoercive situation, where the only moral offense of the researcher was voyeurism. This also suggests that scientists ought to start seriously considering the validity of cognitive studies generally on animals in laboratories. It isn't simply the stress, the unnatural surroundings, etc., but the oppressor/oppressee relationship which may heavily influence the results. Surely such factors would need to be taken into account in the case of studies of human cognition. Where it was discovered that someone was coerced into working math problems, for example, the validity and generality of the results would have to be questioned. Why not with other species?

Returning to the decision cube, we can ask whether it can be used to generate clear decisions. As Driscoll & Bateson point out, their decision procedure cannot literally be a cost/benefit analysis, since the costs and benefits do not involve commensurable sorts of things. This does not show that some sort of informal decision procedure cannot be derived from their decision cube. We do attempt in many spheres to weigh incommensurables against each other. For example, we are willing to risk our own lives in driving various places in order to obtain various benefits. These are informal, personal decisions which obviously vary among individuals. But what is needed here is more than such subjective and individual judgements, and unfortunately Driscoll & Bateson offer us no way of transcending these. There is no basis provided for concluding that *m* amount of knowledge is worth *n* amount of suffering for *x* number of individuals, even assuming we could find metrics for measuring *m* and *n*. This simply underscores the highly informal, qualitative nature of such decisions; the seemingly quantitatively precise picture of the cube notwithstanding. The cube is not really, then, even a set of rules, but rather a general directive to take the three factors of animal pain, concrete probable benefit and quality of research into account in making decisions about animal research. It might exert some influence by suggesting that the knowledge gained must be very important where the pain inflicted upon animals is great. But without some notion of important knowledge or severity of allowable pain this will ultimately be up to individual judgement, and it is unlikely that different researchers would apply the cube in the same way. The likelihood of finding a metric for the importance of knowledge, quite aside from the problem of finding a way of comparing this with degrees of animal suffering, is extremely low indeed, for fairly obvious reasons. And even if a metric for the importance of knowledge could be found, this could not be used to apply to *proposed* research, since the fruitfulness of a research program cannot be known in advance. As Fox (1986), Gallistel (1988) and others have pointed out, serendipity has historically played an important role in science, so that research which promised little of interest outside a narrow theoretical field has contributed vastly to human knowledge and has reaped benefits. Conversely, research dealing with the most pressing human needs may yield nothing if the theoretical background for a breakthrough is not in place. Thus, it is not really possible to use the dimension of "expected benefits." Perhaps the best that can be done along the lines of the proposed metric is to eliminate research which is badly designed and thus cannot yield results, and to eliminate highly invasive and painful research except in the most "pressing" cases, however these might be defined.

While only Driscoll & Bateson's proposal has been examined in detail, the problems encountered in their model are quite general, and surely can be raised against all cost/benefit approaches to this issue. Unlike the humane treatment approach, cost/benefit approaches are simply too vague to offer a clear ethic justifying the use of animals in scientific research. Two key reasons for this unclarity are the lack of an account of the value of scientific research (especially research on animals) and the lack of an account of the value of animal life. The highly controversial issue of the value of scientific research upon animals would take us too far afield but the issue of the value of animal life is unavoidable for a critique of moderate positions. Accordingly, it is to this issue which we finally must turn.

MODERATION AND THE VALUE OF ANIMAL LIFE

Neither humane treatment nor cost/benefit approaches offer explicit accounts of the value of animal life, and to that extent they remain incomplete and unjustified ethics. But certain assumptions about the value of animal life are clearly implicit in both approaches, and these assumptions can be examined to consider the prospects for providing them with justification.

Clearly, animal life is to be assigned a lesser value than human life, according to these views. But, how much less? Presumably, the life of a chimpanzee is worth much less than the life of a human being, in such systems, and the life of a rat or mouse is worth even less than that of the chimpanzee. That something like this is tacitly operating in decisions regarding experimentation can be seen from the widely observed rule that all things being equal, a "lower" (less complex or evolved) organism should be used in preference to a higher one (Russell & Burch 1959; Fox 1986). Obviously evolutionary theory itself does not rate species as less or more valuable, so such ratings must come from some religious or secular normative system.

What defenders of either of the two moderate approaches I have examined (humane treatment and cost/benefit) need is some rational way of assigning value to animal life while still maintaining a large enough value gap between humans and other animals that research on animals can be justified. This value difference must be a difference in kind, rather than degree, in order to justify the difference in kind of treatment provided to human vs. animal subjects of experiments. This is particularly clear in regard to the principle of humane treatment discussed above. But this is also true with regard to the cost/benefit approach. For, this very weighing treats animal life in a way categorically different from human life. No proponents of this approach suggest, for example, that even the most serious, pressing and otherwise unobtainable benefits gained from scientific research can justify invasive and deadly research upon human beings.

What possible bases are there for such categorical value differences? In classical utilitarianism, this sort of ranking isn't done. Jeremy Bentham granted all sentient beings equal consideration. Even though John Stuart Mill considered some sorts of pleasures to be higher than others, the beings themselves are not considered less valuable (since it is the experiences or preferences, not the 'receptacles' who have them, that is important). Thus, cost/benefit analyses such as Driscoll & Bateson's cannot hope to find a foundation in utilitarianism, since there is no calculus for weighing the utilities of unequal beings.

Commonly maintained religious justifications involving the divine origins of human beings and their dominionistic role regarding animals are unpromising, and not simply because they rest upon faith and are thus outside the scientific perspective. In the past, scriptural justifications have been used to motivate discrimination against blacks, women and children, but the scriptural bases of these discriminations have subsequently been undermined. Such a reassessment is currently taking place in regard to the dominionistic perspective on animals in a number of different faiths (Regan 1986).

Secular justifications in terms of important differences between human beings and other species are problematic given that not all humans will have the characteristics. For example, it has been argued in many places that such human characteristics as superior intelligence, linguistic ability, or ability to reason morally explain why we deserve much more serious moral consideration than the rest of the animals. But, for any such characteristic or combination of characteristics, we can find members of our species who lack the characteristic, but who we still wish to offer the same degree of moral consideration as normal humans. For example, profoundly retarded, brain damaged or senile individuals may lack the characteristics of superior intelligence, linguistic ability, and ability to reason morally. Such "marginal case" individuals are nevertheless entitled to the same consideration of their interests as normal humans. This by now familiar objection is often called the "marginal case" argument (for background see Regan 1983; Singer 1990).

Evolutionary theory, our knowledge of genetics, neurophysiology and psychology all point to the continuity among species, and this continuity makes categorical differences in treatment hard to explain. For example, chimpanzees are both psychologically adept and socially complex, and human beings share 98% of their genetic material with them. In fact, recent work in genetics has made it possible to artificially inseminate a female chimpanzee with human sperm (Hollands in press). It is not surprising that this experiment was terminated with abortion of the resultant fetus. For, not only could the researchers involved expect moral outrage from the religious and secular community alike, but the creation of such a hybrid would represent a living refutation to the claim that species differences represent important qualitative boundaries. How would such an individual be classified? Should a half-human primate receive the protection accorded to human subjects, or only the protections of the Federal Animal Welfare Act? What about a 1/4 human, or a 3/4 human?

Of course, chimpanzees may pose special difficulties which can be resolved by extending greater moral concern to them, but this does not show that other species pose the same problems. The marginal case objection considered above naturally leads to the question, "But, where do you draw the line?" This question suggests that a single line must be drawn somewhere, limiting who shall be given rights or serious moral consideration. It may also suggest that when lines are drawn there is always an arbitrary aspect, so that pointing to problematic cases, such as humans who lack the characteristics special to the species, ought not to count against the general validity of a proposed criterion. Both of these points deserve further consideration.

First let us consider the question of the moral significance of species boundaries in relation to the larger question of what makes boundaries either morally significant or morally arbitrary. Why assume that we must explain the relevance of species boundaries in terms of intellectual or other differences? Why not take species boundaries as morally significant in themselves?

In general, when we draw a conventional boundary we do not assume that it alone can legitimately be used to motivate differences in treatment. For example, if a city is divided up into neighborhoods, the neighborhoods create artificial boundaries. If we find that a bank is granting loans in one neighborhood but not another, we consider this "redlining" to be illegitimate. On the other hand, if a tornado hits one neighborhood but not another, then the disaster creates a morally relevant difference between the neighborhoods, rendering government assistance for one neighborhood but not the other fair. But suppose that a very rich person lives in the neighborhood and his house has not been damaged. If this person argued that he should also receive assistance, since he lived in the neighborhood, we would not consider this a strong moral argument (though for the sake of *legal* convenience, he might in fact be treated in the same way). Thus, conventional boundaries cannot legitimize differences in treatment, except where they are correlated with genuinely morally relevant differences.

The notion of species is especially troublesome because, from a scientific point of view, it has changed from a real to a relatively arbitrary boundary. In the old Aristotelian notion, species differences were essential and immutable differences. When Descartes claimed that animals lacked minds and feelings, he was referring to a fundamental metaphysical gulf. Within his framework, no "marginal cases" were possible; animals all lacked minds and humans all possessed them, and the lack of a mind left animals without any moral claim whatever. But the Darwinian conception offers a quite different perspective. As Rachels (1989) argues, modern evolutionary theory does not see species classifications as corresponding to any immutable or essential differences. Rather, according to Darwin and modern evolutionary theory, there are no such fixed essences, but a multitude of organisms that resemble one another in some ways but differ in others. The only reality is the individual. How those individuals are grouped is more or less arbitrary. As Darwin put it:

> I look at the term species, as one arbitrarily given for the sake of convenience to a set of individuals closely resembling each other, and that it does not essentially differ from the term variety, which is given to less distinct and more fluctuating forms. The term variety, again, in comparison with mere individual differences, is also applied arbitrarily, and for mere convenience sake. (Darwin 1859: 52)

Rachels has argued that this recognition of the arbitrary and changeable nature of species boundaries has implications for how we think about moral issues. This is already evident from the fact that it is only within this modern perspective that "marginal case" objections can arise. These objections arise with the recognition that species boundaries do not mark important metaphysical boundaries, and in fact are conventional. Thus, if they are to be used as a basis for differences in value and treatment, they must be correlated with morally significant differences. But if we need to check whether the species boundary correlates appropriately with the morally significant boundary, we might as well just dispense with the species boundary and use the real, morally significant boundary in the first place. This idea accords well with the point made above about the arbitrariness of neighborhood boundaries. I argued in that case that a rich person whose house was not damaged was not morally entitled to disaster relief simply on the basis of living in the same neighborhood with those whose houses were destroyed. Where boundaries are arbitrary, then, we ought to look at the characteristics of individuals themselves to decide how to treat them. This is precisely what Rachels suggests with respect to species boundaries. Rachels argues for this approach with a different sort of "marginal case." Suppose that a chimpanzee (or perhaps one of the 1/4 or 1/2 chimpanzee/human hybrids discussed above) were to learn how to read and speak English. Rachels describes such a possibility:

> And suppose he eventually was able to converse about science, literature, and morals. Finally he wants to attend university classes. Now there might be various arguments about whether to permit this, but suppose someone argued as follows: "Only humans should be allowed to attend these classes. Humans can read, talk, and understand science. Chimps cannot." But this chimp can do those things. "Yes, but normal chimps

> cannot, and that is what matters." Is this a good argument? Regardless of what other arguments might be persuasive, this one is weak. It assumes that we should determine how an individual is to be treated, not on the basis of its qualities, but on the basis of other individuals' qualities. This chimp is not permitted to do something that requires reading, despite the fact that he can read, because other chimps cannot. That seems not only unfair, but irrational. (Rachels 1989: 100)

Thus, the important question in whether to allow a chimp or a human to go to college is simply whether that individual is prepared and capable. Rachels proposes that in general, individuals ought to be treated according to their individual qualities, rather than on the basis of characteristics of groups they may belong to, and he refers to this approach as "Moral Individualism," in order to contrast it with varieties of speciesism.

Whether or not Moral Individualism can itself be defended, Rachel's discussion shows clearly the arbitrariness of species boundaries. This poses a serious challenge to moderate approaches, since the assumption that the species boundary marks an essential and qualitative difference which can motivate profound differences in value and treatment is far from a moderate assumption. This assumption is more consistent with a pre-Darwinian Cartesian view of animals which accords them no moral standing whatsoever. And yet this immoderate assumption is clearly necessary to the two so-called moderate views that I have discussed.

TOWARD A GENUINE MODERATION

It may reasonably be objected at this point that even if species boundaries turn out to be arbitrary, we have not yet arrived at a means of determining which boundaries are genuinely morally significant boundaries. Thus, we still have not answered the question, "where do you draw the line?" Nor does the account offered so far explain how we are to treat individuals. Indeed, one might suppose just on the basis of the argument so far that Moral Individualism could justify invasive research on mentally enfeebled human beings, given that they are similar intellectually to individuals of other species who are routinely used in this way. But Rachels introduces a second principle which shows how an individual's characteristics are to be used to determine what treatment is appropriate. As Rachels argues, relevant differences vary with the different kinds of treatment we have in mind. A difference between individuals that justifies one sort of difference in treatment might be completely irrelevant to justifying another difference in treatment.

> Suppose, for example, the admissions committee of a law school accepts one applicant but rejects another. Asked to justify this, they might explain that the first applicant had a miserable record. Or, to take a different sort of example, suppose a doctor treats two patients differently: he gives one a shot of penicillin, and puts the other's arm in a plaster cast. Again, this can be justified by pointing to a relevant difference between them: the first patient had an infection while the second had a broken arm.
>
> Now, suppose we switch things around. Suppose the law school admissions committee is asked to justify admitting A while rejecting B, and replies that A had an infection but B had a broken arm. Or suppose the doctor is asked to justify giving A a shot of penicillin, while putting B's arm in a cast, and replies that A had better college grades and test scores. Both replies are, of course, silly, for it is clear that what is relevant in the one context is irrelevant in the other. (Rachels 1989: 98)

Rachels expresses this point in a general principle:

> Whether a difference between individuals justifies a difference in treatment depends on the kind of treatment that is in question. A difference that justifies one kind of difference in treatment need not justify another. (Rachels 1989: 99)

This principle is straightforward and rational, and accords well with our intuitions about how we ought to treat each other. Whether someone is intelligent, can do mathematics or speak English is highly relevant if we are considering whether to admit her to an American college. But it is not relevant when we are considering whether to inject her with a debilitating disease. Thus, we have no qualms in not admitting profoundly retarded persons to college, but we have great qualms about using them in medical experiments. To take another example, whether someone is a rational autonomous agent is highly relevant if we are considering some paternalistic action on her behalf. It is less relevant if we are contemplating dripping oven cleaner in her eyes. Since animals differ from humans in myriad ways, this principle licenses many differences in treatment depending on the characteristics of the individual and the type of treatment we are considering. What it does not license, however, is using a single difference between individuals, such as intelligence, to justify wholesale differences in treatment which have nothing to do with the difference in question.

Thus the answer to the question, "where do you draw the line?" is that there is no *one* line which can consistently be drawn, but many, and these lines do not correspond to species boundaries, but vary with the characteristics of individuals and the types of treatments we are contemplating. This approach deserves serious examination, since it offers a well-motivated and clear explanation of the idea of a morally relevant difference, which accounts for the ways in which we treat each other and shows how to extend this analysis to members of other species. Using this approach it is clear how increasing knowledge about characteristics of other species, such as their social and psychological needs, can be integrated into our ethical system.

Clearly, Moral Individualism, in rejecting speciesistic assumptions, is quite different from the versions of moderation examined in this paper. But it does offer another notion of moderation, in which the risks and benefits of research could be shared among individuals in a more equitable way. The consequences of adopting this approach would not be the abolition of all research on animals, but instead the abolition of research which ignores relevant features of research subjects in deciding how to treat them. This would entail radical changes, to be sure. But while these changes would seem radical, relative to the *status quo*, it is worth noting that the *status quo* is not itself moderate, in a very important sense. The assumption that other species are available for our consumption in myriad ways, that they can be destroyed to satisfy our taste for their flesh and our desires for fashion and entertainment, are immoderate assumptions which lead to immoderate consequences in many areas. The destruction of whole species, the annihilation of the rainforests in order to graze cattle for meat consumption, the pollution of water from feedlot runnoff are only a few of the countless consequences of this immoderate assumption. How does one become truly moderate, in such a world of immoderation?

ACKNOWLEDGEMENTS

I wish to thank the editors of this volume, Dale Jamieson and Marc Bekoff, whose criticisms of an earlier draft of this paper were exceptionally helpful. I would also like to thank Lawrence Finsen, as well as an anonymous reviewer, for useful criticisms.

LITERATURE CITED

Adams, Perrie M. 1982. Investigator responsibilities in animal experimentation. In: *Scientific Perspectives on Animal Welfare* (ed. by J. W. Dodds & B. Orlans), pp. 39-43. New York: Academic Press.

Animal Welfare Act: 1982 & 1985 Supplement, United States Code Title 7.

Bekoff, M. 1976. The ethics of experimentation with non-human subjects: Should man judge by vision alone? *The Biologist* 58, 30-31.

Bekoff, M., & Jamieson, D. 1990. Reflective ethology, applied philosophy, and the moral status of animals. *Perspectives in Ethology* 9.

Blumenstyk, G. 1989. With state legislatures as the battleground, scientists and college officials fight Animal-welfare groups. *The Chronicle of Higher Education* 35 No. 30, A1 & A26.

Darwin, C. 1859. *The Origin of Species by Means of Natural Selection.* London: John Murray.

Davis, H. 1989. Theoretical note on the moral development of rats. *Journal of Comparative Psychology* 103, 88-90.

Dresser, R. 1988. Standards for animal research: Looking at the middle. *The Journal of Medicine and Philosophy* 13, 123-143.

Driscoll, J.W. & Bateson, P. 1988. Animals in behavioural research. *Animal Behaviour* 36, 1569- 1574.

Fox, M.A. 1986. *The Case for Animal Experimentation: An Evolutionary and Ethical Perspective.* Berkeley, California: University of California Press.

Frey, R.G. & Patton, W. 1989. Vivisection, morals and medicine: An exchange. In: *Animal Rights and Human Obligations* 2nd Edition (ed. by T. Regan & P. Singer), pp. 223-236. Englewood Cliffs, New Jersey: Prentice-Hall.

Gallistel, C.R. 1981. The case for unrestricted research using animals. *American Psychologist* 36, 4, 357-362.

Goy, R.W. 1982. Policy statement on principles for the ethical uses of animals at the Wisconsin Regional Primate Center. *American Journal of Primatology* 3, 345-347.

Griffin, D.R. 1984. *Animal Thinking.* Cambridge: Harvard, Massachusetts: University Press.

Hollands, C. in press. Trivial and questionable research on animals. In: *Animal Experimentation: The Consensus Changes.* Philadelphia, Pennsylvania: Temple University Press.

King, F.A., Yarbrough, C.J., Anderson, D.C., Gordon, T.P. & Gould, K.G. 1988. Primates. *Science* 240, 1475-1482.

Linden, J.A.M. & Barnes, J. 1988. Animal experimentation in Canada: Legal provisions and policy alternatives. Paper presented at the joint Medical Research Council/National Institutes of Health Workshop on Animals in Research, April 26, 1988, Ottawa, Ontario.

Miller, N.E. 1985. The value of behavioral research on animals. *American Psychologist* 40, 423-440.

National Institutes of Health. 1985. *Guide for the Care and Use of Laboratory Animals.* Washington, D.C.: U. S. Department of Health and Human Services, Public Health Service.

Rachels, J. 1989. Darwin, species and morality. In: *Animal Rights and Human Obligations,* 2nd Edition (ed. by T. Regan & P. Singer), pp. 95-103. Englewood Cliffs, New Jersey: Prentice Hall.

Regan, T. 1983. *The Case for Animal Rights.* Berkeley, California: University of California Press.

______. 1986. (ed.) *Animal Sacrifices.* Philadelphia, Pennsylvania: Temple University Press.

Rollin, B. 1981. *Animal Rights and Human Morality.* Buffalo, New York: Prometheus Books.

Russell, W.M.S. & Burch, R.L. 1959. *The Principles of Humane Experimental Technique.* Springfield, Illinois: Charles Thomas.

Sapontzis, S.F. 1987. *Morals, Reason and Animals.* Philadelphia, Pennsylvania: Temple University Press.

Schweitzer, A. 1929/1989. The ethic of reverence for life. In: *Animal Rights and Human Obligations,* 2nd Edition, (ed. by T. Regan & P. Singer), pp. 32-37. Englewood Cliffs, New Jersey: Prentice Hall.

Singer, P. 1990. *Animal Liberation: A New Ethics for Our Treatment of Animals,* 2nd Edition. New York: New York Review of Books.

Thomas, J.A., Hamm, T.E., Perkins, P.L. & Raffin. 1988. Animal research at Stanford University: Principles, policies and practices. *New England Journal of Medicine* June, 1630-1632.

16. Sympathy, Empathy, and Understanding Animal Feelings—and Feelings for Animals

Michael W. Fox

Should affectional states be considered and included in studies of animal behavior? By affectional states, I mean animal feelings or emotions. A major experimental variable in ethological studies which will be considered is not only the presence of a human observer but also the affectional and perceptual state of the investigator. This is as relevant to this question as the use of "unscientific" subjective terminology. Descriptions and explanations of animal behavior that include affective, emotional and motivational elements could be put to good use by students, farmers, veterinarians, animal caretakers and others. This means that ethological studies need to be framed in a broader dimension that encompasses the feeling-states of animals in identifiable human terms of emotional experience and reaction. In a narrower framework that ignores affectional states we have what amounts to a Cartesian or "mechanomorphic" picture of the animal, which is as limited as one that is overly anthropomorphic and imbued with the anthropocentric projections and expectations of the observer. In this essay I will endeavor to find the middle ground between these two extremes of perceiving, describing and even treating animals either as mechanomorphs or anthropomorphs, recognizing that my "reasonable" position may seem radical to both sides, a reconciliation of which, however, is both reasonable and necessary.

THE OBJECTIFICATION OF ANIMALS

There is an ancient Japanese saying that if you observe an animal very closely, its spirit will enter you. In Zen philosophy, this is analogous to the nondualistic state of consciousness where the observer and the observed become one. From a Western perspective this sense of oneness is often called love. Shorn of sentimentality and possessiveness, this perception of another being - be it human or nonhuman, sentient or nonsentient - is what philosopher Martin Buber (1970) termed "I-Thou." This perceptual and relational state is contrasted with the more common "I-It" state where the observed is objectified, as something other, if not alien, and different, if not inferior. These two states are not mutually exclusive, but when the latter is not embraced by the former, we have a condition which is potentially pathological. As the poet Meridel Le Seur once said, "You can only destroy that which you objectify."

The "scientific method" is based upon a philosophy that is purportedly amoral, value free (which is a value in itself) and objective (which is almost inevitably a subjective consensus). When this "scientific method" is applied to the study of sentient beings we have the possibility of a pathological condition; the perceptual and relational state or attitude created by the aforementioned philosophy of science, in objectifying animals, denies them subjectivity. This makes any study of their subjective/affective states at best limited and at worst can lead to cruel and inhumane treatment.

But we should not place the entire blame upon the scientific method for the avoidance of attributing emotionality or "feelings" to our animal kin. As a culture, we tend to value rationality and intellect over "irrational" emotion; emotionality is distrusted, even feared, repressed and limited. A lack of appreciation and awareness of one's own subjective states can lead to the devaluing and ultimate denial of subjective states in animals. (When I asked a professor of animal science during a public debate before an audience of pig farmers if he felt that pigs had feelings he replied, "We need more research to be certain." A year later in testimony before Congress in opposition of a bill that would allow formula-fed veal calves sufficient room in their narrow crates to enable them to turn around and lie down comfortably, this same person stated, "There is no scientific evidence that the behavioral needs of veal calves are not satisfied" (under existing husbandry conditions).

The denial of emotionality in animals is as much a reflection of our own emotional condition as it is a product of many cultural influences that will be shortly explored. The scientific method is not the only reason why nonhuman animals are too often objectified and "mechanomorphized" by scientists and others. There is also the enculturated taboo of anthropomorphizing animals which has its roots, in part, in the religious and philosophical traditions of Aristotle, St. Augustine and St. Thomas

Aquinas, among others, who set up the body/soul dualism and contended that the animal soul is inferior to the immortal human soul (Thomas 1983). French philosopher René Descartes took this one step farther by denying animals their subjectivity. Devoid of feelings, the cries of animals being vivisected were dismissed by Descartes as being simply the sounds of their body-machines breaking down.

Another historical reason for the objectification of animals and the denial of soul, spirit and a subjective emotional and cognitive world of their own may be linked with the early Christian opposition to "pagan" animism and pantheism (Thomas 1983). The ensoulment and deification of animals was seen as pagan idolatry. Little wonder then that the teachings of Christian leaders like St. Francis of Assisi, who saw animals as brothers and sisters and whose sacramental and fundamentally pantheistic attitude toward the Creation never became part of the mainstream of Judeo-Christian tradition (Fox 1989).

Language is another subtle, and sometimes not so subtle, factor involved in the objectification of animals and in the denial of their individuality and emotionality. Statements such as "Animals *that* or *which...*" (rather than *who*) and "The animal and its world" (rather than his/her or simply "Animals and *their* world") are commonplace. The word animal is used pejoratively in describing human behavior: "The rapist behaved like an animal." (Cognitively and linguistically we also have set up a false duality between "humans and animals." Since humans are animals, it is more correct, for example, to refer to "human and nonhuman animals" and "human and nonhuman primates.")

The objectification of animals is seen in the use of language to "sanitize" subjective terminology and give studies of animal behavior and psychology a more objective scientific aura. There is a plethora of papers that put animals' emotional states such as fear and rage in quotations as though they are not real or cannot be proven. And there is the new-speak double-think rhetoric and psychobabble of terminology; aversive stimulation instead of painful electrocution; avoidance behavior instead of fear and flight, etc.

Language can be used to create distance and in the process limit understanding: where there is an emotional distance, understanding of others' behavior, which involves empathetic attunement to others' affective states and emotional reactions, is shallow and limited. Distance-creating language can also be used as a defense against emotional closeness, effectively cutting off sympathy in order to avoid others' suffering and the associated empathic burden of concern and responsibility.

While a behavioral scientist may be comfortable with using objective, distancing terminology because he/she is only interested in one shallow or limited aspect of the animal's behavioral repertoire, such reductionism is no guarantee of good science, and it is often as much an obstacle to creativity as it is to insuring the humane treatment of research animals.

Linguistic, cognitive and perceptual sets (which arise from a variety of cultural sources as well as from student indoctrination and from the traditions of various scientific disciplines), like habit-fixations and obsessive compulsions, are difficult to break when the individual is set in an objectifying mode and cannot see the world from a different perspective. Ethical blindness is one consequence, as when there is no empathy toward another fellow sentient being and no ability to intuitively and imaginatively see the world through the animal's eyes. The only constraints then against unethical conduct and cruel treatment are legal ones, which are both cumbersome and not well regulated in relation to the protection of animals in society today. But in the absence of sympathy, empathy and compassion as a boundless ethic, legal constraints become increasingly necessary to fill the moral vacuum of human insensitivity and indifference toward other sentient beings.

The origin of ethical sensibility is in empathetic sensitivity and in the absence of the latter, legal and moral codes are needed as a substitute for ethical sensibility. When animals are objectified and mechanomorphized, even though there is ample neurochemical and behavioral evidence of their sentience, we are dealing with a pathological condition that entails more than simply the denial of such evidence. It entails a denial of empathy or fellow-feeling.

This denial does not mean that those who deny animals their subjectivity and emotionality are necessarily cruel, unfeeling and uncaring people. Denial is one way to create distance so that others' distress and the emotional burden associated with the killing and suffering of animals under our dominion are avoided.

SYMPATHY, EMPATHY AND UNDERSTANDING

Barbara McClintock, awarded the Nobel prize for her work on corn plant genetics, attributes her scientific approach, if not her success, to more than observational and analytical skills. A certain intuitive sensitivity or inward openness to living things is also involved as part of her "scientific method." In good biological research she contends that one must have the patience "to hear what the material has to say to you" and the openness to "let it come to you." And it is paramount that one has a "feeling for the organism." (Keller 1983)

Jay McDaniel (1986: 34) summarizes this "feeling for the organism" as follows:

> What is McClintock's 'feeling for the organism?' Judging from her comments and Keller's commentary, it is an appreciative and intuitive apprehension of an organism in three of its aspects. In the first place, it is a feeling for the organism as a *unique individual.* 'No two plants are alike,' McClintock tells Keller, 'they're all different, and as a consequence, you have to know that difference.' McClintock continues: 'I start with the seedling, and I don't want to leave it. I don't feel I really know the story if I don't watch the plant all the way along. So I know every plant in the field, I know them intimately, and I find it a great pleasure to know them.'
>
> In the second place, it is a feeling for the organism as a *mysterious other.* Keller quotes Einstein as saying that science often proceeds from 'a deep longing to understand even a faint reflexion of the reason revealed in the world,' but then she tells us that on this point McClintock may differ. 'McClintock's feeling for the organism is not simply a longing to behold the 'reason revealed in this world.' Rather it is 'a longing to embrace the world in its very being, through reason and beyond.' And how does organism supercede reason? 'For by McClintock, reason - at least in the conventional sense of the word - is not by itself adequate to describe the vast complexity - even mystery - of living forms. Organisms have a life and order of their own that scientists can only partially fathom.' This life of its own is the organism's otherness, and the intuitable and yet ungraspable elusiveness of this life is the organism's mystery.
>
> In the third place, McClintock's feeling for the organism is a sensitivity to the creature as a *fellow*

> *subject.* Keller writes that 'over the years a special kind of sympathetic understanding grew in McClintock, heightening her powers of discernment, until finally, the objects of her study became subjects in their own right.' These objects 'claim from her a special kind of attention that most of us experience only in relation to other persons.' In fact, as Keller explains, the natural objects are not simply objects for McClintock; they are organisms.
>
> Organism is for her a code word - not simply a plant or animal ('Every component of the organism is as much of an organism as every other part') - but the name of a living form, of object-as-subject. With an uncharacteristic lapse into hyperbole, McClintock says: 'Every time I walk on grass I feel sorry because I know the grass is screaming at me.'

To respect the life of an animal and to acknowledge its subjectivity and emotionality comes closest, for the atheist and agnostic, to Albert Schweitzer's (1965) philosophy of reverence for life. The denial of empathy or fellow-feeling is also a denial of Self: the Self that is the co-inherent monad of Buber's "I-Thou" relationship; the Universal Self of Buddhism, Hinduism and the divinity within of Christian sacramentalism and Pantheism (Fox 1989). A deep feeling of kinship with all life and respect for other sentient beings cannot therefore arise from a set of moral and legal codes but from a feeling *for* the animal itself as a fellow sentient being. And it is this way of feeling that determines primarily, I believe, how we perceive and treat animals. (It should also be an integral part of the scientific method according to both McClintock and Lorenz; see below.) To be emotionally, empathetically attuned to an animal is facilitated by early childhood contact with animal companions (pejoratively called "pets") and appropriate parental instruction and example. Such attunement (and at-onement) can help override those cultural and other influences that create distance between human and nonhuman animals later in life. And it its certainly a prerequisite to responsible "pet ownership" as it is to the proper care and handling of farm animals and those kept in zoos and laboratories (Fox 1984, 1986).

Without empathy, animal husbandry becomes animal production and management, and veterinary medicine becomes more of a scientific business serving the animal industries than a healing art helping to repair the clearly broken Covenant of humane and responsible stewardship of the animal kingdom and Earth's creation.

For the ethologist and animal psychologist, the words of Konrad Lorenz are relevant "Before you can study an animal, you must first love it." In response to my foreword in Men and Wolf (Fox 1987), he sent to me the following personal comments:

> I could not disagree less to what you say about our emotional response to animals. Finding an analogy between animal and human behaviour is necessarily the cause of recognizing the subjective experience of animals. It cannot fail to create sympathy and love which is all to the good if one avoids the arrogant belief that knowing that an animal possesses subjective experience like our own allows us insight into the question what an animal feels. But a man who has no respect (and respect is exactly the right word to use) for the animal as an experiencing sentient being, can never gain any understanding of it even from the most objective research...
>
> I think that empathy is just the right word to describe the state of mind while being in contact with an animal; a relationship which implies a profound respect but no pre-conceived expectation.

Lorenz, in his wisdom, sees the objectivity of the scientific method as being without any pre-conceived expectation, as distinct from the more usual interpretation of objectivity as objectifying emotional detachment. In other words, the scientific method entails having an open mind (no pre-conceived expectations) and also an open heart. But the heart is too often closed through rationalization, denial and those cultural and other influences alluded to earlier.

The depth of feeling and empathetic awareness for other living things that Naess (1973) sees us acquiring "with maturity" is expressed in these words of Australian aborigine Bill Neidjie (1985: 51, 60):

> Feeling all these trees, all this country. When this wind blow you can feel it. Same for country...you feel it. You can look, but feeling... that make you.
>
> If you feel sore...headache, sore body, that mean somebody killing tree or grass. You feel because your body [is] in that tree or earth. Nobody can tell you, you got to feel it yourself.

In an essay on this subject of pan-empathy (emotional attunement with the environment and all sentient beings) I made the following distinctions between empathy and sympathy (Fox 1984: 61):

> Empathy is defined variously as: the intellectual identification with or vicarious experiencing of the feelings, thoughts, or attitudes of another (*Random House Dictionary*); the power of projecting one's personality into and so fully understanding the object of contemplation (*Oxford Dictionary*); and the imaginative projection of one's own consciousness into another being (*Webster's Dictionary*).

Sympathy and empathy are distinctly different phenomena. Sympathy is the sharing of another's emotions, especially grief and anguish, involving pity and compassion. Empathy (from the Greek term meaning affection, and a more recent German term *einfuhlung*, which means "a feeling in"), entails the power of *understanding* and imaginatively entering into another's feelings. While the two are not mutually exclusive, empathy implies some level of objective knowledge and therefore a greater accuracy of perception and affect than are seen in sympathy, which, because it is more subjective, may be a less accurate and more intuitive way of perceiving and responding to another's emotions. In our relations with animals (as with each other), sympathetic concern may or may not be misplaced, while empathetic concern, since it includes both objective understanding (of both the animal's nature and our ethical responsibilities) and emotional involvement, is likely to be more accurate and, therefore, less often confounded by anthropomorphic projections.

ANTHROPOMORPHIZING

Because certain biological phenomena are quantifiable and can be objectively identified and measured, the scientific study of animal physiology, pathology, nutrition, ecology, and ethology is possible. There are other natural, biological phenomena, however, which cannot be objectively determined or "tested" empirically and, as such, represent an almost insurmountable barrier to the scientific method *per se.* For example, pain sensitivity, awareness and the entire spectrum of human emotions may be subjective, experiential phenomena which defy objective assessment and analysis. Although neuroanatomical pathways, and physiological, neuroendocrine, and behavioral mechanisms, systems, and processes have been identified in the psychophysiology of sensation, perception, cognition, and emotion, and even though we may know how and why an animal responds in certain ways (to stress, deprivation, disease, and so on), there is no way of scientifically determining what the animal is actually feeling or experiencing subjectively. This "credibility gap" can be closed by arguing that if the psychophysiological and behavioral responses of the animal are analogous (if not homologous) to those associated with various emotional responses and affective states in humans, then it is more likely that the animal is feeling as we might than that it is experiencing some completely different emotion. There may or may not be a close correlation with context and evoking stimuli (see Walker 1983, for detailed evidence of similarities of mental states in humans and animals). Hurnik & Lehman (1982: 135-136) argue that:

> The evidence that we have that an animal is afraid or in pain does not consist in dubious analogies to human behavior. For example, what grounds are available to support the contention that a sheep which sees or smells a wolf feels afraid? We do not say that we know that the sheep is afraid because when human beings are in contact with wolves they feel afraid. Such reasoning would be fallacious and might lead to absurd conclusions. Rather the evidence that the sheep feels fear in the vicinity of the wolf includes observations of physiological and behavioral factors as well as the consideration that fear appears to make a significant contribution to the animal's chance of survival. While it might be suggested that we don't need the hypothesis

that the animal feels in order to explain the animal's behavior in the presence of the wolf - such an explanation can be given without reference to the animal's mental state, we believe that this suggestion is superficial. We believe that reference to the animal's fear is warranted because the best available descriptions and explanations of the sheep's observable behavior make reference to its fear. Reasoning in this way is in accord with sound canons of scientific method; it is not anthropomorphic.

Griffin (1981: 170) writes that:

> The possibility that animals have mental experiences is often dismissed as anthropomorphic because it is held to imply that other species have the same mental experiences as man might have under comparable circumstances. But this widespread view itself contains the questionable assumption that human mental experiences are the only kind that conceivably exist. This belief that mental experiences are a unique attribute of a single species is not only unparsimonious; it is conceited. It seems more likely than not that mental experiences, like many other characters, are widespread, at least among multicellular animals, but differ greatly in nature and complexity.

Without such correlated evidence, we run the risk of projecting our own feelings onto the animal, which may or may not be accurate. Yet there are those who would believe that there are no grounds for such anthropomorphism because they adhere to the Cartesian belief that animals, although they may experience pain, are not aware, in and of themselves as we are, of emotional states. In the light of clear scientific evidence, not of emotional awareness in animals but of analogous neuroanatomical systems and psychophysiological mechanisms and responses to stress in both vertebrate animals and human beings, such Cartesian mechanomorphizing of animals in untenable. There is sufficient documented evidence from stress research, animal psychology, and neurophysiology to support the probability that the subjective emotional world of animals is more similar to the various subjective states of human consciousness than it is different (Fox 1974; Griffin 1982; Walker 1983).

Objective proof of what an animal or human being is subjectively experiencing is impossible, however, and this fact alone demonstrates that "reality," "truth," and "knowledge" are not the exclusive domain of science. It is clearly unwise therefore to rely exclusively upon the scientific method to answer questions of animal consciousness and welfare. The closest approximation is through anthropomorphic correlation of empirically derived observations, for example, of analogous psychophysiological responses to stress in animals that are seen in humans in association with subjective emotional states such as fear and anxiety. Such anthropomorphic correlation between animals and humans is the basis of comparative medicine, physiology, pathology, animal "models" of human disease processes, and so forth. The stressors and other environmental factors triggering such psychophysiological responses may or may not be similar, significant differences being attributable to species, strain, individual, and age-related variables.

The "credibility gap" thus widens when there is a lack of correlation between how the animal responds psychophysiologically to stress, pain, or other environmental stimuli and how a human might under similar circumstances because there would then be a greater element of doubt as to what the animal might be actually feeling. Since animals are likely to suffer in a number of different circumstances in which a human being would not suffer, the anthropomorphic identification/correlation with the animal should not be limited to investigating analogous contexts, stressors, or other emotion-evoking stimuli. Similarly, looking only at analogous behavioral reactions can be no less misleading, since the ethological repertoire of animals is different in many respects from the emotion-correlated behavioral reactions of human beings. Whereas cold may cause a calf or human infant to suffer, a hamster will simply hibernate. Clearly, therefore, a detailed knowledge of species characteristics and environmental requirements is essential, otherwise anthropomorphic concern and care would be inaccurate and could jeopardize the animal's welfare. However, when there are significant species differences in behavior and physiology compared to humans, we are even less able to imagine how they might be feeling by "putting ourselves in their place." A lack of correlation between physiological and behavioral changes in the animal and in human beings adds further to the problem of welfare assessment if no such correlations can be made.

ETHOLOGY, EMPATHY AND PERCEPTION

Empathy is a perceptual and cognitive phenomenon, not simply an anthropomorphic "humanizing" projection: it is analogous to what phenomenologist Merleau-Ponty terms "lateral coexistential knowledge," as distinct from objective, "vertical" (i.e. Cartesian mind over body) knowing and perceiving. Dallery (1978: 74, 76) illustrates this mode of perception as follows:

> This is not the place to summarize Merleau-Ponty's magisterial work, The *Phenomenology of Perception* (1946). For our purposes, it is important to note that perception is described as the complex, always open, temporal 'access' between world and perceiver. It is neither a causal process nor a process distinct from social relations, speech, or understanding (as it would be if perception were a 'thought of seeing'). So in perceiving a snake, for example, I do not simply receive an impression of a sinuous form having a certain mottled pattern; I do not see a cold, indifferent fact, or have a bunch of impressions to which I might or might not endow some value depending on my feelings; I see the *snake*, which is to say that I see its behavior in an environment proper to it and that I 'appropriate' the snake's way of being, the snake's perception of certain things around it. But I am free to regard the snake as an object and admire its beauty, or to loathe its slithering.

There is knowledge and feeling inherent in such empathetic perception. Dallery continues:

> To see the animal moving in its environment is already to 'care' about the animal, since in a way I put myself in its place. I say it is foraging, or mating, or fleeing; I know what it is doing because these are analogues of my behavior...But if beasts have no interior being and are automata, as Descartes held, I cannot 'think in their place.' In fact, I cannot really perceive them. They become real to me only as I add to certain sensations

> meanings that come from my sentiment of intellect. In outline, this is the tendency of modern thought. Perception is relegated either to blind mechanisms (as in skeptical empiricism and objective psychology) or to operations of the mind (as in Cartesianism and Kantianism). For Husserl and Merleau-Ponty, this amounts to canceling out perception and losing the world (at least losing it in and by means of philosophy). Merleau-Ponty then is not speaking metaphorically when he charges both camps in the modern tradition with blindness; he does not mean blindness to things in the environment (loss of the ability to see) but blindness to the world as lived, the world as open to environments of other beasts, as providing the ground of our coexistence of being together.

This I call simply a lack of empathy, which makes us dehumanize ourselves by objectifying the world, the causes of which need careful study.

From the existential phenomenologist's perspective, the difference between detached objectivity and rational empathy can be viewed as follows. Dallery (1978) equates the former with "vertical" Cartesian, hierarchical, instrumental, perceptional knowledge and the latter with "lateral" coexistential knowledge and perception. So where does sympathy fit into this paradigm? Dallery does not answer this question. It lies, I believe, in the "lateral" or coexistential knowledge. And it is easily inhibited by the "vertical" dimension of Cartesian thought and perception. Hence Cartesianism, while not inhibiting rational intellectual development, can impair the expression of sympathy which is a prerequisite for the development of rational empathy and moral maturity.

To conclude: In relation to a person's emotional rapport with an animal, is empathy possible? Sympathetic concern for animals is often judged, sometimes correctly, as being a sentimental, anthropomorphic projection. Sheer subjective sympathy toward an animal, without objective understanding of its behavior and needs, can lead to erroneous assumptions as to its well-being, and to misjudgment of others' treatment of animals as being cruel. Empathy is possible when the "feelings, thought, or attitudes of another" can be vicariously experienced: thus when there is objective knowledge about what an animal's overt behavior signifies, and what emotional states, intentions, and expectations such overt behavior reflects, empathy is possible. Without such objective knowledge, we have sympathy and varying degrees of anthropomorphization. Understanding and sympathy combine to make empathy possible. Ethology - and empathetic ethologists - have a vital role to play, not only in advancing the scientific understanding of animal behavior, but also in improving the care and welfare of animals under humankind's too often ignorant and insensitive dominion. In a spiritual and ecological sense, therefore, I see this as an ethical imperative for ethologists who can make a significant contribution to healing our relationship with the animal kingdom and the rest of Earth's creation.

LITERATURE CITED

Buber, M. 1970. *I and Thou* (Transl. by Walter Kaufmann). New York: Charles Scribner.

Dallery, C. 1978. Thinking and being with beasts. In: *On the Fifth Day: Animal Rights and Human Ethics* (ed. by R.K. Morris & M.W. Fox), pp. 70-92. Washington, D.C.: Aeropolis Press.

Fox, M.W. 1984. *Farm Animals: Husbandry, Behavior and Veterinary Practice.* Baltimore, Maryland: University Park Press.

_____. 1984. Empathy, humaneness and animal welfare. In: *Advances in Animal Welfare Science* (ed. by M.W. Fox & L. Mickley), pp. 61-73. Boston, Massachusetts: Martinus Nijhoff.

_____. 1986. *Laboratory Animal Husbandry. Ethology, Welfare and Experimental Variables* Albany, New York: SUNY Press.

_____. 1987. *Foreword to Man and Wolf* (ed. by H. Frank) Boston, Massachusetts: Dr. W. Junk Publishers.

_____. 1989. *The Life and Teachings of St. Francis of Assisi.* Washington, D.C.: Center for Respect of Life and Environment.

Griffin, D. 1981. *The Question of Animal Awareness.* 2nd Edition. New York: Rockefeller University Press.

Hurnik, F. & Lehman, H. 1982. Unnecessary suffering: Definition and evidence. *International Journal for the Study of Animal Problems* 3, 131-137.

Keller, E. Fox. 1983. *A Feeling for the Organism: The Life and Work of Barbara McClintock.* San Francisco, California: W.H. Freeman & Co.

McDaniel, J. 1986. Christian spirituality as openness to fellow creatures. *Environmental Ethics* 8, 33-46.

Naess, A. 1973. The shallow and the deep, long-range ecology movements: A summary. *Inquiry* 16, 95-100.

Neidjie, B. 1985. *Kakadu Man.* New South Wales, Australia: Mybrood P/L Inc.

Schweitzer, A. 1965. *The Teaching of Reverence for Life.* New York: Holt, Rinehart and Winston.

Thomas, K. 1983. *Man and the Natural World: The History of Modern Sensibility.* New York: Pantheon.

Walker, S. 1983. *Animal Thought* Boston, Massachusetts: Routledge and Kegan Paul.

About the Editors and Authors

Kathleen Akins (Beckman Institute, University of Illinois, Urbana, Illinois 61801) is an Assistant Professor of Philosophy and a faculty member of the Neuroscience Program at the University of Illinois, Champaign-Urbana. She received her Ph.D. in philosophy from the University of Michigan, spending two years in a predoctoral position at the Center for Cognitive Studies at Tufts University. Dr. Akins' main philosophical interest is in the philosophy of mind/brain, focusing upon the ways in which empirical knowledge can solve or change the form of traditional philosophical questions about the mind/brain. Her neuroscience interests have focused upon the principles of sensory processing and neuroethology; at present, she is engaged in neurophysiological research on mammalian visual systems.

Steven N. Austad (Harvard University Biological Laboratories, Harvard University, Cambridge, Massachusetts 02138) received a bachelor's degree in English literature from the University of California, Los Angeles and a Ph.D. in biology from Purdue University. He is currently an Assistant Professor in the Department of Organismic and Evolutionary Biology at Harvard University. During breaks in his academic career, Dr. Austad was a wild animal trainer for the Hollywood film industry, a New York City taxi cab driver, and a pool hustler in small towns around the U.S. His current research concerns the ecology of mammalian aging.

Patrick Bateson (Sub-Department of Animal Behaviour, University of Cambridge, Madingley, Cambridge, England CB3 8AA) is an ethologist, generally interested in the biology of behavior and specifically in behavioral development. He was graduated in zoology from Cambridge in 1960. He received his Ph.D. in animal behavior from Cambridge in 1963. As a Harkness Fellow, he spent two years at the Stanford University Medical Center in California and then returned to Cambridge in 1985. He was Director of the Sub-Department of Animal Behaviour at Cambridge from 1976-1988. Currently he is Professor of

---

For Volumes I and II

Ethology. He has been a Fellow of King's College, Cambridge since 1964 and is now its head. In 1983 he was elected a Fellow of the Royal Society. He co-authored *Measuring Behaviour*, edited *Mate Choice* and coedited *Growing Points in Ethology*, the series *Perspectives in Ethology*, and most recently *The Domestic Cat: The Biology of its Behaviour*.

Marc Bekoff (Department of Environmental, Population, and Organismic Biology, University of Colorado, Boulder, Colorado 80309-0334) is a Professor at the University of Colorado, Boulder. He received his B.A. and Ph.D. degrees at Washington University (St. Louis) and attended the Cornell University Medical College for two years, during which time he developed his long-lasting interests in behavioral and neurobiology. In 1981 he was awarded a John Simon Guggenheim Memorial Fellowship for his work on the social behavior and ecology of coyotes. He has done extensive research on the development of behavior in various canids (wolves, dogs, coyotes) and also has done field work on the behavior of Adélie penguins living at the Cape Crozier Rookery in Antarctica, the social ecology of coyotes in Jackson, Wyoming, and the social behavior and ecology of Evening Grosbeaks in Colorado. Currently he is returning to his earlier interests in the development of social behavior, especially play and communication in canids, and also is studying vigilance (anti-predatory) behavior in grosbeaks. Nonacademic interests include reading spy novels, hiking, and bicycle racing; in 1986 he became the first American to win his age class at the Veteran's (age-graded) Tour de France bicycle race.

Irwin S. Bernstein (Department of Psychology, University of Georgia, Athens, Georgia 30602) is a Research Professor of Psychology at the University of Georgia with an adjunct appointment in Zoology. He is also a Research Professor at the Yerkes Regional Primate Research Center of Emory University. Educated at Cornell University and the University of Chicago, Dr. Bernstein took his first full time professional job at the original Yerkes Laboratories of Primate Biology and has continued his research at Yerkes ever since. His affiliation with the University of Georgia began with a part time appointment in 1968 which became full time in 1971. His research has included laboratory and field studies (Panama, Colombia, Thailand and Malaysia) and has focused on primate social organization. He is probably best known for his studies of primate aggressive behavior.

Andrew Blaustein (Department of Zoology, Oregon State University, Corvalis, Oregon 97331) is Professor of Zoology at Oregon State University. He received his Ph.D. from the University of California at Santa Barbara. He has broad interests in behavior and ecology. His primary research interests are in the development of behavior and the evolution of social systems with an emphasis on amphibians and small mammals.

Karen E. Brakke (Language Research Center, Georgia State University, Atlanta, Georgia 30303) is a graduate student at Georgia State University. She received her BA from Carleton College in Minnesota in 1985 and her M.A. in psychology from Georgia State University in 1989. She has worked on the Language Acquisition Project at the Language Research Center since 1985.

Gordon M. Burghardt (Department of Psychology, University of Tennessee, Knoxville, Tennessee 37996) is a Professor of Psychology at the University of Tennessee at Knoxville with joint appointments in Zoology and Ecology. He currently directs the Ethology (Life Sciences) Graduate Program. He is past President of the Animal Behavior Society; he was also American editor of *Ethology* for six years. He is a Fellow of the American Psychological Association and a U.S. member of the International Ethological Conference Committee. Dr. Burghardt attended the University of Chicago, receiving the A.B. degree in 1963 and the Ph.D. in 1966. Although publishing research articles in several areas, including animal play, behavioral ontogeny, history and theory in ethology, and bear and human behavior, his main interests are in reptile behavior and chemoreception. In addition to laboratory work, Dr. Burghardt has been involved in extensive field studies on green iguana behavior in Panama and Venezuela, and is a Research Associate of the Smithsonian Tropical Research Institute. His books include *The Development of Behavior: Comparative and Evolutionary Aspects* (co-edited with M. Bekoff, Garland STPM, 1978), *Iguanas of the World: Their Behavior, Ecology, and Conservation* (co-edited with A.S. Rand, Noyes, 1982), and *Foundations of Comparative Ethology* (Van Nostrand Reinhold, 1985).

Richard W. Burkhardt, Jr. (Department of History, university of Illinois, Champaign, Illinois 61801) is Professor of History and Departmental Affiliate in Ecology, Ethology, and

Evolution at the University of Illinois at Urbana-Champaign. He is the author of *The Spirit of System: Lamarck and Evolutionary Biology* (Harvard University Press, 1977) and numerous articles on the history of evolutionary theory, hereditary theory, and the study of animal behavior. Currently he is writing a book on the establishment of ethology as a scientific discipline in the twentieth century.

John A. Byers (Department of Biological Sciences, University of Idaho, Moscow, Idaho 83843) is Professor of Biology at the University of Idaho. He received his B.A. at Swarthmore College and his Ph.D. at the University of Colorado, working on the development of behavior in peccaries in Arizona. He has conducted detailed research on play behavior and currently is investigating how natural variation in early experience is associated with subsequent variation in adult social behavior and reproductive success in pronghorn living on the National Bison Range, Montana.

Stephen R.L. Clark (Department of Philosophy, The University of Liverpool, Liverpool, England L69 3BX) is Professor of Philosophy at Liverpool University and author of *Aristotle's Man, The Moral Status of Animals, The Nature of the Beast, From Athens to Jerusalem, The Mysteries of Religion*, and *Civil Peace and Sacred Order*. He is married, with three children.

Robert Coulson (Department of Entomology, Texas A & M University, College Station, Texas 77843) is Professor of Entomology at Texas A&M University. He holds a B.S. degree in Biology from Furman University and M.S. and Ph.D degrees in Entomology from the University of Georgia. His research interests currently center on artificial intelligence applications in natural resource management and the influences of insect populations and communities on forest landscapes. He is co-founder and a principal in the Knowledge Engineering Laboratory (KEL) of the Texas Agricultural Experiment Station.

Roger Crisp (University College, Oxford, England OX1 4BH) is a British Academy Postdoctoral Research Fellow and Honorary Junior Research Fellow at University College, Oxford. He has previously been Lecturer in Philosophy at Hertford, Magdalen and St Anne's Colleges, Oxford. He has published articles in a number of areas of philosophy, edited a book on terrorism, and is currently working on a translation of, and introduction to, Aristotle's *Ethics*.

Lawson Crowe (Department of Philosophy, University of Colorado, Boulder, Colorado 80309) is Professor of Philosophy in the Department of Philosophy and Fellow of the Institute for Behavioral Genetics at the University of Colorado, Boulder. He is former Chancellor of that University, and also served there as Provost, Vice President for Research, and Dean of the Graduate School. His primary field of interest is applied ethics. He has contributed articles to professional journals on ethical issues in medicine, alcohol research, behavior genetics, information technology, and science policy. He teaches undergraduate and graduate courses in ethics, and is a member of the Center for Values and Social Policy of the Department of Philosophy. He is also a member of the Committee for the History and Philosophy of Science.

Thomas J. Daniels (Department of Medical Entomology, new York Medical College, Armonk, New York 10504) is at the New York Medical College where he is studying Lyme's disease. He received his B.A. degree from Seton Hall University, his Master's from Ohio State based on work that he did on the behavior of free-ranging dogs living in Newark, New Jersey, and his Ph.D. from the University of Colorado, for which he studied the social behavior and ecology of feral and free-ranging dogs living in the southwestern United States.

Randolf DiDomenico (Department of Environmental, Population, and Organismic Biology, University of Colorado, Boulder Colorado 80309-0334) is a Postdoctoral Research Associate at the University of Colorado, Boulder. His research interest concerns the philosophical and theoretical assumptions of experimental approaches to understanding the neural basis of behavior. His research has used both lesions and high-speed cinematics to test theoretical predictions on the role of the Mauthner neuron in the escape response of teleost fishes. His background training includes a B.A. in Philosophy and a B.A. in Biology from the University of Colorado. His Ph.D. research was done in Biology at the University of Colorado.

John Dupré (Department of Philosophy, Stanford University, Stanford, California 94305) is an Associate Professor of

Philosophy at Stanford University, where he specializes in the philosophy of biology. He is the editor of *The Latest on the Best: Essays on Evolution and Optimality* (Bradford Books/MIT Press, 1987), and has also written on a variety of topics ranging from metaphysics to feminism. Despairing in the quest for intelligent human life, he has recently been studying attempts to explore the intelligence of other anthropoids. He lives in Palo Alto, California with his wife, the literary critic Regenia Gagnier.

Robert Eaton (Department of Environmental, Population, and Organismic Biology, University of Colorado, Boulder, Colorado 80309-0334) is an Associate Professor of Biology at the University of Colorado, Boulder. His research interest is focused on the development and neuroethology of rapid avoidance responses of teleost fishes for which he employs a technical repertoire of high-speed cinematics, neurophysiology and neuroanatomy. He is editor of the book, *Neural Mechanisms of Startle Behavior*, which reviews a range of organisms from coelenterates to mammals. He received his B.A. and Ph.D. degrees at the University of California, Riverside and his M.A. degree at the University of Oregon. He held a National Research Service Award from the N.I.H. to do postdoctoral work at Scripps Institution of Oceanography, an I.N.S.E.R.M. fellowship at the University of Paris, and was recently a Visiting Scholar at Scripps Institution of Oceanography.

John Fentress (Department of Psychology, Dalhousie University, Halifax, Nova Scotia, Canada B3H 4J1) is Professor of Psychology and Biology, Dalhousie University, Nova Scotia, Canada. As an undergraduate at Amherst College he majored in psychology with a minor in biology, at which time he became attracted to research in ethology. He received his Ph.D. from Cambridge University in 1965 under the guidance of W.H. Thorpe and R.A. Hinde. This was followed by postdoctoral research with R.A. Doty at the Center for Brain Research, University of Rochester. He taught previously at the University of Oregon in Eugene. He has served as President of the Animal Behavior Society, as well as Secretary of the International Society for Neuroethology, and Group Chairman for the Life Sciences, Natural Sciences and Engineering Research Council of Canada. Fentress' research interests focus upon analyses of patterned behavior as adaptive processes. Wolves, laboratory rodents, and most recently children, have contributed to his investigations.

Susan Isen Finsen (Department of Philosophy, California State University, San Bernadino, California 92373) received her B.A. in psychology from Reed College, and her Ph.D. in History and Philosophy of Science from Indiana University. She currently teaches at California State University, San Bernadino. Her work in philosophy of science has focused upon the foundations of psycholinguistics and evolutionary theory. Her recent research focuses on ethics and animals, with particular focus upon the ethics of animal research. She is coauthoring a book with Lawrence Finsen on the animal rights movement.

John Andrew Fisher (Department of Philosophy, University of Colorado, Boulder, Colorado 80309) is Associate Professor of Philosophy and Chair of the Philosophy Department at the University of Colorado, Boulder. He received his Ph.D. in philosophy from the University of Minnesota in 1971. Dr. Fisher has published in the areas of epistemology, aesthetics, and philosophy of language. Concerning animals he has published "Taking Sympathy Seriously: A Defense of Our Moral Psychology Toward Animals" (*Environmental Ethics*, 1987, 9, 197-215). His interest in the explanation of animal behavior stems from two sources. One is an interest in the explanation of human behavior, especially linguistic behavior - his dissertation was on Chomsky, Wittgenstein and the controversial claim that ordinary speakers know and follow rules of grammar. The other source is a fascination with the other-minds problem and its extension to nonhuman animals.

L. Joseph Folse (Wildlife and Fisheries, Texas A & M University, College Station, Texas 77843) is Associate Professor of Wildlife and Fisheries Sciences at Texas A&M University, where he develops and coordinates the use of innovative approaches for modeling ecological systems. His focus is on the interface between individual animals and their environments, using techniques from artificial intelligence and software engineering to solve modeling problems intractable with traditional programming techniques. His current interest is in developing computer programming environments to support simulation of complex systems with event-driven interactions and dynamic structures. He has also done ecological field studies with grassland bird communities in East Africa and Texas. He holds a B.S. in Physics from the University of Texas, and M.S. and Ph.D.

degrees in Wildlife and Fisheries Sciences from Texas A&M university.

Michael W. Fox (Humane Society of the United States, 2100 L Street NW, Washington, DC 20037) joined the Humane Society of the United States (HSUS) in 1976 and has produced numerous publications and developed several technical research programs which applied scientific methods to the investigation of the many uses of animals, notably laboratory, companion and farm animals. He currently serves as Vice President for the HSUS, heading the division of Bioethics and Farm Animals. In addition, he was named Director of the Center for Respect of Life and Environment in 1987, a new division of the HSUS. Dr. Fox has authored over thirty books, is a contributing editor to *McCall's* magazine, and has a nationwide syndicated newspaper column, "Ask Your Animal doctor." He is also a consulting veterinarian, and gives lectures, seminars and presentations both in the U.S. and abroad on a variety of topics related to animal welfare, behavior, and conservation. Dr. Fox has a veterinary degree from London's Royal Veterinary College; also a Ph.D. in medicine, and a D.Sc. in ethology/animal behavior both from London University, England. He is profiled in *Who's Who in America* and *Who's Who in the World.*

Bennett Galef, Jr. (Department of Psychology, McMaster University, Hamilton, Ontario, Canada L8X 4K1) was born and reared in new York City, attended Princeton University (A.B., 1962), and received both his M.A. (1965) and Ph.D. (1968) from the University of Pennsylvania. He has worked throughout his academic career at McMaster University in Hamilton, Ontario where he is now Professor of Psychology and Associate Member of the Department of Biology. Dr. Galef was co-founder and, for six years, co-organizer of the Winter Animal Behavior Conferences. He serves on the editorial boards of several journals, is a fellow of several academic societies, and contributes articles, book chapters, and book reviews to scholarly publications. When not at his desk or in his laboratory, he is frequently to be found on either the ski slopes or trout streams of Rocky Mountain states. This chapter was written while Dr. Galef was on sabbatical leave at the University of Colorado in Boulder trying to decide whether he was vacationing or at work.

John L. Gittleman (Department of Zoology, University of Tennessee, Knoxville, Tennessee 37996) is an Assistant Professor in the Department of Zoology at the University of Tennessee. He received the D. Phil. from the University of Sussex (England) and then completed postdoctoral appointments at the National Zoo (Smithsonian Institution) and the University of Tennessee. Dr. Gittleman has a broad interest in mammalian behavior, ecology, morphology, and evolution. He is best known for his comparative research on the behavioral ecology of carnivores, including their social organization, life history patterns, and allometry. He has also worked on communal care and reproductive energetics of mammals. More recently, his research has focused on methodological aspects of the comparative approach in behavioral and evolutionary studies.

Donald R. Griffin (Concord Field Station, Harvard University, Old Causeway Rd., Bedford Massachusetts 07130) is an Associate of the Museum of Comparative Zoology at Harvard University. He grew up in Westchester County, New York as an enthusiastic naturalist. As an undergraduate and graduate student at Harvard he studied migration and homing in bats and eventually demonstrated that the seven gram *Myotis lucifugus* could survive for more than 20 years, and that females may reproduce when 12 years old. Together with Robert Galambos he also discovered that bats avoid obstacles by hearing echoes of ultrasonic orientation sounds, and later he found echolocation to be a versatile mode of perception that is used to locate and capture flying insects. From 1942 to 1945 Dr. Griffin participated in applied research on communication equipment, cold weather clothing, and infrared viewing devices. He taught comparative physiology from 1946 to 1953 at Cornell, elementary zoology and neuroethology at Harvard from 1953 to 1965, and from 1965 until his retirement in 1986 he was Professor at The Rockefeller University where he established a program of research in animal behavior jointly sponsored by Rockefeller and the New York Zoological Society. From 1987 to 1989 he was a visiting lecturer at Princeton. Long dissatisfied with the prevailing behavioristic approaches to animal behavior, Dr. Griffin has advocated a cognitive ethology in which cognition and mental experiences are recognized as potentially important attributes of animals that are difficult but not impossible to investigate scientifically.

Lori Gruen (Department of Philosophy, University of Colorado, Boulder, Colorado 80309) is currently working on her Ph.D. at the University of Colorado, Boulder, where she is affiliated with the Center for Values and Social Policy. Co-author of *Animal Liberation: A Graphic Guide,* she has also published articles on ethical issues relating to women, animals, and the environment.

Louis M. Herman (Kewalo Basin Marine Mammal Laboratory, University of Hawaii at Manoa, Honolulu, Hawaii 98614) is a Professor of Psychology at the University of Hawaii and Director of the Kewalo Basin Marine Mammal Laboratory. His interests are in laboratory and field studies of communication and cognition in dolphins and whales. Dr. Herman received his Ph.D. from the Pennsylvania State University in Experimental Psychology.

Dale Jamieson (Department of Philosophy, University of Colorado, Boulder, Colorado 80309) is Associate Professor of Philosophy and Director of the Center for Values and Social Policy at the University of Colorado, Boulder. He is also an Adjunct Scientist in the Environmental and Societal Impacts Group at the National Center for Atmospheric Research. His early work was in philosophy of language, and he has written widely on ethical issues concerning our treatment of animals. Recently he has begun to work in philosophy of science, focusing on behavioral biology and atmospheric science.

Walter D. Koenig (Hastings Reservation, Star Route Box 80, Carmel Valley, California 93924) is Associate Research Zoologist at Hastings Reservation, Carmel Valley, California, and Adjunct Associate Professor in the Department of Zoology, University of California, Berkeley. He obtained his Ph.D. from the University of California, Berkeley in 1978 and has been studying the social behavior of acorn woodpeckers since 1974.

Donald E. Kroodsma (Department of Zoology, University of Massachusetts, Amherst, Massachusetts 01003) is currently Professor of Biology at the University of Massachusetts. He received his B.A. from Hope College and his Ph.D. from Oregon State University where he was a Woodrow Wilson Fellow. From 1972-74 he was a postdoctoral fellow at Rockefeller University, where he remained until 1980 as an Assistant Professor and Associate Director of the Field Research Center. He has been at the University of Massachusetts since 1980. His main interests are in animal communication, especially the ecology, evolution, ontogeny and neural control of vocal behavior in birds.

Steven L. Lima (Department of Life Sciences, Indiana State University, Terre Haute, Indiana 47809) received his Ph.D. degree in 1985 at the University of Rochester. He has had postdoctoral experience at Simon Fraser University and the University of Arizona and currently is an Assistant Professor at Indiana State University. Dr. Lima's research interests are in the fields of foraging ecology, predation risk, and vigilance behavior. He is an active contributor to the theoretical and empirical literature on these and other subjects.

Merry E. Makela (Department of Entomology, Texas A & M University, College Station, Texas 77843) is an Assistant Research Scientist in the Department of Entomology at Texas A&M University where she develops behavioral based simulation models of insect population dynamics and genetics. She received her B.A. in 1968 and her Ph.D. in 1975, in Zoology and Genetics respectively, both from the University of Texas at Austin. During a two year postdoctoral stay at the Center for Demographic and Population Genetics at the University of Texas Health Science Center in Houston she studied the causes of high blood pressure among Aymara Indians in Northern Chili. Then, as a Research Assistant Professor at the School of Public Health she studied the epidemiology of central nervous system trauma in the Houston/Galveston metropolitan area and participated in a nation-wide clinical trial to evaluate the efficacy of high dose barbiturates in traumatically comatose patients. In 1984 she took a position with General Electric working on the NASA Spacelab Life Sciences program studying the causes of space motion sickness. She moved to Texas A&M in 1986 to work with Dr. Nicholas Stone on a variety of analytical and simulation studies related to insect pest management. She has since developed methods to simulate the behavior of organisms using an object-oriented programming paradigm and is currently working on a population genetic and dynamic model to predict the spread and impact of Africanized honey bees in Texas.

Gail R. Michener (Department of Biological Sciences, University of Lethbridge, Lethbridge, Alberta, Canada T1K 3M4) is Associate Professor of Biological Sciences at the University of Lethbridge in Alberta, Canada. She was born in Britain and educated in Australia (B.Sc. from the University of Adelaide, 1967) and Canada (Ph.D. from the University of Saskatchewan, 1972). She was a Lecturer in Zoology at the University of Cape Coast, Ghana, West Africa and a Walton Killam Post-doctoral Fellow at the University of Alberta, Canada, before joining the University of Lethbridge as an NSERC University Research Fellow in 1981. She has been studying the behavioral ecology of Richardson's ground squirrels in Canada since 1969. Her other research interests including nesting success of black-billed magpies and natural history of *Dianthidium* bees. She co-edited *The Biology of Ground-Dwelling Squirrels* (1984, University of Nebraska Press) with J. Murie, served as an Associate Editor for the *Journal of Mammalogy* from 1983 to 1987, and has been an Executive Member of the Animal Behavior Society (1985-1988) and the Canadian Council on Animal Care (1987-1990).

Robert W. Mitchell (Department of Psychology, Memphis State University, Memphis, Tennessee 38152) is currently doing post-doctoral research at Memphis State University. He received his Bachelor's degree at Purdue University, his Master's degree at the University of Hawaii, and his Ph.D. at Clark University, where he studied dog-human play. He has long been interested in deception in nonhumans as evidence for mental representation, and has co-edited with Nicholas Thompson a book on this topic. His main current interests are in the areas of development and cognition, and in the conceptual clarification of the application of common sense concepts (such as deception, imitation, play, and mirror-self-recognition) to various organisms.

Sandra D. Mitchell (Department of Philosophy, University of California, La Jolla, California 92093) is an Assistant Professor in the Department of Philosophy and a member of the interdisciplinary Science Studies Program at the University of California, San Diego. Her training is in history and philosophy of science. She received a B.A. in Philosophy from Pitzer College in Claremont, California, an M.Sc. in Philosophy, Logic, and Scientific Method from the London School of Economics and a Ph.D. in History and Philosophy of Science from the University of Pittsburgh. Prior to her appointment at UCSD she taught at the Ohio State University. Her research interests include general models of scientific explanation, the structure of evolutionary theories both in biology and the social sciences, and the role of presuppositions in prescribing and proscribing the questions which scientists ask. Her publications in these areas are: (1987) Competing units of selection?: A Case of symbiosis *Philosophy of Science* 54, 351-67; (1987) Can evolution adapt to cultural selection? *Philosophy of Science Association* 2, 87-96; (1989) The causal background of functional explanation. *International Studies in the Philosophy of Science* 3, 213-229.

Palmer Morrel-Samuels (Kewalo Basin Marine Mammal Laboratory, University of Hawaii at Manoa, Honolulu, Hawaii 98614) is a research associate at the Kewalo Basin Marine Mammal Laboratory. He has special interests in gestural communication. Dr. Morrel-Samuels received his Ph.D. from Columbia University in Experimental Social Psychology.

Ronald L. Mumme (Department of Biology, Memphis State University, Memphis, Tennessee 38152) is Assistant Professor at Memphis State University. He obtained his Ph.D. from the University of California, Berkeley in 1984 and is currently conducting an experimental study of helping behavior in Florida scrub jays at Archbold Biological Station.

Jane M. Packard (Wildlife and Fisheries, Sciences, Texas A&M University, Nagle Hall, College Station, Texas 77843) is Assistant Professor of Wildlife and Fisheries Sciences at Texas A&M University. As a research assistant for Konrad Lorenz at the Max Planck Institüt für Verhaltensphysiologie in 1972-1973, she gained appreciation of cultural differences in styles of thinking used in explanation and prediction of animal behavior. During doctoral work at the University of Minnesota and postdoctoral work at the University of Florida, she recognized the need to incorporate behavioral decision rules in individually-based models of animal movements and population dynamics. She has pursued this goal to address questions of sustainable management of large vertebrate populations, working in collaboration with coauthors who are knowledgeable about the applications of artificial intelligence to simulation modeling.

Michael Philips (Department of Philosophy, Portland State University, Portland, Oregon 97207) received his bachelor's degree from the University of California, Riverside and his Ph.D. from Johns Hopkins University. He is currently a Visiting Professor of Philosophy at the University of British Columbia while on leave from his permanent position as Professor of

Philosophy at Portland State University. Besides his philosophical endeavors, Dr. Philips is a playwright, commentator for Oregon Public Radio, and an avid juggler.

Richard Porter (Peabody College, Vanderbilt University, Nashville, Tennessee 37203) is Professor of Psychology at Vanderbilt University. After receiving his Ph.D. in psychology at Wayne State University in 1970, he spent two years at Leicester University as a post-doctoral fellow. His primary research interests are the development of social behavior and social olfaction in mammals, including humans.

Bernard E. Rollin (Department of Philosophy, Colorado State University, Ft. Collins, Colorado 80523) is a Professor of Philosophy, Professor of Physiology and Biophysics, and Director of Bioethical Planning at Colorado State University. He has published extensively on the moral status of animals, animal pain, animal consciousness, veterinary and human medical ethics, genetic engineering and other issues in bioethics. Rollin has lectured all over the world on bioethics, is a consultant to many governments, and is a principal architect of recent U.S. federal legislation governing the treatment of laboratory animals. His most recent book is *The Unheeded Cry: Animal Consciousness, Animal Pain, and Science* (Oxford University Press, 1989).

Alexander Rosenberg (Department of Philosophy, University of California, Riverside, California 92521) is Professor and Chair of the Philosophy Department at the University of California, Riverside. He is the author of *Sociobiology and the Preemption of Social Science* (Johns Hopkins, 1981), *The Structure of Biological Science* (Cambridge, 1984), and *Philosophy of Social Science* (Westview and Oxford, 1988). He has held a Guggenheim fellowship and also works in metaphysics and the philosophy of psychology.

Michael L. Rosenzweig (Department of Ecology and Evolutionary Biology, University of Arizona, Tucson, Arizona 85721) was the founding head of the Department of Ecology and Evolutionary Biology at The University of Arizona, where he currently is a Professor. He is an evolutionary ecologist. He took his Ph.D. (in 1966 at the University of Pennsylvania) with R.H. MacArthur and W. John Smith. He has worked on a variety of theoretical and empirical research problems in ecology and evolution, especially optimal habitat use, predation and competitive relationships in desert animals. He is the founder and Editor-in-Chief of the journal *Evolutionary Ecology*.

Sue Savage-Rumbaugh (Language Research Center, Georgia State University, Atlanta, Georgia 30303) is currently the principal investigator of the National Institute of Child Health and Human Development funded project, "Language Acquisition in *Pan paniscus*" at the Language Research Center. She is also co-investigator, with Dr. Duane Rumbaugh, of the "Cognitive project", a study of numerical and mapping abilities in apes, and was the principal investigator for the "Animal Model Project" (1980-1985), both also funded by the National Institute of Child Health and Human Development. Dr. Savage-Rumbaugh earned her B.A. in psychology from Southwest Missouri University in 1970. She received her M.S. in psychology from the University of Oklahoma in 1972, and her Ph.D. in psychology from the same university in 1975. Dr. Savage-Rumbaugh has been a faculty member of Emory University in Atlanta since 1976, and a faculty member of Georgia State University in Atlanta since 1982.

W. John Smith (Department of Biology, University of Pennsylvania, Philadelphia, Pennsylvania 19104) is Professor of Biology at The University of Pennsylvania. He was born in Toronto, Canada and received his Bachelor's degree with honors from Carleton University and his Ph.D. from Harvard in 1961, working with Ernst Mayr. Dr. Smith has done field work throughout much of the New World, and currently is working in the East-Central and Southwestern United States and in Venezuela and Panama. He is currently performing experimental studies on singing behavior - how different ways of singing render different interactional behavior of the singer possible. Dr. Smith has published numerous papers dealing with communication and is the author of *The Behavior of Communications: An Ethological Approach* (Harvard University Press, 1977).

Sarah Stebbins (Department of Philosophy, Columbia University, New York, New York 10027) teaches philosophy at Columbia University. She has written on ape language research as well as on topics in the philosophy of language and the philosophy of logic. She is an amateur dog trainer and exhibitor. Dr. Stebbins taught at the Linguistics Society of America Summer Institute in 1987 and was a member of the Mind and Brain Research Group at the Center for Interdisciplinary Research (ZiF) at the University of Bielefeld, West Germany.

Nicholas D. Stone (Department of Entomology, Virginia Polytechnic Institute and State University, Blacksburg, Virginia 24060) is Associate Professor of Entomology at Virginia Polytechnic Institute and State University where he works on systems science and artificial intelligence approaches to pest management and insect population ecology. He received his A.B. in Biology from Harvard University in 1980 and his Ph.D. in Entomology from the University of California at Berkeley in 1984. From 1984 to 1988 he was Assistant Professor in the Department of Entomology at Texas A&M University where he was a cofounder of the Knowledge Engineering Laboratory in the Texas Agricultural Experiment Station. Since moving to VPI & SU, he has helped to create the Center for Computer-Aided Decision-Making in the College of Agricultural and Life Sciences. His current research interests are the integration of simulation modeling with knowledge-based systems, the use of artificial intelligence techniques to model animal behavior, and farm-level planning systems.

Randy Thornhill (Department of Biology, University of New Mexico, Albuquerque, New Mexico 87131) is Professor of Biology at the University of New Mexico, Albuquerque, New Mexico. He received his B.S. (Zoology) and M.S. (Entomology) from Auburn University and his Ph.D. (Zoology) degree from the University of Michigan. His research interests include the diversity of insect mating systems, sexual selection, and evolutionary methodology. This research has let to numerous research articles and a book, *The Evolution of Insect Mating Systems* (Harvard University Press, 1983). His current research focuses primarily on tests of hypotheses for how sexual selection has worked to design elaborate male secondary sexual traits and how natural selection has worked to design female mate choice in red jungle fowl, on the role of good genes sexual selection in scorpionflies, on the evolution of psychological pain in humans, and on the evolution of men's sexual psychology. He has received a Harry Frank Guggenheim award and the Senior Distinguished Scientist Award of the Alexander von Humboldt Foundation.

Hugh Wilder (Department of Philosophy, College of Charleston, Charleston, South Carolina 29401) is a Professor in the

Philosophy Department at the College of Charleston, South Carolina. His research interests are in philosophy of mind, philosophy of language, and aesthetics. With Judith de Luce, he co-edited *Language in Primates* (Springer-Verlag, 1983).

Thomas G. Wynn (Department of Anthropology, University of Colorado, Colorado Springs, Colorado 80933) received his Ph.D. degree at the University of Illinois in 1977. He is currently Professor and Chairman of the Department of Anthropology at the University of Colorado at Colorado Springs. His research interests include the evolution of human intelligence, the evolution of tool behavior, Paleolithic archaeology, and prehistory of plains in Colorado. He has recently published *The Evolution of Spatial Competence* (1989, the University of Illinois Press) and has written numerous articles for professional journals, including the *Journal of Human Evolution, World Archaeology, Man,* and *American Anthropologist.*

Index